Essential Organic Chemistry

Second Edition

Paula Yurkanis Bruice

University of California, Santa Barbara

Prentice Hall

New York   Boston   San Francisco
London   Toronto   Sydney   Tokyo   Singapore   Madrid
Mexico City   Munich   Paris   Cape Town   Hong Kong   Montreal

Library of Congress Cataloging-in-Publication Data

Bruice, Paula Yurkanis
  Essential organic chemistry/Paula Yurkanis Bruice.—2nd ed.
    p. cm.
  Includes bibliographical references and index.
  ISBN 0-321-59695-1
  1. Chemistry, Organic—Textbooks. I. Title.

QD251.3.B777 2010
547—dc22   2008050453

Acquisitions Editor: Dawn Giovanniello
Assistant Editor: Jessica Neumann
Editor in Chief, Science: Nicole Folchetti
Marketing Manager: Elizabeth Averbeck
Editorial Assistant: Kristen Wallerius
Marketing Assistant: Keri Parcells
Managing Editor, Chemistry and Geosciences: Gina M. Cheselka
Project Manager, Science: Beth Sweeten
Senior Operations Supervisor: Alan Fischer
Art Director: Suzanne Behnke
Associate Media Producer: Kristin Mayo
Art Editor: Connie Long
Art Studio: Imagineering
Photo Researcher: Yvonne Gerin
Spectra: Reproduced by permission of Aldrich Chemical Co.
Interior Designer: Joseph Sengotta
Cover Designer: Suzanne Behnke
Cover Photo: Biwa Inc/Photonica/Getty Images
Production Services/Composition: Preparé, Inc.

To Meghan, Kenton, and Alec

with love and immense respect;

and to Tom, my best friend

Printed in the United States of America
10 9 8 7 6 5 4 3 2

ISBN-10: 0-321-59695-1
ISBN-13: 978-0-321-59695-6

Prentice Hall
is an imprint of

www.pearsonhighered.com

Interest Boxes

Brief Contents

Contents

iv

5 The Reactions of Alkenes and Alkynes: An Introduction to Multistep Synthesis   107

6 Isomers and Stereochemistry   143

7 Delocalized Electrons and Their Effect on Stability, Reactivity, and pK_a: Ultraviolet and Visible Spectroscopy   169

8 Aromaticity: Reactions of Benzene and Substituted Benzenes   197

Boxed Features: The Structure of DNA: Watson, Crick, Franklin, and Wilkins   534   ■   Sickle Cell Anemia   544   ■   Antibiotics that Act by Inhibiting Translation   545   ■   DNA Fingerprinting   549   ■   Resisting Herbicides   549

21.10   The Economics of Drugs • Government Regulations   566
Summary   567   ■   Problems   567
Boxed Features: Drug Safety   559   ■   Orphan Drugs   567

Appendices

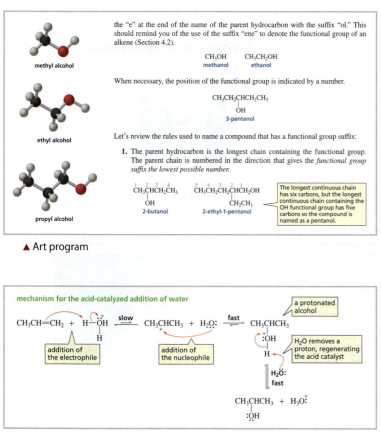

the "e" at the end of the name of the parent hydrocarbon with the suffix "ol." This should remind you of the use of the suffix "ene" to denote the functional group of an alkene (Section 4.2).

CH₃OH   CH₃CH₂OH
methanol   ethanol

When necessary, the position of the functional group is indicated by a number.

CH₃CH₂CHCH₂CH₃
|
OH
3-pentanol

Let's review the rules used to name a compound that has a functional group suffix:

1. The parent hydrocarbon is the longest chain containing the functional group. The parent chain is numbered in the direction that gives the *functional group suffix the lowest possible number.*

CH₃CHCH₂CH₃
|
OH
2-butanol

CH₃CH₂CH₂CHCH₂OH
|
CH₂CH₃
2-ethyl-1-pentanol

The longest continuous chain has six carbons, but the longest continuous chain containing the OH functional group has five carbons so the compound is named as a pentanol.

▲ Art program

mechanism for the acid-catalyzed addition of water

CH₃CH=CH₂ + H—ÖH —slow→ CH₃CHCH₃ + H₂Ö: —fast→ CH₃CHCH₃

addition of the electrophile

addition of the nucleophile

a protonated alcohol

H₂O removes a proton, regenerating the acid catalyst

CH₃CHCH₃ + H₃O⁺

▲ Mechanism

Art Program: Rich in Three-Dimensional, Computer-Generated Structures

This edition continues to present energy-minimized, **three-dimensional structures** throughout the text that give students an appreciation of the three-dimensional shapes of organic molecules. Color in the illustration program is used not simply for show but to highlight and organize the information. Colors are employed in consistent ways (for example, mechanism arrows are always red), but there is no need for a student to memorize a color palette.

Companion Website

WWW icons in the margins identify 3-D molecules, movies, and interactive animations on the Companion Website (http://www.chemplace.com) that are pertinent to the material being discussed. For a description of the Website's content, see The Chemistry Place for Essential Organic Chemistry under Resources for Students.

Changes to This Edition

Responses from instructors and students have led to adjustments in the coverage and distribution of certain topics and encouraged expansion of the book's most successful pedagogical features.

Content and Organization

The chapter on isomers and stereochemistry has been moved closer to the front of the book. The chapter on radicals has been deleted; however, the discussion of radicals in biological systems is retained and is incorporated with the radical chemistry found in Chapter 5. A discussion of catalysis has been added to Chapter 5, and mass spectrometry has been added to the chapter on "Determining the Structures of Organic Compounds." There are new sections that will allow students to apply what they have learned about reactivity to synthetic problems: for example, using substitution reactions to synthesize organic compounds, and using orientation effects to synthesize disubstituted benzenes. There is also new material on solvent effects, and the reactions of carbohydrates in basic solutions. Mechanistic details are expanded to facilitate students' understanding. Lastly, much of the book has been rewritten with the goal of making the material easier to understand.

Pedagogical Elements

The mechanisms in this edition are introduced by bold green titles and are presented in the form of bulleted steps to make the individual steps stand out more clearly. The mechanisms are not boxed; students should see mechanisms as being central to the understanding of the discipline and not something that is set aside or that they can come back to later. This edition also includes more voice boxes to aid student learning, more interest boxes, and many new problems.

LIST OF RESOURCES

For Students

Study Guide and Solutions Manual (0321592581) by Paula Yurkanis Bruice. This *Study Guide and Solutions Manual* contains complete and detailed explanations of the solutions to the problems in the text. It also contains exercises on "drawing curved arrows" and on "drawing resonance contributors." Each chapter contains a Practice Test, with the answers found at the back of the book.

The Chemistry Place for Essential Organic Chemistry, 2e (http://www.chemplace.com) Built to complement *Essential Organic Chemistry* as part of an integrated course package and identified by WWW icons in the margins, the easy-to-use Companion Website features the following modules for each chapter:

- **Interactive Tutorial and Animation Galleries** highlight central concepts and illustrate key mechanisms.

- **Molecule Galleries** feature hundreds of 3-D molecular models of compounds. Students can rotate and compare models, change their representation, and examine electrostatic potential map surfaces—a unique feature of learning organic chemistry on the Web.

- **Practice Exercises and Quizzes** allow students to test their understanding of the material. Each question includes a hint with a cross-reference to a text reading and detailed feedback. Although I had never been a fan of multiple-choice questions, the quality of these questions, written by a team of authors, has changed my mind.

Molecular Modeling Workbook (0131410407) Features SpartanView™ and SpartanBuild™ software. This workbook includes a software tutorial and numerous challenging exercises students can tackle to solve problems involving structure building and analysis using the tools included in the two pieces of Spartan software. Available free when packaged with the text. Contact your Pearson Prentice Hall representative for details.

Prentice Hall Molecular Model Kit (02055081363) This best-selling model kit allows students to build space-filling and ball-and-stick models of common organic molecules. It allows accurate depiction of double and triple bonds, including heteroatomic molecules (which some model kits cannot handle well).

Prentice Hall Framework Molecular Model Kit (0133300765) This model kit allows students to build scale models with precise interatomic distances and bond angles. This is the most accurate model kit available.

For Instructors

Online Test Item File (0321602668) by Debbie Beard, Mississippi State University. Includes a selection of more than 1200 multiple-choice, short answer, and essay test questions.

Online Instructor Resource Center (0321633466) This resource features almost all the art from the text, including tables, in PowerPoint™ format. Also included are Lecture PowerPoints, Classroom Response PowerPoints and Problem PowerPoints for each chapter.

ACKNOWLEDGMENTS

I am enormously grateful to the following reviewers who made this book a reality. The value of their work cannot be overstated.

Second Edition Reviewers

Deborah Booth, *University of Southern Mississippi*
Paul Buonora, *California State University–Long Beach*
Tom Chang, *Utah State University*
Dana Chatellier, *University of Delaware*
Amy Deveau, *University of New England*
J. Brent Friesen, *Dominican University*
Anne Gorden, *Auburn University*
Christine Hermann, *University of Radford*
Scott Lewis, *James Madison University*
Cynthia McGowan, *Merrimack College*
Keith Mead, *Mississippi State University*
Amy Pollock, *Michigan State University*

Manuscript Reviewers

Ardeshir Azadnia, *Michigan State University*
Debbie Beard, *Mississippi State University*
J. Phillip Bowen, *University of North Carolina–Greensboro*
Tim Burch, *Milwaukee Area Technical College*
Dana Chatellier, *University of Delaware*
Michelle Chatellier, *University of Delaware*
Long Chiang, *University of Massachusetts–Lowell*
Jan Dekker, *Reedley College*
Olga Dolgounitcheva, *Kansas State University*
John Droske, *University of Wisconsin–Stevens Point*
Eric Enholm, *University of Florida*
Gregory Friestad, *University of Vermont*
Wesley Fritz, *College of Dupage*
Robert Gooden, *Southern University*
Michael Groziak, *California State University–Hayward*
Steve Holmgren, *Montana State University*
Robert Hudson, *University of Western Ontario*
Richard Johnson, *University of New Hampshire*
Alan Kennan, *Colorado State University*
Spencer Knapp, *Rutgers University*
Mike Nuckols, *North Carolina State University*
Ed Parish, *Auburn University*
Mark W. Peczuh, *University of Connecticut*
Suzanne Purrington, *North Carolina State University*
Charles Rose, *University of Nevada–Reno*
Preet Saluja, *Triton College*
Joseph Sloop, *United States Military Academy*
Robert Swindell, *University of Arkansas*

About the Author

Paula Bruice with Zeus and Abigail

Paula Yurkanis Bruice was raised primarily in Massachusetts. After graduating from the Girls' Latin School in Boston, she received an A.B. from Mount Holyoke College and a Ph.D. in chemistry from the University of Virginia. She was awarded an NIH postdoctoral fellowship for study in the Department of Biochemistry at the University of Virginia Medical School and held a postdoctoral appointment in the Department of Pharmacology at Yale Medical School.

She has been a member of the faculty at the University of California, Santa Barbara, since 1972, where she has received the Associated Students Teacher of the Year Award, the Academic Senate Distinguished Teaching Award, two Mortar Board Professor of the Year Awards, and the UCSB Alumni Association Teaching Award. Her research interests center on the mechanism and catalysis of organic reactions, particularly those of biological significance. Paula has a daughter and a son who are physicians and a son who is a lawyer. Her main hobbies are reading mystery/suspense novels and enjoying her pets (two dogs, two cats, and two parrots).

CHAPTER 1

Electronic Structure and Covalent Bonding

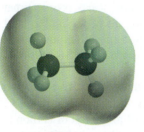

Ethane

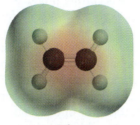

Ethene

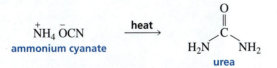

Ethyne

To stay alive, early humans must have been able to tell the difference between two kinds of materials in their world. "You can live on roots and berries," they might have said, "but you can't live on dirt. You can stay warm by burning tree branches, but you can't burn rocks."

By the early eighteenth century, scientists thought they had grasped the nature of that difference. Compounds derived from living sources were believed to contain an unmeasurable vital force—the essence of life. Because they came from organisms, they were called "organic" compounds. Compounds derived from minerals—those lacking that vital force—were "inorganic."

Because chemists could not create life in the laboratory, they assumed they could not create compounds that had a vital force. Since this was their mind-set, you can imagine how surprised chemists were in 1828 when Friedrich Wöhler produced urea—a compound known to be excreted by mammals—by heating ammonium cyanate, an inorganic mineral.

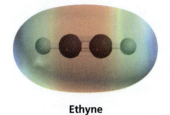

$$\overset{+}{N}H_4 \; \overset{-}{O}CN \quad \xrightarrow{\text{heat}} \quad$$
ammonium cyanate

urea

For the first time, an "organic" compound had been obtained from something other than a living organism and certainly without the aid of any kind of vital force. Clearly, chemists needed a new definition for "organic compounds." **Organic compounds** are now defined as *compounds that contain carbon*.

Why is an entire branch of chemistry devoted to the study of carbon-containing compounds? We study organic chemistry because just about all of the molecules that

make life possible—proteins, enzymes, vitamins, lipids, carbohydrates, and nucleic acids—contain carbon; thus, the chemical reactions that take place in living systems, including our own bodies, are reactions of organic compounds. Most of the compounds found in nature—those we rely on for food, medicine, clothing (cotton, wool, silk), and energy (natural gas, petroleum)—are organic compounds as well.

Organic compounds are not, however, limited to those found in nature. Chemists have learned to synthesize millions of organic compounds never found in nature, including synthetic fabrics, plastics, synthetic rubber, medicines, and even things like photographic film and Super Glue. Many of these synthetic compounds prevent shortages of naturally occurring products. For example, it has been estimated that if synthetic materials were not available for clothing, all of the arable land in the United States would have to be used for the production of cotton and wool just to provide enough material to clothe us. Currently, there are about 16 million known organic compounds, and many more are possible.

What makes carbon so special? Why are there so many carbon-containing compounds? The answer lies in carbon's position in the periodic table. Carbon is in the center of the second row of elements. The atoms to the left of carbon have a tendency to give up electrons, whereas the atoms to the right have a tendency to accept electrons (Section 1.3).

the second row of the periodic table

Because carbon is in the middle, it neither readily gives up nor readily accepts electrons. Instead, it shares electrons. Carbon can share electrons with several different kinds of atoms, and it can also share electrons with other carbon atoms. Consequently, carbon is able to form millions of stable compounds with a wide range of chemical properties simply by sharing electrons.

When we study organic chemistry, we study how organic compounds react. When an organic compound reacts, some existing bonds break and some new bonds form. Bonds form when two atoms share electrons, and bonds break when two atoms no longer share electrons. How readily a bond forms and how easily it breaks depend on the particular electrons that are shared, which, in turn, depend on the atoms to which the electrons belong. So if we are going to start our study of organic chemistry at the beginning, we must start with an understanding of the structure of an atom—what electrons an atom has and where they are located.

NATURAL VERSUS SYNTHETIC

It is a popular belief that natural substances—those made in nature—are superior to synthetic ones—those made in the laboratory. Yet when a chemist synthesizes a compound, such as penicillin, it is exactly the same in all respects as the compound synthesized in nature. Sometimes chemists can improve on nature. For example, chemists have synthesized analogs of morphine—compounds with structures similar to but not identical to that of morphine—that have painkilling effects like morphine but, unlike morphine, are not habit forming. Chemists have synthesized analogs of penicillin that do not produce the allergic responses that a significant fraction of the population experiences from naturally produced penicillin, or that do not have the bacterial resistance of the naturally produced antibiotic.

A field of poppies growing in Afghanistan. Commercial morphine is obtained from opium, the juice obtained from this species of poppy.

1.1 THE STRUCTURE OF AN ATOM

An atom consists of a tiny dense nucleus surrounded by electrons that are spread throughout a relatively large volume of space around the nucleus. The nucleus contains *positively charged protons* and *neutral neutrons*, so it is positively charged. The electrons are *negatively charged*. Because the amount of positive charge on a proton equals the amount of negative charge on an electron, a neutral atom has an equal number of protons and electrons. Atoms can gain electrons and thereby become negatively charged, or they can lose electrons and become positively charged. However, the number of protons in an atom does not change.

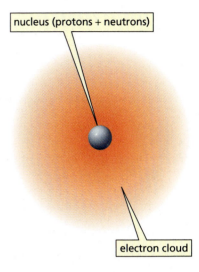

nucleus (protons + neutrons)

electron cloud

Protons and neutrons have approximately the same mass and are about 1800 times more massive than an electron. This means that most of the *mass* of an atom is in its nucleus. However, most of the *volume* of an atom is occupied by its electrons, and that is where our focus will be because it is the electrons that form chemical bonds.

The **atomic number** of an atom equals the *number of protons* in its nucleus. The atomic number is also the number of electrons that surround the nucleus of a neutral atom. For example, the atomic number of carbon is 6, which means that a neutral carbon atom has six protons and six electrons.

The **mass number** of an atom is the *sum of its protons and neutrons*. All carbon atoms have the same atomic number because they all have the same number of protons. They do not all have the same mass number because they do not all have the same number of neutrons. For example, 98.89% of naturally occurring carbon atoms have six neutrons—giving them a mass number of 12—and 1.11% have seven neutrons—giving them a mass number of 13. These two different kinds of carbon atoms (^{12}C and ^{13}C) are called isotopes. **Isotopes** have the same atomic number (that is, the same number of protons), but different mass numbers because they have different numbers of neutrons.

Naturally occurring carbon also contains a trace amount of ^{14}C, which has six protons and eight neutrons. This isotope of carbon is radioactive, decaying with a half-life of 5730 years. (The *half-life* is the time it takes for one-half of the nuclei to decay.) As long as a plant or an animal is alive, it takes in as much ^{14}C as it excretes or exhales. When it dies, it no longer takes in ^{14}C, so the ^{14}C in the organism slowly decreases. Therefore, the age of an organic substance can be determined by its ^{14}C content.

The **atomic weight** (or **atomic mass**) of a naturally occurring element is the *average mass of its atoms*. For example, carbon has an atomic weight of 12.011 atomic mass units. The **molecular weight** of a compound is the *sum of the atomic weights* of all the atoms in the molecule.

PROBLEM 1 ♦

Oxygen has three isotopes with mass numbers of 16, 17, and 18. The atomic number of oxygen is eight. How many protons and neutrons does each of the isotopes have?

1.2 HOW THE ELECTRONS IN AN ATOM ARE DISTRIBUTED

The electrons in an atom can be thought of as occupying a set of shells that surround the nucleus. The way in which the electrons are distributed in these shells is based on a theory developed by Einstein. The first shell is the smallest and the one closest to the nucleus; the second shell is larger and extends farther from the nucleus; and the third and higher numbered shells extend even farther out. Each shell consists of subshells known as **atomic orbitals**. The first shell has only an *s* atomic orbital; the second shell consists of *s* and *p* atomic orbitals; and the third shell consists of *s*, *p*, and *d* atomic orbitals (Table 1.1).

Table 1.1   Distribution of Electrons in the First Three Shells That Surround the Nucleus

	First shell	Second shell	Third shell
Atomic orbitals	s	s, p	s, p, d
Number of atomic orbitals	1	1, 3	1, 3, 5
Maximum number of electrons	2	8	18

Each shell contains one s orbital. The second and higher shells—in addition to their s orbital—each contain three p orbitals. The three p orbitals have the same energy. The third and higher shells—in addition to their s and p orbitals—also contain five d orbitals. Because an orbital can contain no more than two electrons (see below), the first shell, with only one atomic orbital, can contain no more than two electrons. The second shell, with four atomic orbitals—one s and three p—can have a total of eight electrons. Eighteen electrons can occupy the nine atomic orbitals—one s, three p, and five d—of the third shell.

An important point to remember is that *the closer the atomic orbital is to the nucleus, the lower is its energy.* Because the s orbital in the first shell (called a $1s$ orbital) is closer to the nucleus than is the s orbital in the second shell (called a $2s$ orbital), the $1s$ orbital is lower in energy. Comparing orbitals in the same shell, we see that an s orbital is lower in energy than a p orbital, and a p orbital is lower in energy than a d orbital.

> **The closer the orbital is to the nucleus, the lower is its energy.**

Relative energies of atomic orbitals: $1s < 2s < 2p < 3s < 3p < 3d$

The **electronic configuration** of an atom describes what orbitals the electrons occupy. The following three rules are used to determine an atom's electronic configuration:

1. An electron always goes into the available orbital with the lowest energy.

2. No more than two electrons can occupy each orbital, and the two electrons must be of opposite spin. (Notice in Table 1.2 that spin in one direction is designated by ↑, and spin in the opposite direction by ↓.)

From these first two rules, we can assign electrons to atomic orbitals for atoms that contain one, two, three, four, or five electrons. The single electron of a hydrogen atom occupies a $1s$ orbital, the second electron of a helium atom fills the $1s$ orbital, the third electron of a lithium atom occupies a $2s$ orbital, the fourth electron of a beryllium atom fills the $2s$ orbital, and the fifth electron of a boron atom occupies one of the $2p$ orbitals. (The subscripts x, y, and z distinguish the three $2p$ orbitals.) Because the three

Table 1.2   The Electronic Configurations of the Smallest Atoms

Atom	Name of element	Atomic number	$1s$	$2s$	$2p_x$	$2p_y$	$2p_z$	$3s$
H	Hydrogen	1	↑					
He	Helium	2	↑↓					
Li	Lithium	3	↑↓	↑				
Be	Beryllium	4	↑↓	↑↓				
B	Boron	5	↑↓	↑↓	↑			
C	Carbon	6	↑↓	↑↓	↑	↑		
N	Nitrogen	7	↑↓	↑↓	↑	↑	↑	
O	Oxygen	8	↑↓	↑↓	↑↓	↑	↑	
F	Fluorine	9	↑↓	↑↓	↑↓	↑↓	↑	
Ne	Neon	10	↑↓	↑↓	↑↓	↑↓	↑↓	
Na	Sodium	11	↑↓	↑↓	↑↓	↑↓	↑↓	↑

ALBERT EINSTEIN

Albert Einstein (1879–1955) was born in Germany. When he was in high school, his father's business failed and his family moved to Milan, Italy. Although Einstein wanted to join his family in Italy, he had to stay behind because German law required compulsory military service after high school. To help him, his high school mathematics teacher wrote a letter saying that Einstein could have a nervous breakdown without his family and also that there was nothing left to teach him. Eventually, Einstein was asked to leave the school because of his disruptive behavior. Popular folklore says he left because of poor grades in Latin and Greek, but his grades in those subjects were fine.

Einstein was visiting the United States when Hitler came to power, so he accepted a position at the Institute for Advanced Study in Princeton, becoming a U.S. citizen in 1940. Although a lifelong pacifist, he wrote a letter to President Roosevelt warning of ominous advances in German nuclear research. This led to the creation of the Manhattan Project, which developed the atomic bomb and tested it in New Mexico in 1945.

2p orbitals have the same energy, the electron can be put into any one of them. Before we can continue to atoms containing six or more electrons, we need the third rule:

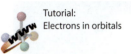
Tutorial:
Electrons in orbitals

3. When there are two or more orbitals with the same energy, an electron will occupy an empty orbital before it will pair up with another electron.

The sixth electron of a carbon atom, therefore, goes into an empty 2p orbital, rather than pairing up with the electron already occupying a 2p orbital (Table 1.2). There is one more empty 2p orbital, so that is where the seventh electron of a nitrogen atom goes. The eighth electron of an oxygen atom pairs up with an electron occupying a 2p orbital rather than going into a higher energy 3s orbital.

Electrons in inner shells (those below the outermost shell) are called **core electrons**. Electrons in the outermost shell are called **valence electrons**. Carbon, for example, has two core electrons and four valence electrons (Table 1.2).

Lithium and sodium each have one valence electron. Elements in the same column of the periodic table have the same number of valence electrons. Because the number of valence electrons is the major factor determining an element's chemical properties, elements in the same column of the **periodic table** have similar chemical properties. (You can find a periodic table inside the back cover of this book.) Thus, the chemical behavior of an element depends on its electronic configuration.

PROBLEM 2 ◆

How many valence electrons do the following atoms have?
a. carbon   **b.** nitrogen   **c.** oxygen   **d.** fluorine

---

PROBLEM 3 ◆

Table 1.2 shows that lithium and sodium each have one valence electron. Find potassium (K) in the periodic table and predict how many valence electrons it has.

PROBLEM-SOLVING STRATEGY

Write the ground-state electronic configuration for chlorine.

The periodic table in the back of the book shows that chlorine has 17 electrons. Now we need to assign the electrons to orbitals using the rules that determine an atom's electronic configuration. Two electrons are in the 1s orbital, two are in the 2s orbital, six are in the 2p orbitals, and two are in the 3s orbital. This accounts for 12 of the 17 electrons. The remaining five electrons are in the 3p orbitals. Therefore, the electronic configuration is written as: $1s^2, 2s^2, 2p^6, 3s^2, 3p^5$

Now continue on to Problem 4.

Shown is a bronze sculpture of **Einstein** *on the grounds of the National Academy of Sciences in Washington, D.C. It measures 21 feet from the top of the head to the tip of the feet and weighs 7000 pounds. In his left hand, Einstein holds the mathematical equations that represent his three most important contributions to science: the photoelectric effect, the equivalency of energy and matter, and the theory of relativity. At his feet is a map of the sky.*

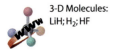

3-D Molecules:
LiH; H$_2$; HF

Electrostatic potential maps (often simply called potential maps) are models that show how charge is distributed in the molecule under the map. The potential maps for LiH, H$_2$, and HF are shown below.

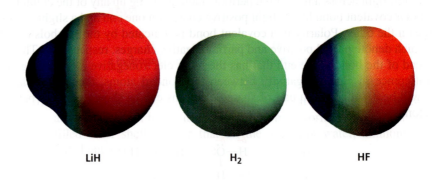

The colors indicate the distribution of charge in the molecule: red signifies electron-rich areas (negative charge); blue signifies electron-deficient areas (positive charge); and green signifies no charge. For example, the potential map for LiH indicates that the hydrogen atom is more negatively charged than the lithium atom. By comparing the three maps, we can tell that the hydrogen in LiH is more negatively charged than a hydrogen in H$_2$, and the hydrogen in HF is more positively charged than a hydrogen in H$_2$.

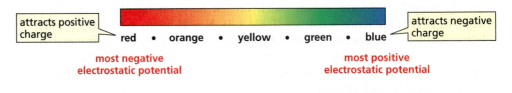

PROBLEM 9 ◆

After examining the potential maps for LiH, HF, and H$_2$, answer the following questions:
a. Which compounds are polar?
b. Which compound has the most positively charged hydrogen?

1.4 HOW THE STRUCTURE OF A COMPOUND IS REPRESENTED

First we will see how compounds are drawn using Lewis structures. Then we will look at the representations of structures that are used more commonly for organic compounds.

Lewis Structures

The chemical symbols we have been using, in which the valence electrons are represented as dots, are called **Lewis structures** (or electron dot structures) after G. N. Lewis (Section 1.3). Lewis structures are useful because they show us which atoms are bonded together and tell us whether any atoms *possess lone-pair electrons* or have a *formal charge*, two concepts we describe below. The Lewis structures for H$_2$O, H$_3$O$^+$, HO$^-$, and H$_2$O$_2$ are

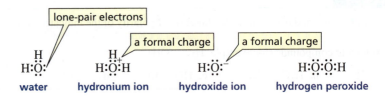

When you draw a Lewis structure, make sure that hydrogen atoms are surrounded by no more than two electrons and that C, O, N, and halogen (F, Cl, Br, I) atoms are surrounded by no more than eight electrons, in accordance with the octet rule. Valence electrons not used in bonding are called **nonbonding electrons, lone-pair electrons**, or simply **lone pairs**.

Movie:
Formal charge

Once you have all the atoms and the electrons in place, you must examine each atom to see whether a formal charge should be assigned to it. A **formal charge** is the *difference* between the number of valence electrons an atom has when it is not bonded to any other atoms and the number of electrons it "owns" when it is bonded. An atom "owns" all of its lone-pair electrons and half of its bonding (shared) electrons.

formal charge = number of valence electrons − (number of lone-pair electrons + 1/2 number of bonding electrons)

For example, an oxygen atom has six valence electrons (Table 1.2). In water (H_2O), oxygen "owns" six electrons (four lone-pair electrons and half of the four bonding electrons). Because the number of electrons it "owns" is equal to the number of its valence electrons ($6 - 6 = 0$), the oxygen atom in water does not have a formal charge. The oxygen atom in the hydronium ion (H_3O^+) "owns" five electrons: two lone-pair electrons plus three (half of six) bonding electrons. Because the number of electrons it "owns" is one less than the number of its valence electrons, its formal charge is $+1$ ($6 - 5 = 1$). The oxygen atom in hydroxide ion HO^- "owns" seven electrons: six lone-pair electrons plus one (half of two) bonding electron. Because it "owns" one more electron than the number of its valence electrons, its formal charge is -1 ($6 - 7 = -1$).

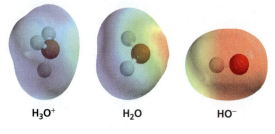

H_3O^+ H_2O HO^-

PROBLEM 10 ♦

The formal charge does not necessarily indicate that the atom has greater or less electron density than other atoms in the molecule without formal charges. You can see this by examining the potential maps for H_2O, H_3O^+, and HO^-.

a. Which atom bears the formal negative charge in the hydroxide ion?
b. Which atom has the greater electron density in the hydroxide ion?
c. Which atom bears the formal positive charge in the hydronium ion?
d. Which atom has the least electron density in the hydronium ion?

Nitrogen has five valence electrons (Table 1.2). Prove to yourself that the appropriate formal charges have been assigned to the nitrogen atoms in the following Lewis structures:

ammonia   ammonium ion   amide anion   hydrazine

Carbon has four valence electrons. Take a moment to understand why the carbon atoms in the following Lewis structures have the indicated formal charges:

methane   methyl cation   methyl anion   methyl radical   ethane
                 a carbocation   a carbanion

A species containing a positively charged carbon atom is called a **carbocation**, and a species containing a negatively charged carbon atom is called a **carbanion**. (Recall that a *cation* is a positively charged ion and an *anion* is a negatively charged ion.) A species containing an atom with a single unpaired electron is called a **radical** (often called a **free radical**).

Hydrogen has one valence electron, and each halogen (F, Cl, Br, I) has seven valence electrons, so the following species have the indicated formal charges:

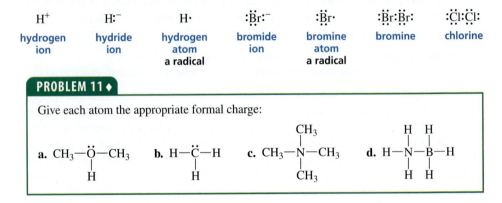

PROBLEM 11 ◆

Give each atom the appropriate formal charge:

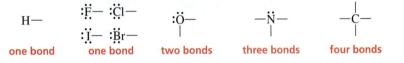

In studying the molecules in this section, notice that when the atoms do not bear a formal charge or an unpaired electron, hydrogen and the halogens always have *one* covalent bond, oxygen always has *two* covalent bonds, nitrogen always has *three* covalent bonds, and carbon has *four* covalent bonds. Atoms that have more bonds or fewer bonds than the number required for a neutral atom will have either a formal charge or an unpaired electron. These numbers are very important to remember when you are first drawing structures of organic compounds because they provide a quick way to recognize when you have made a mistake.

$$H- \qquad :\ddot{F}- \quad :\ddot{C}l- \qquad :\ddot{O}- \qquad -\ddot{N}- \qquad -\overset{|}{\underset{|}{C}}-$$
$$\qquad :\ddot{I}- \quad :\ddot{B}r-$$

one bond   one bond   two bonds   three bonds   four bonds

In the following Lewis structures, notice that each atom has a filled outer shell. Also notice that since none of the molecules has a formal charge or an unpaired electron, C forms four bonds, N forms three bonds, O forms two bonds, and H and Br each form one bond.

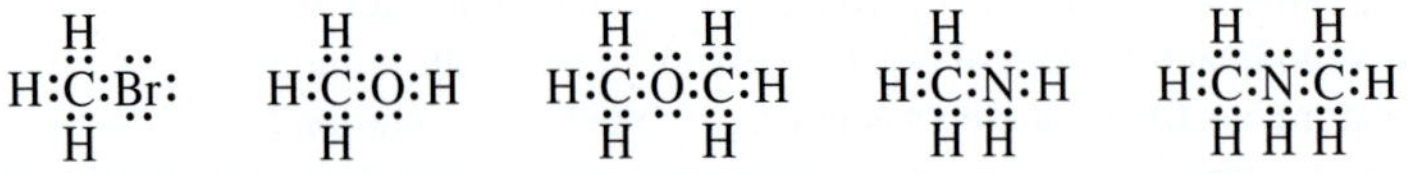

Because a pair of shared electrons can be shown as a line between two atoms, compare the preceding structures with the following ones:

PROBLEM 12 ◆

Draw the Lewis structure for each of the following:

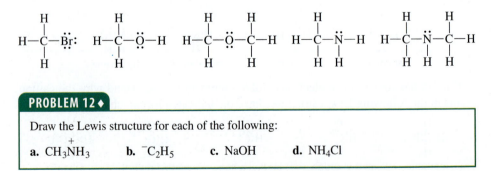

Kekulé Structures

In **Kekulé structures**, the bonding electrons are drawn as lines and the lone-pair electrons are usually left out entirely, unless they are needed to draw attention to some chemical property of the molecule. (Although lone-pair electrons are not shown, you should remember that neutral nitrogen, oxygen, and halogen atoms always have them: one pair in the case of nitrogen, two pairs in the case of oxygen, and three pairs in the case of a halogen.)

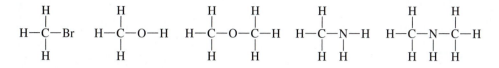

Condensed Structures

Frequently, structures are simplified by omitting some (or all) of the covalent bonds and listing atoms bonded to a particular carbon (or nitrogen or oxygen) next to it with subscripts as necessary. These structures are called **condensed structures**. Compare the following examples with the Kekulé structures shown above:

$$CH_3Br \qquad CH_3OH \qquad CH_3OCH_3 \qquad CH_3NH_2 \qquad CH_3NHCH_3$$

You can find more examples of condensed structures and the conventions commonly used to create them in Table 1.4.

PROBLEM 13 ◆

Draw the lone-pair electrons that are not shown in the following structures:

a. $CH_3CH_2NH_2$ **c.** CH_3CH_2OH **e.** CH_3CH_2Cl

b. CH_3NHCH_3 **d.** CH_3OCH_3 **f.** $HONH_2$

PROBLEM 14 ◆

Draw condensed structures for the compounds represented by the following models (black = C, white = H, red = O, blue = N, green = Cl):

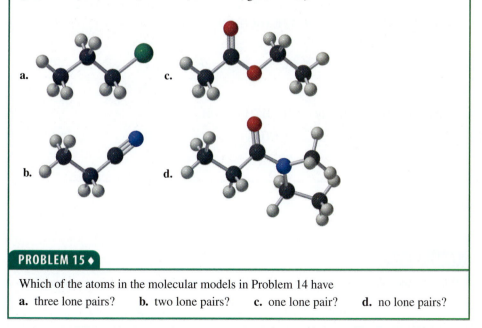

a.

b.

c.

d.

PROBLEM 15 ◆

Which of the atoms in the molecular models in Problem 14 have

a. three lone pairs?   **b.** two lone pairs?   **c.** one lone pair?   **d.** no lone pairs?

Table 1.4   Kekulé and Condensed Structures

Kekulé structure	Condensed structures

Atoms bonded to a carbon are shown to the right of the carbon. Atoms other than H can be shown hanging from the carbon.

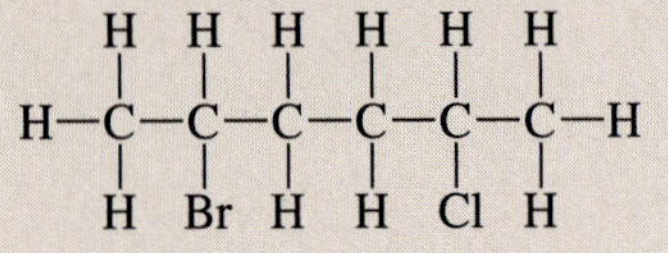

$CH_3CHBrCH_2CH_2CHClCH_3$ or $CH_3CHCH_2CH_2CHCH_3$
                                            | |
                                            Br   Cl

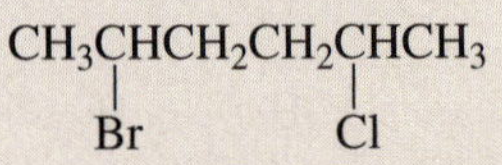

Repeating CH_2 groups can be shown in parentheses.

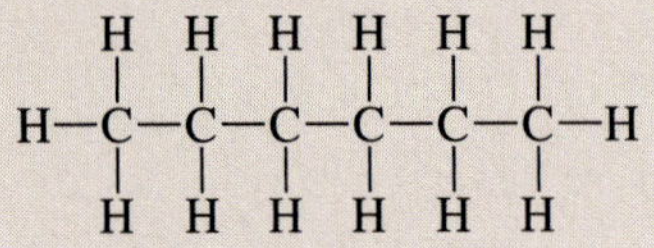

$CH_3CH_2CH_2CH_2CH_2CH_3$ or $CH_3(CH_2)_4CH_3$

Groups bonded to a carbon can be shown (in parentheses) to the right of the carbon, or hanging from the carbon.

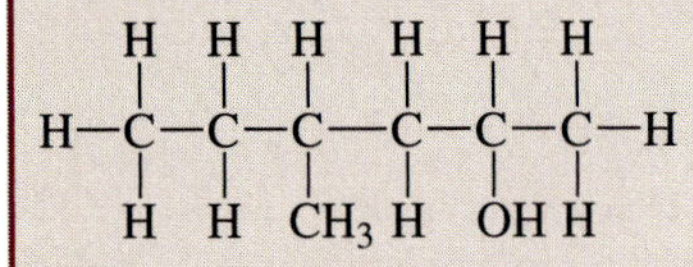

$CH_3CH_2CH(CH_3)CH_2CH(OH)CH_3$ or $CH_3CH_2CHCH_2CHCH_3$
                                                    | |
                                                    CH_3 OH

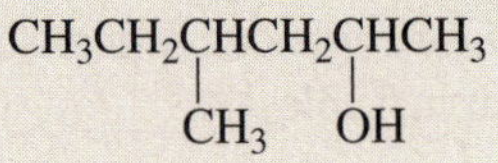

Groups bonded to the far-right carbon are not put in parentheses.

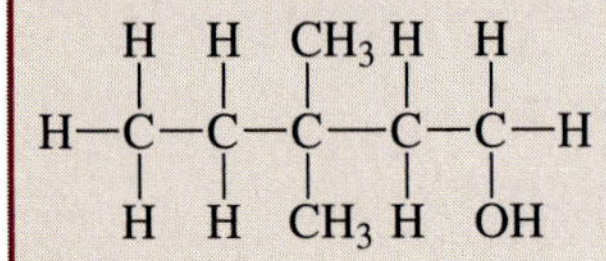

$CH_3CH_2C(CH_3)_2CH_2CH_2OH$ or $CH_3CH_2CCH_2CH_2OH$
                                                     | (with CH₃ above and below)

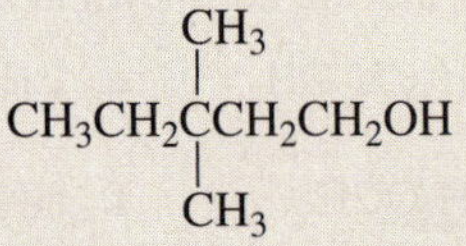

Two or more identical groups considered bonded to the "first" atom on the left can be shown (in parentheses) to the left of that atom, or hanging from the atom.

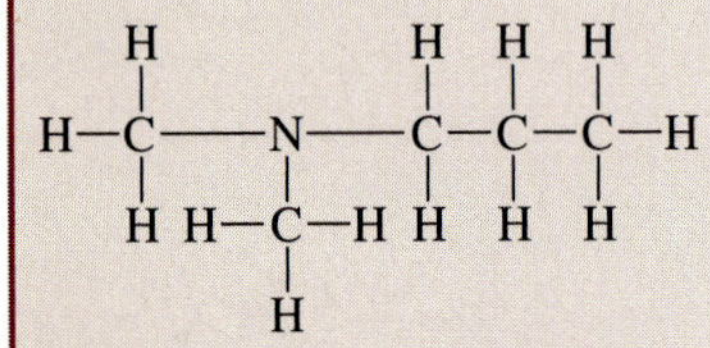

$(CH_3)_2NCH_2CH_2CH_3$ or $CH_3NCH_2CH_2CH_3$
                                            |
                                            CH_3

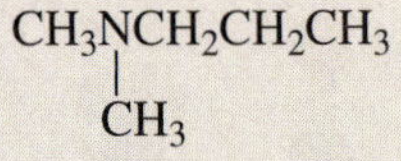

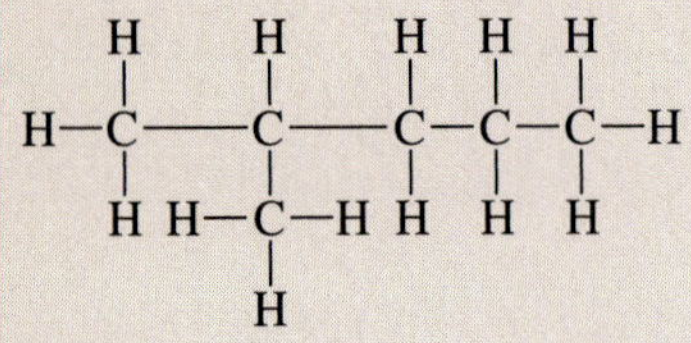

$(CH_3)_2CHCH_2CH_2CH_3$ or $CH_3CHCH_2CH_2CH_3$
                                            |
                                            CH_3

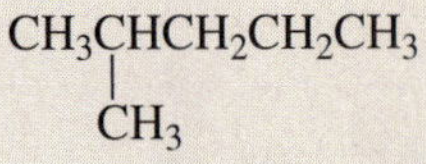

PROBLEM 16

Which of the following molecular formulas are not possible for an organic compound?

C_2H_6 C_2H_7 C_3H_9 C_3H_8 C_4H_{10}

PROBLEM 17

a. Draw two Lewis structures for C_2H_6O.
b. Draw three Lewis structures for C_3H_8O.
(*Hint:* The two Lewis structures in part a are **constitutional isomers**; they have the same atoms, but differ in the way the atoms are connected; see page 143. The three Lewis structures in part b are also constitutional isomers.)

PROBLEM 18

Expand the following condensed structures to show the covalent bonds and lone pairs:
a. $CH_3NH(CH_2)_2CH_3$ **c.** $(CH_3)_3COH$
b. $(CH_3)_2CHCl$ **d.** $(CH_3)_3C(CH_2)_3CH(CH_3)_2$

1.5 ATOMIC ORBITALS

We have seen that electrons are distributed into different atomic orbitals (Table 1.2). An **orbital** is a three-dimensional region around the nucleus where an electron is most likely to be found. But what does an orbital look like? The *s* orbital is a sphere with the nucleus at its center. Thus, when we say that an electron occupies a 1*s* orbital, we mean that there is a greater than 90% probability that the electron is in the space defined by the sphere.

An orbital tells us the volume of space around the nucleus where an electron is most likely to be found.

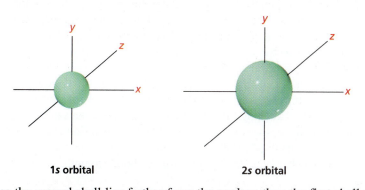

1s orbital   **2s orbital**

Because the second shell lies farther from the nucleus than the first shell (Section 1.2), the average distance from the nucleus is greater for an electron in a 2*s* orbital than for an electron in a 1*s* orbital. A 2*s* orbital, therefore, is represented by a larger sphere. Because of the greater size of the 2*s* orbital, its average electron density is less than the average electron density of a 1*s* orbital.

Unlike *s* orbitals, which resemble spheres, a *p* orbital has two lobes. Generally, the lobes are depicted as teardrop-shaped, but computer-generated representations reveal that they are shaped more like doorknobs. In Section 1.2, we saw that the second and higher numbered shells each contain three *p* orbitals, and the three *p* orbitals have the same energy. The p_x orbital is symmetrical about the *x*-axis, the p_y orbital is symmetrical about the *y*-axis, and the p_z orbital is symmetrical about the *z*-axis. This means that each *p* orbital is perpendicular to the other two *p* orbitals. The energy of a 2*p* orbital is greater than that of a 2*s* orbital because the average location of an electron in a 2*p* orbital is farther away from the nucleus.

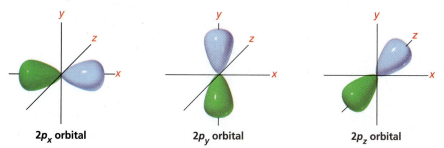

$2p_x$ orbital   **$2p_y$ orbital**   **$2p_z$ orbital**

**computer-generated
2*p* orbital**

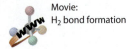

Movie:
H₂ bond formation

1.6 HOW ATOMS FORM COVALENT BONDS

How do atoms form covalent bonds in order to form molecules? Let's look first at the bonding in a hydrogen molecule (H_2). The covalent bond is formed when the $1s$ orbital of one hydrogen atom overlaps the $1s$ orbital of a second hydrogen atom. The covalent bond that is formed when the two orbitals overlap is called a **sigma (σ) bond**.

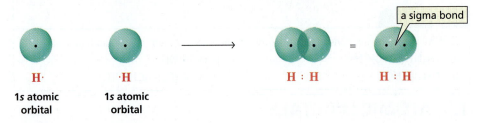

Why do atoms form covalent bonds? As the two orbitals start to overlap to form the covalent bond, energy is released (and stability increases) because the electron in each atom is attracted both to its own nucleus and to the positively charged nucleus of the other atom (Figure 1.2). Thus atoms form covalent bonds because the covalently bonded atoms are more stable than the individual atoms. The attraction of the negatively charged electrons for the positively charged nuclei is what holds the atoms together. The more the orbitals overlap, the more the energy decreases until the atoms are so close together that their positively charged nuclei start to repel each other. This repulsion causes a large increase in energy. Maximum stability (that is, minimum energy) is achieved when the nuclei are a certain distance apart. This distance is the **bond length** of the new covalent bond. The length of the H—H bond is 0.74 Å.

As Figure 1.2 shows, energy is released when a covalent bond forms. When the H—H bond forms, 105 kcal/mol or 439 kJ/mol of energy is released (1 kcal = 4.184 kJ).* Breaking the bond requires precisely the same amount of energy. Thus, the **bond strength**—also called the **bond dissociation energy**—is the energy required to break the bond, or the energy released when the bond is formed. Every covalent bond has a characteristic bond length and bond strength.

Maximum stability corresponds to minimum energy.

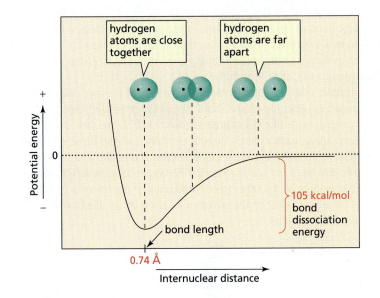

▶ **Figure 1.2**
The change in energy that occurs as two $1s$ atomic orbitals approach each other. The internuclear distance at minimum energy is the length of the H—H covalent bond.

---

*Joules are the Système International (SI) units for energy, although many chemists use calories. We will use both in this book.

1.7 HOW SINGLE BONDS ARE FORMED IN ORGANIC COMPOUNDS

We will begin the discussion of bonding in organic compounds by looking at the bonding in methane, a compound with only one carbon atom. Then we will examine the bonding in ethane, a compound with two carbons attached by a carbon–carbon *single bond*.

The Bonds in Methane

Methane (CH_4) has four covalent C—H bonds. Because all four bonds have the same length and all the bond angles are the same (109.5°), we can conclude that the four C—H bonds in methane are identical. Four different ways to represent a methane molecule are shown here.

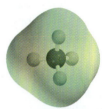

3-D Molecule: Methane

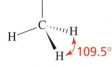

perspective formula of methane

ball-and-stick model of methane

space-filling model of methane

electrostatic potential map for methane

In a **perspective formula**, bonds in the plane of the paper are drawn as solid lines, bonds protruding out of the plane of the paper toward the viewer are drawn as solid wedges, and those protruding back from the plane of the paper away from the viewer are drawn as hatched wedges.

The potential map of methane shows that neither carbon nor hydrogen carries much of a charge: there are neither red areas, representing partially negatively charged atoms, nor blue areas, representing partially positively charged atoms. (Compare this map with the potential map for water on page 11). The absence of partially charged atoms is due to the similar electronegativities of carbon and hydrogen, which cause them to share their bonding electrons relatively equally. Methane is therefore a **nonpolar molecule**.

You may be surprised to learn that carbon forms four covalent bonds since you know that carbon has only two unpaired electrons in its electronic configuration (Table 1.2). But if carbon formed only two covalent bonds, it would not complete its octet. We therefore need to come up with an explanation that accounts for carbon's forming four covalent bonds.

If one of the electrons in carbon's $2s$ orbital were promoted into the empty $2p$ orbital, the new electronic configuration would have four unpaired electrons; thus, four covalent bonds could be formed.

If carbon used an s orbital and three p orbitals to form these four bonds, the bond formed with the s orbital would be different from the three bonds formed with p orbitals. What could account for the fact that the four C—H bonds in methane are identical if they are made using one s and three p orbitals? The answer is that carbon uses *hybrid orbitals*.

Hybrid orbitals are mixed orbitals that result from combining atomic orbitals. The concept of combining orbitals was first proposed by Linus Pauling in 1931. If the one

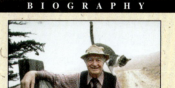

s and three *p* orbitals of the second shell are all combined and then apportioned into four equal orbitals, each of the four resulting orbitals will be one part *s* and three parts *p*. Therefore, each orbital has 25% *s* character and 75% *p* character. This type of mixed orbital is called an sp^3 (stated "*s-p*-three" not "*s-p*-cubed") orbital. (The superscript 3 means that three *p* orbitals were mixed with one *s* orbital to form the hybrid orbitals.) Each of the four sp^3 orbitals has the same energy.

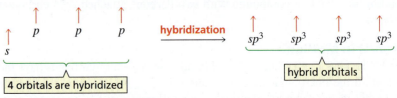

Like a *p* orbital, an sp^3 orbital has two lobes. Unlike the lobes of a *p* orbital, the two lobes of an sp^3 orbital are not the same size (Figure 1.3). The larger lobe of the sp^3 orbital is used in covalent bond formation.

▶ **Figure 1.3**
An *s* orbital and three *p* orbitals hybridize to form four sp^3 orbitals. An sp^3 orbital is more stable than a *p* orbital, but not as stable as an *s* orbital.

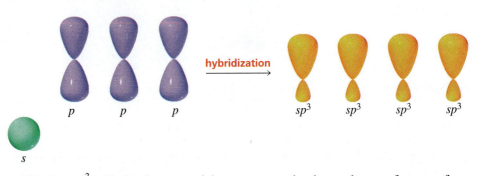

The four sp^3 orbitals adopt a spatial arrangement that keeps them as far away from each other as possible (Figure 1.4a). They do this because electrons repel each other, and moving the orbitals as far from each other as possible minimizes the repulsion. When four orbitals move as far from each other as possible, they point toward the corners of a regular tetrahedron (a pyramid with four faces, each an equilateral triangle). Each of the four C—H bonds in methane is formed from overlap of an sp^3 orbital of carbon with the *s* orbital of a hydrogen (Figure 1.4b). This explains why the four C—H bonds are identical.

▶ **Figure 1.4**
(a) The four sp^3 orbitals are directed toward the corners of a tetrahedron, causing each bond angle to be 109.5°. (b) An orbital picture of methane, showing the overlap of each sp^3 orbital of carbon with the *s* orbital of a hydrogen. (For clarity, the smaller lobes of the sp^3 orbitals are not shown.)

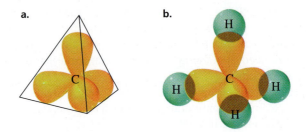

The angle between any two bonds that point from the center to the corners of a tetrahedron are 109.5°. The bond angles in methane therefore are 109.5°. This is called a **tetrahedral bond angle**. A carbon, such as the one in methane, that forms covalent bonds using four equivalent sp^3 orbitals is called a **tetrahedral carbon**.

Electron pairs stay as far from each other as possible.

Hybrid orbitals may appear to have been contrived just to make things fit—and that is exactly the case. Nevertheless, they give us a very good picture of the bonding in organic compounds.

> **Note to the student**
> It is important to understand what molecules look like in three dimensions. Therefore be sure to visit the textbook's website (*http://www.chemplace.com*) and look at the three-dimensional representations of molecules that can be found in the molecule gallery prepared for each chapter.

The Bonds in Ethane

The two carbon atoms in ethane (CH_3CH_3) are tetrahedral. Each carbon uses four sp^3 orbitals to form four covalent bonds:

ethane

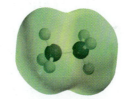 3-D Molecule: Ethane

One sp^3 orbital of one carbon overlaps an sp^3 orbital of the other carbon to form the C—C bond. Each of the remaining three sp^3 orbitals of each carbon overlaps the s orbital of a hydrogen to form a C—H bond. Thus, the C—C bond is formed by sp^3–sp^3 overlap, and each C—H bond is formed by sp^3–s overlap (Figure 1.5).

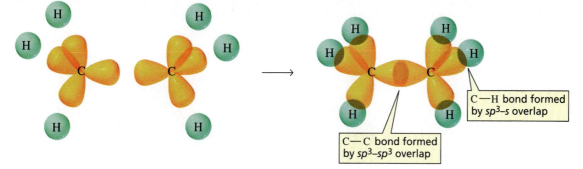

C—H bond formed by sp^3–s overlap

C—C bond formed by sp^3–sp^3 overlap

▲ **Figure 1.5**
An orbital picture of ethane. The C—C bond is formed by sp^3–sp^3 overlap, and each C—H bond is formed by sp^3–s overlap. (The smaller lobes of the sp^3 orbitals are not shown.)

Each of the bond angles in ethane is nearly the tetrahedral bond angle of 109.5°, and the length of the C—C bond is 1.54 Å. Ethane, like methane, is a nonpolar molecule.

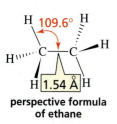

perspective formula of ethane   **ball-and-stick model of ethane**   **space-filling model of ethane**   **electrostatic potential map for ethane**

All the bonds in methane and ethane are sigma (σ) bonds. We will see that all **single bonds** found in organic compounds are *sigma bonds*.

PROBLEM 19 ♦

What orbitals are used to form the 10 covalent bonds in propane ($CH_3CH_2CH_3$)?

1.8 HOW A DOUBLE BOND IS FORMED: THE BONDS IN ETHENE

Each of the carbon atoms in ethene (also called ethylene) forms four bonds, but each is bonded to only three atoms:

ethene
ethylene

To bond to three atoms, each carbon hybridizes three atomic orbitals: an *s* orbital and two of the *p* orbitals). Because three orbitals are hybridized, three hybrid orbitals are formed. These are called sp^2 orbitals. After hybridization, each carbon atom has three identical sp^2 orbitals and one *p* orbital:

The axes of the three sp^2 orbitals lie in a plane (Figure 1.6a). To minimize electron repulsion, the three orbitals need to get as far from each other as possible. Therefore the bond angles are all close to 120°. The unhybridized *p* orbital is perpendicular to the plane defined by the axes of the sp^2 orbitals (Figure 1.6b).

▶ **Figure 1.6**
(a) The three sp^2 orbitals lie in a plane.
(b) The unhybridized *p* orbital is perpendicular to the plane. (The smaller lobes of the sp^2 orbitals are not shown.)

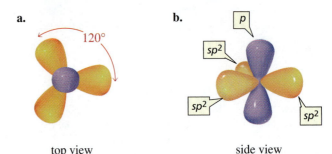

The carbons in ethene are held together by two bonds. Two bonds connecting two atoms is called a **double bond**. The two carbon–carbon bonds in the double bond are not identical. One of them results from the overlap of an sp^2 orbital of one carbon with an sp^2 orbital of the other carbon; this is a sigma (σ) bond. Each carbon uses its other two sp^2 orbitals to overlap the *s* orbital of a hydrogen to form the C—H bonds (Figure 1.7a). The second carbon–carbon bond results from side-to-side overlap of the two unhybridized *p* orbitals (Figure 1.7b). Side-to-side overlap of *p* orbitals forms a **pi (π) bond**. Thus, one of the bonds in a double bond is a σ bond, and the other is a π bond. All the C—H bonds are σ bonds.

Side-to-side overlap of two *p* atomic orbitals forms a π bond.

The two *p* orbitals that overlap to form the π bond must be parallel to each other for maximum overlap to occur. This forces the triangle formed by one carbon and two hydrogens to lie in the same plane as the triangle formed by the other carbon and two

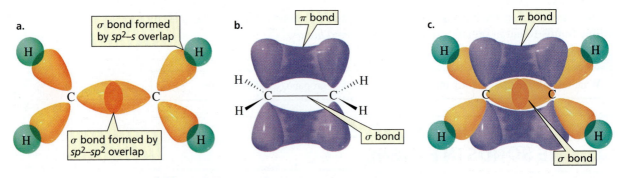

▲ **Figure 1.7**
(a) One C—C bond in ethene is a σ bond formed by sp^2–sp^2 overlap, and the C—H bonds are formed by sp^2–*s* overlap.
(b) The second C—C bond is a π bond formed by side-to-side overlap of a *p* orbital of one carbon with a *p* orbital of the other carbon.
(c) There is an accumulation of electron density above and below the plane containing the two carbons and four hydrogens.

hydrogens. As a result, all six atoms of ethene lie in the same plane, and the electrons in the *p* orbitals occupy a volume of space above and below the plane (Figure 1.7c). The potential map for ethene shows that it is a nonpolar molecule with a slight accumulation of negative charge (the pale orange area) above the two carbons. (If you could turn the potential map over to show the hidden side, a similar accumulation of negative charge would be found there.)

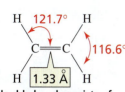

a double bond consists of
one σ bond and one π bond

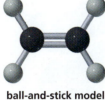

ball-and-stick model
of ethene

space-filling model
of ethene

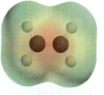

electrostatic potential map
for ethene

Four electrons hold the carbons together in a carbon–carbon double bond; only two electrons hold the carbons together in a carbon–carbon single bond. This means that a carbon–carbon double bond is stronger and shorter than a carbon–carbon single bond.

DIAMOND AND GRAPHITE: SUBSTANCES THAT CONTAIN ONLY CARBON ATOMS

Diamond is the hardest of all substances. Graphite, in contrast, is a slippery, soft solid most familiar to us as the "lead" in pencils. Both materials, in spite of their very different physical properties, contain only carbon atoms. The two substances differ solely in the nature of the bonds holding the carbon atoms together. Diamond consists of a rigid three-dimensional network of atoms, with each carbon bonded to four other carbons via sp^3 orbitals. The carbon atoms in graphite, on the other hand, are sp^2 hybridized, so each bonds to only three other carbons. This planar arrangement causes the atoms in graphite to lie in flat, layered sheets that can shear off of neighboring sheets. When you write with a pencil, sheets of carbon atoms shear off to leave a thin trail of graphite.

1.9 HOW A TRIPLE BOND IS FORMED: THE BONDS IN ETHYNE

The carbon atoms in ethyne (also called acetylene) are each bonded to only two atoms—a hydrogen and another carbon:

$$\text{H}-\text{C}\equiv\text{C}-\text{H}$$

ethyne
acetylene

Because each carbon forms covalent bonds with two atoms, only two orbitals (an *s* and a *p*) are hybridized. Two identical *sp* orbitals result. Each carbon atom in ethyne, therefore, has two hybridized *sp* orbitals and two unhybridized *p* orbitals (Figure 1.8).

To minimize electron repulsion, the two *sp* orbitals point in opposite directions (Figure 1.8).

The carbons in ethyne are held together by three bonds. Three bonds connecting two atoms is called a **triple bond**. One of the *sp* orbitals of one carbon in ethyne overlaps an *sp* orbital of the other carbon to form a carbon–carbon σ bond. The other *sp* orbital of each carbon overlaps the *s* orbital of a hydrogen to form a C—H σ bond

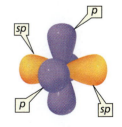

▲ **Figure 1.8**
The two *sp* orbitals are oriented 180° away from each other, perpendicular to the two unhybridized *p* orbitals. (The smaller lobes of the *sp* orbitals are not shown.)

(Figure 1.9a). Because the two *sp* orbitals point in opposite directions, the bond angles are 180°. The two unhybridized *p* orbitals are perpendicular to each other, and both are perpendicular to the *sp* orbitals. Each of the unhybridized *p* orbitals engages in side-to-side overlap with a parallel *p* orbital on the other carbon, with the result that two π bonds are formed (Figure 1.9b).

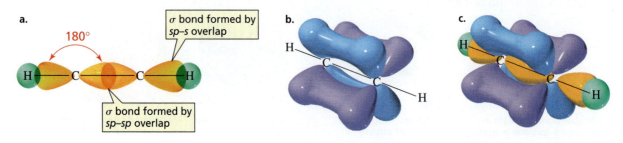

a.

180°

σ bond formed by
sp–s overlap

σ bond formed by
sp–sp overlap

b.

c.

▲ **Figure 1.9**
(a) The C—C σ bond in ethyne is formed by *sp–sp* overlap, and the C—H bonds are formed by *sp–s* overlap. The carbon atoms and the atoms bonded to them are in a straight line.
(b) The two π bonds are formed by side-to-side overlap of the *p* orbitals of one carbon with the *p* orbitals of the other carbon.
(c) The triple bond has an electron-dense region above and below and in front of and in back of the internuclear axis of the molecule.

A **triple bond** therefore consists of one σ bond and two π bonds. Because the two unhybridized *p* orbitals on each carbon are perpendicular to each other, there is a region of high electron density above and below, *and* in front of and in back of, the internuclear axis of the molecule (Figure 1.9c). The potential map for ethyne shows that negative charge accumulates in a cylinder that wraps around the egg-shaped molecule.

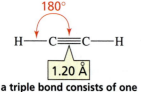

180°

H—C≡C—H

1.20 Å

a triple bond consists of one
σ bond and two π bonds

ball-and-stick model
of ethyne

space-filling model
of ethyne

electrostatic potential map
for ethyne

Because the two carbon atoms in a triple bond are held together by six electrons, a triple bond is stronger and shorter than a double bond.

3-D Molecule:
Ethyne

PROBLEM 20 **SOLVED**

For each of the following species:
a. Draw its Lewis structure.
b. Describe the orbitals used by each carbon atom in bonding and indicate the approximate bond angles.
   1. HCOH   **2.** CCl_4 **3.** HCN

Solution to 20a   Because HCOH is neutral, we know that each H forms one bond, the oxygen forms two bonds, and the carbon forms four bonds. Our first attempt at a Lewis structure (drawing the atoms in the order given by the Kekulé structure) shows that carbon is the only atom that does not form the needed number of bonds.

H—C—O—H

If we place a double bond between the carbon and the oxygen and move an H, all the atoms end up with the correct number of bonds. All the C and H atoms have filled outer shells but lone-pair electrons need to be added to give the oxygen atom a filled outer shell. When we check to see if any atom needs to be assigned a formal charge, we find that none of them does.

:O:
‖
H—C—H

Solution to 20b   Because the carbon atom forms a double bond, we know that carbon uses sp^2 orbitals (as it does in ethene) to bond to the two hydrogens and the oxygen. It uses its "leftover" p orbital to form the second bond to oxygen. Because carbon is sp^2 hybridized, the bond angles are approximately 120°.

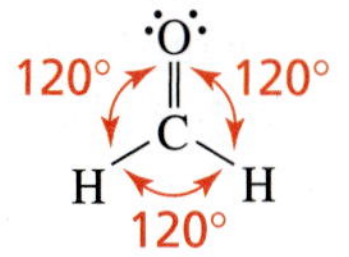

1.10 BONDING IN THE METHYL CATION, THE METHYL RADICAL, AND THE METHYL ANION

Not all carbon atoms form four bonds. A carbon with a positive charge, a negative charge, or an unpaired electron forms only three bonds. Now we will see what orbitals carbon uses when it forms three bonds.

The Methyl Cation ($^+CH_3$)

The positively charged carbon in the methyl cation is bonded to three atoms, so it hybridizes three orbitals—an s orbital and two p orbitals. Therefore, it forms its three covalent bonds using sp^2 orbitals. Its unhybridized p orbital remains empty. The positively charged carbon and the three atoms bonded to it lie in a plane. The p orbital stands perpendicular to the plane.

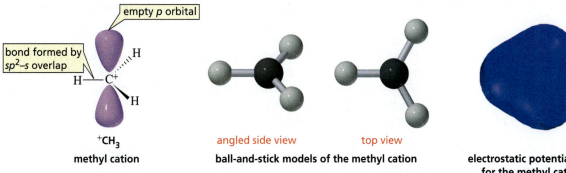

$^+CH_3$
methyl cation

angled side view   top view

ball-and-stick models of the methyl cation

electrostatic potential map for the methyl cation

The Methyl Radical ($\cdot CH_3$)

The carbon atom in the methyl radical is also sp^2 hybridized. The methyl radical differs by one unpaired electron from the methyl cation. That electron is in the p orbital. Notice the similarity in the ball-and-stick models of the methyl cation and the methyl radical. The potential maps, however, are quite different because of the additional electron in the methyl radical.

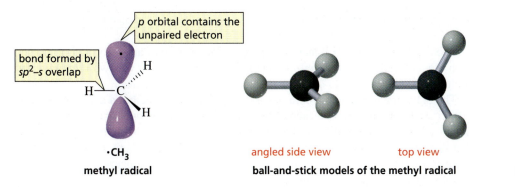

$\cdot CH_3$
methyl radical

angled side view   top view

ball-and-stick models of the methyl radical

electrostatic potential map for the methyl radical

The Methyl Anion ($\bar{\cdot}\text{CH}_3$)

The negatively charged carbon in the methyl anion has three pairs of bonding electrons and one lone pair. The four pairs of electrons are farthest apart when the four orbitals containing the bonding and lone-pair electrons point toward the corners of a tetrahedron. In other words, a negatively charged carbon is sp^3 hybridized. In the methyl anion, three of carbon's sp^3 orbitals each overlap the s orbital of a hydrogen, and the fourth sp^3 orbital holds the lone pair.

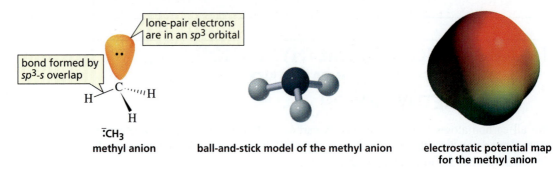

$\bar{\cdot}\text{CH}_3$
methyl anion   **ball-and-stick model of the methyl anion**   **electrostatic potential map for the methyl anion**

Take a moment to compare the potential maps for the methyl cation, the methyl radical, and the methyl anion.

1.11 THE BONDS IN WATER

water

3-D Molecule:
Water

The bond angles in a molecule indicate which orbitals are used in bond formation.

The oxygen atom in water (H_2O) forms two covalent bonds and has two lone pairs.

Because the electronic configuration of oxygen shows that it has two unpaired electrons (Table 1.2), oxygen does not need to promote an electron to form the number (two) of covalent bonds required to complete its octet. If we assume that oxygen uses p orbitals to form the two O—H bonds, as predicted by oxygen's electronic configuration, we would expect a bond angle of about 90° because the two p orbitals are at right angles to each other. However, the experimentally observed bond angle is 104.5°. In addition, we would expect the lone pairs to be chemically different because one pair would be in an s orbital and the other would be in a p orbital. The lone pairs, however, are known to be identical.

To explain the observed bond angle and the fact that the lone pairs are identical, oxygen must use hybrid orbitals to form covalent bonds—just as carbon does. The s orbital and the three p orbitals must hybridize to produce four identical sp^3 orbitals.

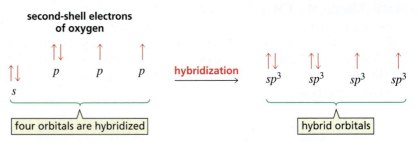

Each of the two O—H bonds is formed by the overlap of an sp^3 orbital of oxygen with the s orbital of a hydrogen. A lone pair occupies each of the two remaining sp^3 orbitals.

The bond angle in water (104.5°) is a little smaller than the bond angle in methane (109.5°) presumably because each of the lone pairs is held by only one nucleus, which makes a lone pair more diffuse than a bonding pair that is held by two nuclei and is therefore relatively confined between them. Consequently, there is more electron repulsion between lone pairs, causing the O—H bonds to squeeze closer together, thereby decreasing the bond angle.

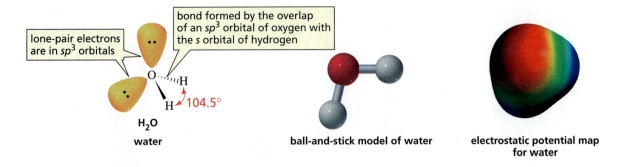

lone-pair electrons are in sp^3 orbitals

bond formed by the overlap of an sp^3 orbital of oxygen with the s orbital of hydrogen

104.5°

H_2O
water

ball-and-stick model of water

electrostatic potential map for water

Compare the potential map for water with that for methane. Water is a polar molecule; methane is nonpolar.

PROBLEM 21 ◆

The bond angles in H_3O^+ are greater than _____ and less than _____.

WATER—A UNIQUE COMPOUND

Water is the most abundant compound found in living organisms. Its unique properties have allowed life to originate and evolve. Its high heat of fusion (the heat required to convert a solid to a liquid) protects organisms from freezing at low temperatures because a lot of heat must be removed from water to freeze it. Its high heat capacity (the heat required to raise the temperature of a substance a given amount) minimizes temperature changes in organisms, and its high heat of vaporization (the heat required to convert a liquid to a gas) allows animals to cool themselves with a minimal loss of body fluid. Because liquid water is denser than ice, ice formed on the surface of water floats and insulates the water below. That is why oceans and lakes don't freeze from the bottom up. It is also why plants and aquatic animals can survive when the ocean or lake they live in freezes.

1.12 THE BONDS IN AMMONIA AND IN THE AMMONIUM ION

The nitrogen atom in ammonia (NH_3) forms three covalent bonds and has one lone pair.

Because nitrogen has three unpaired electrons in its electronic configuration (Table 1.2), it can form three covalent bonds without having to promote an electron. The experimentally observed bond angles in NH_3 are 107.3°, indicating that nitrogen also uses hybrid orbitals when it forms covalent bonds. Like carbon and oxygen, the one s and three p orbitals of the second shell of nitrogen hybridize to form four identical sp^3 orbitals:

$$H—\ddot{N}—H$$
$$|$$
$$H$$

ammonia

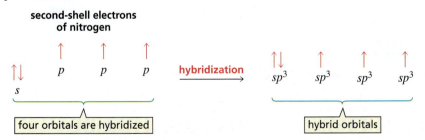

second-shell electrons of nitrogen

p p p

s

four orbitals are hybridized

hybridization

sp^3 sp^3 sp^3 sp^3

hybrid orbitals

Each of the N—H bonds in NH_3 is formed by the overlap of an sp^3 orbital of nitrogen with the s orbital of a hydrogen. The single lone pair occupies an sp^3 orbital. The bond angle (107.3°) is smaller than the tetrahedral bond angle (109.5°) because of the relatively diffuse lone pair. Notice that the bond angles in NH_3 (107.3°) are larger than

the bond angles in H_2O (104.5°) because nitrogen has only one lone pair, whereas oxygen has two lone pairs.

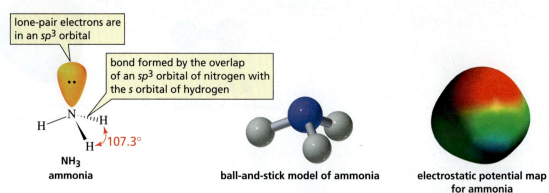

lone-pair electrons are in an sp^3 orbital

bond formed by the overlap of an sp^3 orbital of nitrogen with the s orbital of hydrogen

107.3°

NH_3
ammonia

ball-and-stick model of ammonia

electrostatic potential map for ammonia

Because the ammonium ion ($^+NH_4$) has four identical N—H bonds and no lone pairs, all the bond angles are 109.5°, just like the bond angles in methane.

109.5°

$^+NH_4$
ammonium ion

ball-and-stick model of the ammonium ion

electrostatic potential map for the ammonium ion

3-D Molecule:
Ammonia

PROBLEM 22 ♦

According to the potential map for the ammonium ion, which atom(s) has (have) the least electron density?

PROBLEM 23 ♦

Compare the potential maps for methane, ammonia, and water. Which is the most polar molecule? Which is the least polar?

electrostatic potential map for methane

electrostatic potential map for ammonia

electrostatic potential map for water

PROBLEM 24 ♦

Predict the approximate bond angles in the methyl carbanion.

1.13 THE BOND IN A HYDROGEN HALIDE

Fluorine, chlorine, bromine, and iodine are known as the halogens; HF, HCl, HBr, and HI are called hydrogen halides. Bond angles will not help us determine the orbitals that form the hydrogen halide bond, as they did with other molecules, because a

hydrogen halide has only one bond and therefore no bond angle. We do know, however, that a halogen's three lone pairs are identical and that lone pairs position themselves to minimize electron repulsion (Section 1.7). Both of these facts suggest that a halogen's three lone pairs are in sp^3 orbitals. Therefore, we will assume that a hydrogen–halogen bond is formed by the overlap of an sp^3 orbital of the halogen with the s orbital of hydrogen.

hydrogen fluoride

hydrogen chloride

hydrogen bromide

hydrogen iodide

H—F̈:
hydrogen fluoride   **ball-and-stick model of hydrogen fluoride**   **electrostatic potential map for hydrogen fluoride**

In the case of fluorine, the sp^3 orbital used in bond formation belongs to the second shell of electrons. In chlorine, the sp^3 orbital belongs to the third shell of electrons. Because the average distance from the nucleus is greater for an electron in the third shell than for an electron in the second shell, the average electron density is less in a $3sp^3$ orbital than in a $2sp^3$ orbital. This means that the electron density in the region where the s orbital of hydrogen overlaps the sp^3 orbital of the halogen decreases as the size of the halogen increases (Figure 1.10). Therefore, the hydrogen–halogen bond becomes longer and weaker as the size (atomic weight) of the halogen increases (Table 1.5).

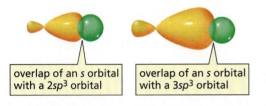

overlap of an s orbital with a $2sp^3$ orbital   overlap of an s orbital with a $3sp^3$ orbital

◀ **Figure 1.10**
There is greater electron density in the region of overlap of an s orbital with a $2sp^3$ orbital than in the region of overlap of an s orbital with a $3sp^3$ orbital.

The greater the electron density in the region of orbital overlap, the stronger is the bond.

The shorter the bond, the stronger it is.

Table 1.5 Hydrogen–Halogen Bond Lengths and Bond Strengths

Hydrogen halide	Bond length (Å)	Bond strength kcal/mol	kJ/mol
H—F	0.917	136	571
H—Cl	1.2746	103	432
H—Br	1.4145	87	366
H—I	1.6090	71	298

PROBLEM 25 ♦

a. Predict the relative lengths and strengths of the bonds in Cl_2 and Br_2.
b. Predict the relative lengths and strengths of the bonds in HF, HCl, and HBr.

PROBLEM 26 ♦

a. Which bond is longer?   **b.** Which bond is stronger?

   1. C—Cl or C—Br   **2.** C—C or C—H   **3.** H—Cl or H—H

1.14 SUMMARY: HYBRIDIZATION, BOND LENGTHS, BOND STRENGTHS, AND BOND ANGLES

All single bonds found in organic compounds are sigma bonds.

The hybridization of a C, O, or N is $sp^{(3 \text{ minus the number of } \pi \text{ bonds})}$

All single bonds are σ bonds. All double bonds are composed of one σ bond and one π bond. All triple bonds are composed of one σ bond and two π bonds. The easiest way to determine the hybridization of a carbon, oxygen, or nitrogen atom is to look at the number of π bonds it forms: if it forms no π bonds, it is sp^3 hybridized; if it forms one π bond, it is sp^2 hybridized; if it forms two π bonds, it is sp hybridized. The exceptions are carbocations and carbon radicals, which are sp^2 hybridized—not because they form a π bond, but because they have an empty or a half-filled p orbital (Section 1.10).

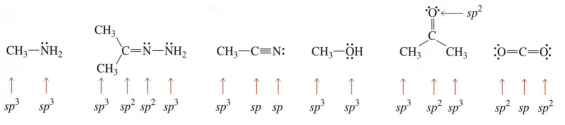

In comparing the lengths and strengths of carbon–carbon single, double, and triple bonds, we see that the more bonds holding two carbon atoms together, the shorter and stronger is the carbon–carbon bond (Table 1.6): Triple bonds are shorter and stronger than double bonds, which are shorter and stronger than single bonds.

A double bond (a σ bond plus a π bond) is stronger than a single bond (a σ bond), but it is not twice as strong. We can conclude, therefore, that a π bond is weaker than a σ bond.

A π bond is weaker than a σ bond.

You may wonder how an electron "knows" what orbital it should go into. In fact, electrons know nothing about orbitals. They simply arrange themselves around atoms in the most stable manner possible. It is chemists who use the concept of orbitals to explain this arrangement.

PROBLEM 27 ◆

Which of the bonds of a carbon–carbon double bond has more effective orbital–orbital overlap, the σ bond or the π bond?

Table 1.6 Comparison of the Bond Angles and the Lengths and Strengths of the Carbon–Carbon Bonds in Ethane, Ethene, and Ethyne

Molecule	Hybridization of carbon	Bond angles	Length of C—C bond (Å)	Strength of C—C bond (kcal/mol)	(kJ/mol)
ethane	sp^3	109.5°	1.54	90	377
ethene	sp^2	120°	1.33	174	728
H—C≡C—H ethyne	sp	180°	1.20	231	967

PROBLEM 28 ◆

a. What is the hybridization of each of the carbon atoms in the following compound?

$$CH_3CHCH=CHCH_2C\equiv CCH_3$$
$$|$$
$$CH_3$$

b. What is the hybridization of each of the carbon, oxygen, and nitrogen atoms in the following compounds?

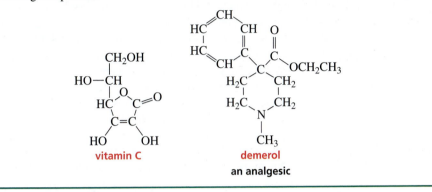

vitamin C demerol
an analgesic

PROBLEM-SOLVING STRATEGY

Predict the approximate bond angle of the C—N—H bond in $(CH_3)_2NH$.

First we need to determine the hybridization of the central atom. Because the nitrogen atom forms only single bonds, we know it is sp^3 hybridized. Next we look to see if there are lone pairs that will affect the bond angle. A neutral nitrogen has one lone pair. Based on these observations, we can predict that the C—N—H bond angle will be about $107.3°$, the same as the H—N—H bond angle in NH_3, another compound with a neutral sp^3 hybridized nitrogen.

Now continue on to Problem 29.

PROBLEM 29 ◆

Predict the approximate bond angles:
a. the C—N—C bond angle in $(CH_3)_2\overset{+}{N}H_2$
b. the C—N—C bond angle in $CH_3CH_2NH_2$
c. the H—C—N bond angle in $(CH_3)_2NH$
d. the C—O—C bond angle in CH_3OCH_3

PROBLEM 30

Describe the orbitals used in bonding and the approximate bond angles in the following compounds. (*Hint:* See Table 1.6.)

a. CH_3OH **b.** $HONH_2$ **c.** $HCOOH$ **d.** N_2

SUMMARY

Organic compounds are compounds that contain carbon. The **atomic number** of an atom is the number of protons in its nucleus. The **mass number** of an atom is the sum of its protons and neutrons. **Isotopes** have the same atomic number but different mass numbers.

An **atomic orbital** indicates where there is a high probability of finding an electron. The closer the atomic orbital is to the nucleus, the lower is its energy. Electrons are assigned to atomic orbitals according to three rules: an electron goes into the available orbital with the lowest energy; no more

than two electrons can be in an orbital; and an electron will occupy an empty orbital before pairing up with an electron in an orbital with the same energy.

The **octet rule** states that an atom will give up, accept, or share electrons in order to fill its outer shell or attain an outer shell with eight electrons. The **electronic configuration** of an atom describes the orbitals occupied by the atom's electrons. Electrons in inner shells are called **core electrons**; electrons in the outermost shell are called **valence electrons**. **Lone-pair electrons** are valence electrons that are not used in bonding. A **bond** formed as a result of the attraction of opposite charges is called an **ionic bond**; a bond formed as a result of sharing electrons is called a **covalent bond**. A **polar covalent bond** is a covalent bond between atoms with different **electronegativities**.

Lewis structures indicate which atoms are bonded together and show **lone-pair electrons** and **formal charges**. A **carbocation** has a positively charged carbon, a

carbanion has a negatively charged carbon, and a **radical** has an unpaired electron.

Bond strength is measured by the **bond dissociation energy**. A σ bond is stronger than a π bond. All **single bonds** in organic compounds are **sigma (σ) bonds**. A **double bond** consists of one σ bond and one π bond, and a **triple bond** consists of one σ bond and two π bonds. Carbon–carbon triple bonds are shorter and stronger than carbon–carbon double bonds, which are shorter and stronger than carbon–carbon single bonds. To form four bonds, carbon promotes an electron from a $2s$ to a $2p$ orbital. C, N, O, and the halogens form bonds using **hybrid orbitals**. The **hybridization** of C, N, or O depends on the number of π bonds the atom forms: no π bonds means that the atom is sp^3 **hybridized**, one π bond indicates that it is sp^2 **hybridized**, and two π bonds signifies that it is sp **hybridized**. Exceptions are carbocations and carbon radicals, which are sp^2 **hybridized**. Bonding and lone-pair electrons around an atom are positioned as far apart as possible.

PROBLEMS

31. Draw a Lewis structure for each of the following species:
   a. H_2CO_3 **b.** CO_3^{2-} **c.** H_2CO **d.** CH_3NH_2 **e.** CO_2 **f.** N_2H_4

32. How many valence electrons do the following atoms have?
   a. carbon and silicon   **b.** nitrogen and phosphorus   **c.** neon and argon   **d.** magnesium and calcium

33. For each of the following species, give the hybridization of the central atom and indicate the bond angles:
   a. NH_3 **b.** $^+NH_4$ **c.** $^-CH_3$ **d.** $C(CH_3)_4$ **e.** $\cdot CH_3$ **f.** $^+CH_3$ **g.** HCN   **h.** H_3O^+

34. Use the symbols $\delta+$ and $\delta-$ to show the direction of polarity of the indicated bond in each of the following compounds:
   a. F—Br   **b.** H_3C—Cl   **c.** H_3C—MgBr   **d.** H_2N—OH

35. Draw the condensed structure of a compound that contains only carbon and hydrogen atoms and that has
   a. three sp^3 carbons.
   b. one sp^3 carbon and two sp^2 carbons.
   c. two sp^3 carbons and two sp carbons.

36. Predict the approximate bond angles:
   a. the H—C—O bond angle in CH_3OH **b.** the C—O—H bond angle in CH_3OH
   c. the H—C—H bond angle in $H_2C{=}O$ **d.** the C—C—N bond angle in $CH_3C{\equiv}N$

37. Give each atom the appropriate formal charge:
   a. H:Ö: **b.** H:Ö·   **c.** H—N̈—H   **d.** H—C̈—H

38. Write the electronic configuration for the following species (carbon's electronic configuration is written as $1s^2\,2s^2\,2p^2$):
   a. Ca   **b.** Ca^{2+} **c.** Ar   **d.** Mg^{2+}

39. Only one of the following formulas describes a compound that exists. Fix the other formulas so they also describe compounds that exist.
   a. $CH_3CH_3CH_3$ **c.** $(CH_3)_2CCH_3$ **e.** $CH_3CH_2CH_2$

   b. CH_5 **d.** $(CH_3)_2CHCH_2CH_3$ **f.** $CH_3CHCH_2CH_3$

40. List the bonds in order of decreasing polarity (that is, list the most polar bond first).
   a. C—O, C—F, C—N   **b.** C—Cl, C—I, C—Br   **c.** H—O, H—N, H—C   **d.** C—H, C—C, C—N

41. Write the Kekulé structure for each of the following compounds:
a. CH₃CHO
b. CH₃OCH₃
c. CH₃COOH
d. (CH₃)₃COH
e. CH₃CH(OH)CH₂CN
f. (CH₃)₂CHCH(CH₃)CH₂C(CH₃)₃

42. Assign the missing formal charges.

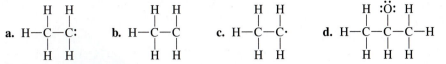

43. Draw the missing lone pairs and assign the missing formal charges.

a. H—C—O—H (with H's)
b. H—C—O—H
c. H—C—O
d. H—C—N—H

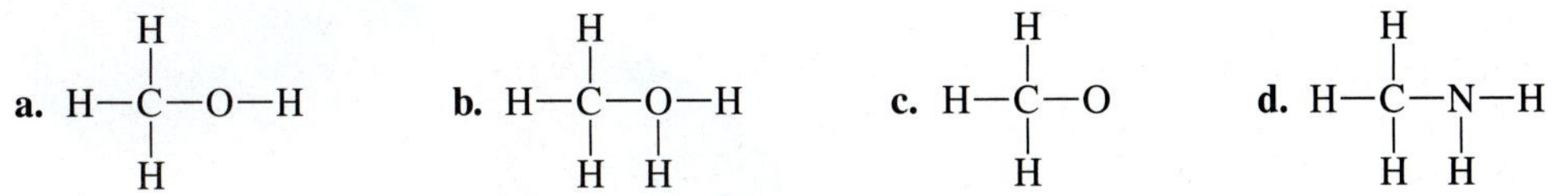

44. Account for the difference in the shape and color of the potential maps for ammonia and the ammonium ion in Section 1.12.

45. What is the hybridization of the indicated atom in each of the following compounds?

a. CH₃CH=CH₂
b. CH₃CCH₃ (with O)
c. CH₃CH₂OH
d. CH₃C≡N
e. CH₃CH=NCH₃
f. CH₃OCH₂CH₃

46. a. Which of the indicated bonds in each compound is shorter?
b. Indicate the hybridization of the C, O, and N atoms in each of the compounds.

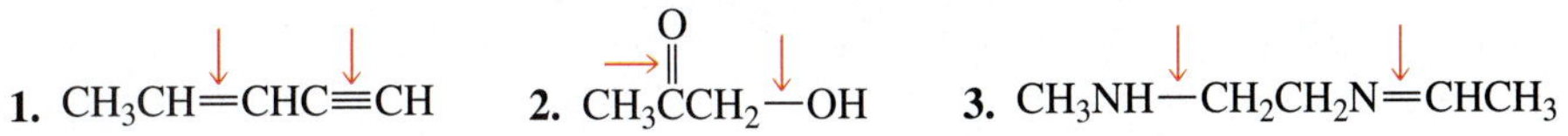

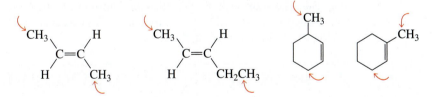

47. Which of the following have a tetrahedral geometry?

H₂O   H₃O⁺   ⁺CH₃   NH₃   ⁺NH₄   ⁻CH₃

48. Do the two *sp*² hybridized carbons and the two indicated atoms lie in the same plane?

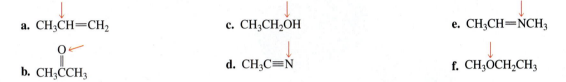

49. For each of the following compounds, indicate the hybridization of each carbon atom and give the approximate values of all the bond angles:
a. CH₃C≡CH
b. CH₃CH=CH₂
c. CH₃CH₂CH₃
d. CH₂=CH—CH=CH₂

50. Sodium methoxide (CH₃ONa) has both ionic and covalent bonds. Which bond is ionic? How many covalent bonds does it have?

51. a. Why is a H—H bond (0.74 Å) shorter than a C—C bond (1.54 Å)?
b. Predict the length of a C—H bond.

52. Explain why the following compound is not stable:

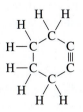

CHAPTER 2

Acids and Bases

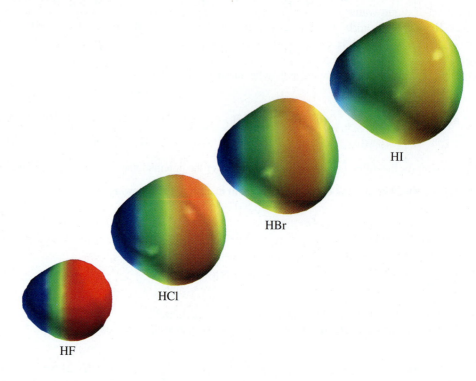

HI

HBr

HCl

HF

Early chemists called any compound that tasted sour an acid (from *acidus*, Latin for "sour"). Some familiar acids are citric acid (found in lemons and other citrus fruits), acetic acid (found in vinegar), and hydrochloric acid (found in stomach acid—the sour taste associated with vomiting). Compounds that neutralize acids were called bases, or alkaline compounds. Glass cleaners and solutions designed to unclog drains are familiar alkaline solutions.

2.1 AN INTRODUCTION TO ACIDS AND BASES

Brønsted and Lowry defined an **acid** as a species that donates a proton, and a **base** as a species that accepts a proton. (Remember that positively charged hydrogen ions are called protons.) In the reaction shown below, hydrogen chloride (HCl) is an acid because it donates a proton to water, and water is a base because it accepts a proton from HCl. Water can accept a proton because it has two lone pairs, either of which can form a covalent bond with a proton. In the reverse reaction, H_3O^+ is an acid because it donates a proton to Cl^-, and Cl^- is a base because it accepts a proton from H_3O^+. The reaction of an acid with a base is called an **acid–base reaction**. Both an acid and a base must be present in an acid–base reaction, because an acid cannot donate a proton unless a base is present to accept it.

$$\text{H}\ddot{\text{C}}\text{l:} \quad + \quad \text{H}_2\ddot{\text{O}}: \quad \rightleftharpoons \quad :\ddot{\text{C}}\text{l:}^- \quad + \quad \text{H}_3\ddot{\text{O}}^+$$

an acid   a base   a base   an acid

Notice that according to the Brønsted–Lowry definitions, *any species that has a hydrogen can potentially act as an acid, and any compound that has a lone pair can potentially act as a base.*

When a compound loses a proton, the resulting species is called its **conjugate base**. Thus, Cl^- is the conjugate base of HCl, and H_2O is the conjugate base of H_3O^+. When a compound accepts a proton, the resulting species is called its **conjugate acid**. Thus, HCl is the conjugate acid of Cl^-, and H_3O^+ is the conjugate acid of H_2O.

In a reaction between ammonia and water, ammonia (NH_3) is a base because it accepts a proton, and water is an acid because it donates a proton. Thus, HO^- is the conjugate base of H_2O, and $^+NH_4$ is the conjugate acid of NH_3. In the reverse reaction, ammonium ion ($^+NH_4$) is an acid because it donates a proton, and hydroxide ion (HO^-) is a base because it accepts a proton.

$$\overset{\cdot\cdot}{N}H_3 \quad + \quad H_2\overset{\cdot\cdot}{O}\text{:} \quad \rightleftharpoons \quad {}^+NH_4 \quad + \quad H\overset{\cdot\cdot}{\underset{\cdot\cdot}{O}}\text{:}^-$$

a base   an acid   an acid   a base

Notice that water can behave as either an acid or a base. It can behave as an acid because it has a proton that it can donate, but it can also behave as a base because it has a lone pair that can accept a proton. In Section 2.2, we will see how we can predict that water acts as a base in the first reaction and as an acid in the second reaction.

Acidity is a measure of the tendency of a compound to give up a proton. **Basicity** is a measure of a compound's affinity for a proton. A strong acid is one that has a strong tendency to give up its proton. This means that its conjugate base must be weak because it has little affinity for the proton. A weak acid has little tendency to give up its proton, indicating that its conjugate base is strong because it has a high affinity for the proton. Thus, the following important relationship exists between an acid and its conjugate base: *the stronger the acid, the weaker is its conjugate base*. For example, since HBr is a stronger acid than HCl, we know that Br^- is a weaker base than Cl^-.

PROBLEM 1 ◆

a. Draw the conjugate acid of each of the following:
   1. NH_3 **2.** Cl^- **3.** HO^- **4.** H_2O

b. Draw the conjugate base of each of the following:
   1. NH_3 **2.** HBr   **3.** HNO_3 **4.** H_2O

PROBLEM 2

a. Write an equation showing CH_3OH reacting as an acid with NH_3 and an equation showing CH_3OH reacting as a base with HCl.

b. Write an equation showing NH_3 reacting as an acid with HO^- and an equation showing NH_3 reacting as a base with HBr.

2.2 pK_a AND pH

When a strong acid such as hydrogen chloride is dissolved in water, almost all the molecules dissociate (break into ions), which means that the *products* are favored at equilibrium—the equilibrium lies to the right. When a much weaker acid, such as acetic acid, is dissolved in water, very few molecules dissociate, so the *reactants* are favored at equilibrium—the equilibrium lies to the left. Two half-headed arrows are used to designate equilibrium reactions. A longer arrow is drawn toward the species favored at equilibrium.

$$H\overset{\cdot\cdot}{\underset{\cdot\cdot}{C}}l\text{:} \quad + \quad H_2\overset{\cdot\cdot}{O}\text{:} \quad \rightleftharpoons \quad H_3\overset{\cdot\cdot}{O}^+ \quad + \quad \text{:}\overset{\cdot\cdot}{\underset{\cdot\cdot}{C}}l\text{:}^-$$

hydrogen chloride

$$H_3C \overset{\displaystyle :O:}{\underset{\displaystyle \overset{\|}{C}}{\diagdown}} \overset{\cdot\cdot}{O}H \quad + \quad H_2\overset{\cdot\cdot}{O}\text{:} \quad \rightleftharpoons \quad H_3\overset{\cdot\cdot}{O}^+ \quad + \quad H_3C \overset{\displaystyle :O:}{\underset{\displaystyle \overset{\|}{C}}{\diagdown}} \overset{\cdot\cdot}{\underset{\cdot\cdot}{O}}\text{:}^-$$

acetic acid

The degree to which an acid (HA) dissociates in an aqueous solution is indicated by the **acid dissociation constant, K_a**. Brackets are used to indicate the concentration in moles/liter, that is, the molarity (M).

$$K_a = \frac{[H_3O^+][A^-]}{[HA]}$$

The larger the acid dissociation constant, the stronger is the acid—that is, the greater is its tendency to give up a proton. Hydrogen chloride, with an acid dissociation constant of 10^7, is a stronger acid than acetic acid, with an acid dissociation constant of only 1.74×10^{-5}. For convenience, the strength of an acid is generally indicated by its **pK_a** value rather than by its K_a value, where

$$pK_a = -\log K_a$$

The pK_a of hydrogen chloride is -7, and the pK_a of acetic acid, a much weaker acid, is 4.76. Notice that the stronger the acid, the smaller is its pK_a value.

The stronger the acid, the smaller is its pK_a value.

very strong acids	$pK_a < 1$
moderately strong acids	$pK_a = 1-3$
weak acids	$pK_a = 3-5$
very weak acids	$pK_a = 5-15$
extremely weak acids	$pK_a > 15$

The concentration of positively charged hydrogen ions in the solution is indicated by **pH**. The concentration can be written as $[H^+]$ or, because a hydrogen ion in water is solvated, as $[H_3O^+]$.

$$pH = -\log[H_3O^+]$$

The lower the pH, the more acidic is the solution. Acidic solutions have pH values less than 7; basic solutions have pH values greater than 7. The pH values of some commonly encountered solutions are shown in the margin. The pH of a solution can be changed simply by adding acid or base to the solution. Do not confuse pH and pK_a: the pH scale is used to describe the acidity of a *solution*; the pK_a is characteristic of a particular *compound*, much like a melting point or a boiling point—it indicates the tendency of the compound to give up its proton.

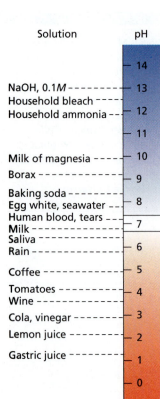

Solution — pH

14
NaOH, 0.1M -------- 13
Household bleach ----
Household ammonia -- 12

11

Milk of magnesia ----- 10
Borax ------------- 9
Baking soda -------- 8
Egg white, seawater --
Human blood, tears --- 7
Milk ---------------
Saliva -------------
Rain --------------- 6
Coffee ------------- 5
Tomatoes ---------- 4
Wine -------------
Cola, vinegar ------- 3
Lemon juice -------- 2
Gastric juice ------- 1

0

PROBLEM 3 ◆

a. Which is a stronger acid, one with a pK_a of 5.2 or one with a pK_a of 5.8?
b. Which is a stronger acid, one with an acid dissociation constant of 3.4×10^{-3} or one with an acid dissociation constant of 2.1×10^{-4}?

PROBLEM-SOLVING STRATEGY

Vitamin C has a pK_a value of 4.17. What is its K_a value?

You will need a calculator to answer this question. Remembering that $pK_a = -\log K_a$:

1. Enter the pK_a value on your calculator.
2. Multiply it by -1.
3. Determine the inverse log by pressing the key labeled 10^x.

You should find that vitamin C has a K_a value of 6.76×10^{-5}.

Now continue on to Problem 4.

PROBLEM 4 ◆

Butyric acid, the compound responsible for the unpleasant odor and taste of sour milk, has a pK_a value of 4.82. What is its K_a value? Is it a stronger or a weaker acid than vitamin C?

ACID RAIN

Rain is mildly acidic (pH = 5.5) because when the CO_2 in the air reacts with water, a weak acid—carbonic acid ($pK_a = 6.4$)—is formed.

$$CO_2 + H_2O \rightleftharpoons H_2CO_3$$
carbonic acid

In some parts of the world, rain has been found to be much more acidic—with pH values as low as 4.3. Acid rain is formed when sulfur dioxide and nitrogen oxides are produced, because when these gases react with water, strong acids—sulfuric acid ($pK_a = -5.0$) and nitric acid ($pK_a = -1.3$)—are formed. Burning fossil fuels for the generation of electric power is the factor most responsible for forming these acid-producing gases.

Acid rain has many deleterious effects. It can destroy aquatic life in lakes and streams; it can make soil so acidic that crops cannot grow; and it can cause the deterioration of paint and building materials, including monuments and statues that are part of our cultural heritage. Marble—a form of calcium carbonate—decays because acid reacts with CO_3^{2-} to form

carbonic acid, which decomposes to CO_2 and H_2O, the reverse of the reaction shown to the left.

$$CO_3^{2-} \underset{\text{H}^+}{\rightleftharpoons} HCO_3^- \underset{\text{H}^+}{\rightleftharpoons} H_2CO_3 \rightleftharpoons CO_2 + H_2O$$

photo taken in 1935   photo taken in 1994

Statue of George Washington in Washington Square Park in Greenwich Village, New York.

PROBLEM 5

Antacids are compounds that neutralize stomach acid. Write the equations that show how Milk of Magnesia, Alka-Seltzer, and Tums remove excess acid.
a. Milk of Magnesia: $Mg(OH)_2$
b. Alka-Seltzer: $KHCO_3$ and $NaHCO_3$
c. Tums: $CaCO_3$

PROBLEM 6 ◆

Are the following body fluids acidic or basic?
a. bile (pH = 8.4)   **b.** urine (pH = 5.9)   **c.** spinal fluid (pH = 7.4)

2.3 ORGANIC ACIDS AND BASES

The most common organic acids are carboxylic acids—compounds that have a COOH group. Acetic acid and formic acid are examples of carboxylic acids. Carboxylic acids have pK_a values ranging from about 3 to 5. (They are weak acids.) The pK_a values of a wide variety of organic compounds are given in Appendix II.

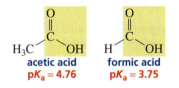

acetic acid   **formic acid**
$pK_a = 4.76$ $pK_a = 3.75$

3-D Molecule:
Acetic acid

The carboxyl group of a carboxylic acid can be represented in several different ways.

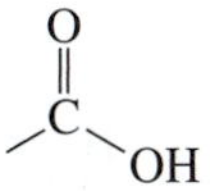

a carboxyl group   —COOH   —CO₂H

carboxyl groups are frequently shown in abbreviated forms

Alcohols—compounds that have an OH group—are much weaker acids than carboxylic acids, with pK_a values close to 16. Methyl alcohol and ethyl alcohol are examples of alcohols.

$$CH_3\underline{OH} \qquad CH_3CH_2\underline{OH}$$
methyl alcohol   **ethyl alcohol**
pK_a = 15.5   **pK_a = 15.9**

We have seen that water can behave both as an acid and as a base. An alcohol behaves similarly; it can behave as an acid and donate a proton, or as a base and accept a proton.

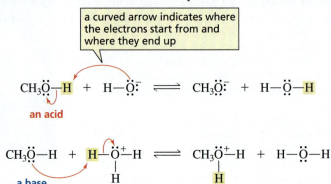

A carboxylic acid also can behave as an acid and donate a proton, or as a base and accept a proton.

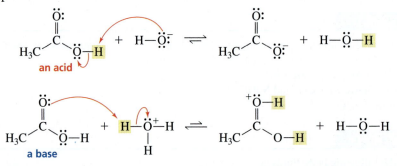

A *protonated* compound is a compound that has gained an additional proton. Protonated alcohols and protonated carboxylic acids are very strong acids. For example, protonated methyl alcohol has a pK_a of −2.5, protonated ethyl alcohol has a pK_a of −2.4, and protonated acetic acid has a pK_a of −6.1. Notice that the sp^2 oxygen of the carboxylic acid is the one that is protonated. We will see why this is so in Section 11.9.

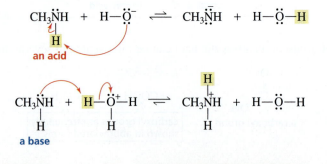

A compound with an NH_2 group is an amine. An amine can behave as an acid and donate a proton, or as a base and accept a proton.

$$CH_3\overset{..}{N}H + H\!-\!\overset{..}{\underset{..}{O}}{}^- \rightleftharpoons CH_3\overset{-}{\overset{..}{N}}H + H\!-\!\overset{..}{\underset{..}{O}}\!-\!H$$
    an acid

$$CH_3\overset{..}{N}H + H\!-\!\overset{+}{\underset{|}{O}}\!-\!H \rightleftharpoons CH_3\overset{+}{\overset{H}{N}}H + H\!-\!\overset{..}{\underset{..}{O}}\!-\!H$$
    a base

Amines, however, have such high pK_a values that they rarely behave as acids. Ammonia also has a high pK_a value.

$$CH_3NH_2 \qquad\qquad NH_3$$

methylamine   ammonia

$pK_a = 40$ $pK_a = 36$

Amines are much more likely to act as bases. In fact, amines are the most common organic bases. Instead of talking about the strength of a base in terms of its pK_b value, it is easier to talk about the strength of its conjugate acid as indicated by its pK_a value, remembering that the stronger the acid, the weaker is its conjugate base. For example, protonated methylamine is a stronger acid than protonated ethylamine, which means that methylamine is a weaker base than ethylamine. Notice that the pK_a values of protonated amines are about 10 to 11.

$$CH_3\overset{+}{N}H_3 \qquad\qquad CH_3CH_2\overset{+}{N}H_3$$

protonated methylamine   protonated ethylamine

$pK_a = 10.7$ $pK_a = 11.0$

It is important to know the approximate pK_a values of the various classes of compounds we have discussed. An easy way to remember them is in units of five, as shown in Table 2.1. (R is used when the particular carboxylic acid, alcohol, or amine is not specified.) Protonated alcohols, protonated carboxylic acids, and protonated water have pK_a values less than 0, carboxylic acids have pK_a values of about 5, protonated amines have pK_a values of about 10, and alcohols and water have pK_a values of about 15. These values are also listed inside the back cover of this book for easy reference.

Be sure to learn the approximate pK_a values given in Table 2.1.

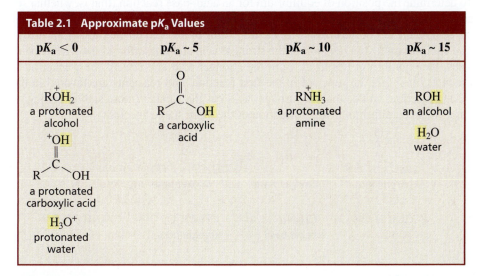

Table 2.1 Approximate pK_a Values

$pK_a < 0$	$pK_a \sim 5$	$pK_a \sim 10$	$pK_a \sim 15$
$R\overset{+}{O}H_2$ a protonated alcohol	$\underset{\text{a carboxylic acid}}{R-\overset{O}{\overset{\|}{C}}-OH}$	$R\overset{+}{N}H_3$ a protonated amine	ROH an alcohol H_2O water
$R-\overset{\overset{+}{O}H}{\overset{\|}{C}}-OH$ a protonated carboxylic acid			
H_3O^+ protonated water			

PROBLEM 7 ◆

a. Which is a stronger base, CH_3COO^- or $HCOO^-$? (The pK_a of CH_3COOH is 4.8; the pK_a of $HCOOH$ is 3.8.)

b. Which is a stronger base, HO^- or $^-NH_2$? (The pK_a of H_2O is 15.7; the pK_a of NH_3 is 36.)

c. Which is a stronger base, H_2O or CH_3OH? (The pK_a of H_3O^+ is -1.7; the pK_a of $CH_3\overset{+}{O}H_2$ is -2.5.)

PROBLEM 8 ◆

Using the pK_a values in Section 2.3, rank the following species in order of decreasing base strength (that is, list the strongest base first):

$$CH_3NH_2 \quad CH_3NH^- \quad CH_3OH \quad CH_3O^- \quad CH_3\overset{O}{\overset{\|}{C}}O^-$$

2.4 HOW TO PREDICT THE OUTCOME OF AN ACID–BASE REACTION

Now let's see how we can predict that water will act as a base in the first reaction in Section 2.1 and as an acid in the second reaction. To determine which of the two reactants of the first reaction will be the acid, we need to compare their pK_a values: the pK_a of hydrogen chloride is −7, and the pK_a of water is 15.7. Because hydrogen chloride is the stronger acid, it will donate a proton to water. Water, therefore, is a base in this reaction. When we compare the pK_a values of the two reactants of the second reaction, we see that the pK_a of ammonia is 36 and the pK_a of water is 15.7. In this case, water is the stronger acid, so it donates a proton to ammonia. Water, therefore, is an acid in this reaction.

PROBLEM 9 ♦

Using the pK_a values in Section 2.3, predict the products of the following reaction:

$$CH_3NH_2 \quad + \quad CH_3OH \quad \rightleftharpoons$$

2.5 HOW TO DETERMINE THE POSITION OF EQUILIBRIUM

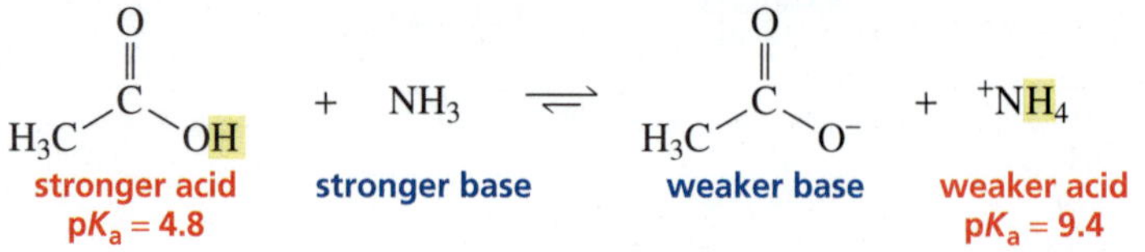

Strong reacts to form weak.

To determine the position of equilibrium for an acid–base reaction (that is, whether reactants or products are favored at equilibrium), we need to compare the pK_a value of the acid on the left of the arrow with the pK_a value of the acid on the right of the arrow. The equilibrium favors *reaction* of the stronger acid and *formation* of the weaker acid. Thus, the equilibrium lies away from the stronger acid and toward the weaker acid. Products, therefore, are favored in the first reaction, and reactants are favored in the second reaction. Notice that the stronger acid has the weaker (more stable) conjugate base, so the equilibrium favors formation of the more stable species.

$$CH_3CH_2OH \quad + \quad CH_3NH_2 \quad \rightleftharpoons \quad CH_3CH_2O^- \quad + \quad CH_3\overset{+}{N}H_3$$

weaker acid pK_a = 15.9	weaker base	stronger base	stronger acid pK_a = 10.7

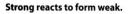

PROBLEM 10

a. For each of the acid–base reactions in Section 2.3, compare the pK_a values of the acids on either side of the equilibrium arrows and convince yourself that the position of equilibrium is in the direction indicated. (The pK_a values you need can be found in Section 2.3 or in Problem 7.)

b. Do the same thing for the equilibria in Section 2.1. (The pK_a of $^+NH_4$ is 9.4.)

2.6 HOW THE STRUCTURE OF AN ACID AFFECTS ITS pK_a

The weaker the base, the stronger is its conjugate acid.

Stable bases are weak bases.

The more stable the base, the stronger is its conjugate acid.

The strength of an acid is determined by the stability of the conjugate base that is formed when the acid gives up its proton: the more stable the base, the stronger is its conjugate acid. A stable base is a base that readily bears the electrons it formerly shared with a proton. In other words, stable bases are weak bases—they don't share

their electrons well. That is why we can say, *the weaker the base, the stronger is its conjugate acid* or, *the more stable the base, the stronger is its conjugate acid.*

Two factors that affect the stability of a base are its *size* and its *electronegativity*. The elements in the second row of the periodic table are all about the same size, but they have very different electronegativities, which increase across the row from left to right. Therefore, of the atoms shown, carbon is the least electronegative and fluorine is the most electronegative.

relative electronegativities: C < N < O < F

> most
> electronegative

If we look at the acids formed by attaching hydrogens to these elements, we see that the most acidic compound is the one that has its hydrogen attached to the most electronegative atom. Thus, HF is the strongest acid and methane is the weakest acid (Table 2.2).

relative acidities: CH_4 < NH_3 < H_2O < HF

> strongest
> acid

If we look at the stabilities of the conjugate bases of these acids, we find that they too increase from left to right because the more electronegative the atom, the better it can bear its negative charge. Thus, we see that the strongest acid has the most stable conjugate base.

relative stabilities: $^-CH_3$ < $^-NH_2$ < HO^- < F^-

> most
> stable

We therefore can conclude that *when the atoms are similar in size, the strongest acid will have its hydrogen attached to the most electronegative atom.*

The effect that the electronegativity of the atom bonded to a hydrogen has on the acidity of that hydrogen can be appreciated when the pK_a values of alcohols and amines are compared. Because oxygen is more electronegative than nitrogen, an alcohol is more acidic than an amine.

When atoms are similar in size, the strongest acid will have its hydrogen attached to the most electronegative atom.

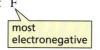

CH_3OH CH_3NH_2
methyl alcohol   **methylamine**
pK_a = 15.5   **pK_a = 40**

Similarly, a protonated alcohol is more acidic than a protonated amine.

$CH_3\overset{+}{O}H_2$ $CH_3\overset{+}{N}H_3$
protonated methyl alcohol   **protonated methylamine**
pK_a = −2.5   **pK_a = 10.7**

Table 2.2 The pK_a Values of Some Simple Acids			
CH_4	**NH_3**	**H_2O**	**HF**
pK_a = ~60	pK_a = 36	pK_a = 15.7	pK_a = 3.2
		H_2S	**HCl**
		pK_a = 7.0	pK_a = −7
			HBr
			pK_a = −9
			HI
			pK_a = −10

In comparing atoms that are very different in size, the *size* of the atom is much more important than its *electronegativity* in determining how well it bears its negative charge. For example, as we proceed down a column in the periodic table, the atoms get larger and their electronegativity decreases. However the stability of the bases increases down the column, so the strength of their conjugate acid *increases*. Thus, HI is the strongest acid of the hydrogen halides, even though iodine is the least electronegative of the halogens. Therefore, *when atoms are very different in size, the strongest acid will have its hydrogen attached to the largest atom.*

When atoms are very different in size, the strongest acid will have its hydrogen attached to the largest atom.

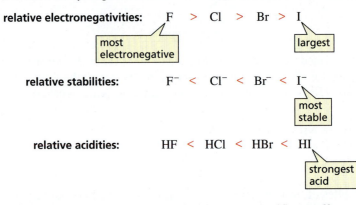

Why does the size of an atom have such a significant effect on the stability of the base that it more than overcomes the difference in electronegativity? The valence electrons of F^- are in a $2sp^3$ orbital, the valence electrons of Cl^- are in a $3sp^3$ orbital, those of Br^- are in a $4sp^3$ orbital, and those of I^- are in a $5sp^3$ orbital. The volume of space occupied by a $3sp^3$ orbital is significantly greater than the volume of space occupied by a $2sp^3$ orbital because a $3sp^3$ orbital extends out farther from the nucleus. Because its negative charge is spread over a larger volume of space, Cl^- is more stable than F^-.

Thus, as the halide ion increases in size, its stability increases because its negative charge is spread over a larger volume of space (its electron density decreases). Therefore, HI is the strongest acid of the hydrogen halides because I^- is the most stable halide ion, even though iodine is the least electronegative of the halogens (Table 2.2). The potential maps illustrate the large difference in size of the halogens:

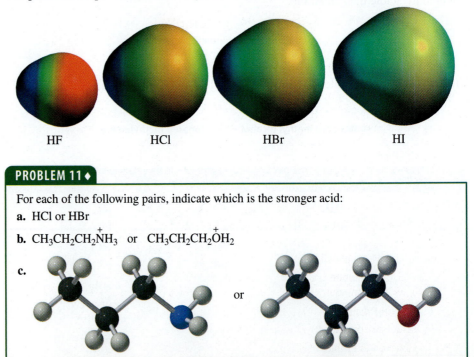

HF   HCl   HBr   HI

PROBLEM 11 ◆

For each of the following pairs, indicate which is the stronger acid:
a. HCl or HBr

b. $CH_3CH_2CH_2\overset{+}{N}H_3$ or $CH_3CH_2CH_2\overset{+}{O}H_2$

c.   or

PROBLEM 12 ◆

a. Which of the halide ions (F⁻, Cl⁻, Br⁻, I⁻) is the strongest base?
b. Which is the weakest base?

PROBLEM 13 ◆

a. Which is more electronegative, oxygen or sulfur?
b. Which is a stronger acid, H_2O or H_2S?
c. Which is a stronger acid, CH_3OH or CH_3SH?
d. Which of the following is a stronger acid?

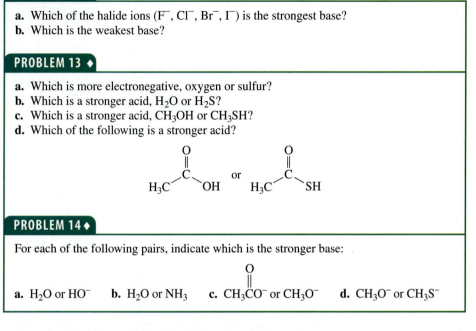

PROBLEM 14 ◆

For each of the following pairs, indicate which is the stronger base:

a. H_2O or HO^- **b.** H_2O or NH_3 **c.** CH_3CO^- or CH_3O^- **d.** CH_3O^- or CH_3S^-

2.7 HOW pH AFFECTS THE STRUCTURE OF AN ORGANIC COMPOUND

Whether an acid will lose a proton in an aqueous solution depends on both the pK_a of the acid and the pH of the solution. *A compound will exist primarily in its acidic form (with its proton) in solutions that are more acidic than the pK_a value of the group that undergoes dissociation. It will exist primarily in its basic form (without its proton) in solutions that are more basic than the pK_a value of the group that undergoes dissociation.* When the pH of a solution equals the pK_a value of the group that undergoes dissociation, the concentration of the compound in its acidic form will equal the concentration of the compound in its basic form.

> A compound will exist primarily in its acidic form if the pH of the solution is less than the compound's pK_a.
>
> A compound will exist primarily in its basic form if the pH of the solution is greater than the compound's pK_a.

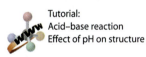

Tutorial:
Acid–base reaction
Effect of pH on structure

$$\begin{array}{ccc}
\text{acidic form} & & \text{basic form} \\
RCOOH & \rightleftharpoons & RCOO^- + H^+ \\
R\overset{+}{N}H_3 & \rightleftharpoons & RNH_2 + H^+
\end{array}$$

PROBLEM-SOLVING STRATEGY

Write the form in which the following compounds will predominate in a solution with a pH = 5.5:

a. CH_3CH_2OH ($pK_a = 15.9$)   **c.** $CH_3\overset{+}{N}H_3$ ($pK_a = 11.0$)

b. $CH_3CH_2\overset{+}{O}H_2$ ($pK_a = -2.5$)

To answer this kind of question, we need to compare the pH of the solution with the pK_a value of the compound's dissociable proton.

a. The pH of the solution is more acidic (5.5) than the pK_a value of CH_3CH_2OH (15.9). Therefore, the compound will exist primarily as CH_3CH_2OH (with its proton).

b. The pH of the solution is more basic (5.5) than the pK_a value of $CH_3CH_2\overset{+}{O}H_2$ (−2.5). Therefore, the compound will exist primarily as CH_3CH_2OH (without its proton).

c. The pH of the solution is more acidic (5.5) than the pK_a value of $CH_3\overset{+}{N}H_3$ (10.7). Therefore, the compound will exist primarily as $CH_3\overset{+}{N}H_3$ (with its proton).

Now continue on to Problem 15.

PROBLEM 15 ♦

For each of the following compounds, shown in their acidic forms, write the form that will predominate in a solution with a pH = 5.5:

a. CH_3COOH ($pK_a = 4.76$)

b. $CH_3CH_2\overset{+}{N}H_3$ ($pK_a = 11.0$)

c. H_3O^+ ($pK_a = -1.7$)

d. HBr ($pK_a = -9$)

e. $^+NH_4$ ($pK_a = 9.4$)

f. $HC{\equiv}N$ ($pK_a = 9.1$)

g. HNO_2 ($pK_a = 3.4$)

h. HNO_3 ($pK_a = -1.3$)

PROBLEM 16 ♦ *SOLVED*

a. Indicate whether a carboxylic acid (RCOOH) with a pK_a of 4.5 will have more charged molecules or more neutral molecules in a solution with the following pH value:

1. pH = 1   **3.** pH = 5   **5.** pH = 10
2. pH = 3   **4.** pH = 7   **6.** pH = 13

b. Answer the same question for a protonated amine ($R\overset{+}{N}H_3$) with a pK_a of 9.

c. Answer the same question for an alcohol (ROH) with a pK_a of 15.

Solution to 16 a1. First determine whether more molecules will be in the acidic form or the basic form; if the pH is less than the pK_a, more molecules will be in the acidic form; if the pH is greater than the pK_a, more molecules will be in the basic form. For 16a1, the pH = 1 and the pK_a = 4.5, so more molecules will be in the acidic form. Now determine whether the acidic form is charged or neutral. The acidic form of a carboxylic acid is neutral, so there will be more neutral molecules in the solution.

PROBLEM 17 ♦

A naturally occurring amino acid such as alanine has both a carboxylic acid group and an amine group. The pK_a values of the two groups are shown.

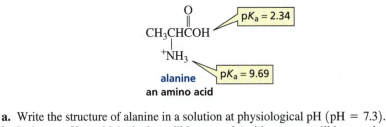

a. Write the structure of alanine in a solution at physiological pH (pH = 7.3).

b. Is there a pH at which alanine will be neutral (neither group will have a charge)?

2.8 BUFFER SOLUTIONS

A solution containing a weak acid (HA) and its conjugate base (A⁻) is called a **buffer solution**. A buffer solution will maintain nearly constant pH when small amounts of acid or base are added to it, because the weak acid can donate a proton to any HO⁻ added to the solution, and its conjugate base can accept any H⁺ that is added to the solution.

can donate an H⁺ to HO⁻

$$HA + HO^- \longrightarrow A^- + H_2O$$

$$A^- + H_3O^+ \longrightarrow HA + H_2O$$

can accept an H⁺ from H_3O^+

BLOOD IS A BUFFERED SOLUTION

Blood is the fluid that transports oxygen to all the cells of the human body. The normal pH of human blood is 7.3 to 7.4. Death will result if this pH decreases to a value less than ~ 6.8 or increases to a value greater than ~ 8.0 for even a few seconds.

Oxygen is carried to cells by a protein in the blood called hemoglobin (HbH$^+$). When hemoglobin binds O_2, hemoglobin loses a proton, which would make the blood more acidic if it did not contain a buffer to maintain its pH.

$$HbH^+ + O_2 \rightleftharpoons HbO_2 + H^+$$

A carbonic acid/bicarbonate (H_2CO_3/HCO_3^-) buffer controls the pH of blood. An important feature of this buffer is that carbonic acid decomposes to CO_2 and H_2O:

$$CO_2 + H_2O \rightleftharpoons \underset{\text{carbonic acid}}{H_2CO_3} \rightleftharpoons \underset{\text{bicarbonate}}{HCO_3^-} + H^+$$

During exercise our metabolism speeds up, producing large amounts of CO_2. The increased concentration of CO_2 shifts the equilibrium between carbonic acid and bicarbonate to the right, which increases the concentration of H$^+$. Significant amounts of lactic acid are also produced during exercise, and this further increases the concentration of H$^+$. Receptors in the brain respond to the increased concentration of H$^+$ by triggering a reflex that increases the rate of breathing. Hemoglobin then releases more oxygen to the cells, and more CO_2 is eliminated by exhalation. Both processes decrease the concentration of H$^+$ in the blood by shifting both equilibria to the left.

Thus, any disorder that decreases the rate and depth of ventilation, such as emphysema, will decrease the pH of the blood—a condition called acidosis. In contrast, any excessive increase in the rate and depth of ventilation, such as hyperventilation due to anxiety, will increase the pH of blood—a condition called alkalosis.

PROBLEM 18

Write the equation that shows how a buffer made by dissolving CH$_3$COOH and CH$_3$COO$^-$Na$^+$ in water prevents the pH of a solution from changing when

a. a small amount of H$^+$ is added to the solution.
b. a small amount of HO$^-$ is added to the solution.

2.9 LEWIS ACIDS AND BASES

In 1923, G. N. Lewis (page 6) offered new definitions for the terms *acid* and *base*. He defined an acid as a species that *accepts a share in an electron pair* and a base as a species that *donates a share in an electron pair*. All proton-donating acids fit the Lewis definition because all proton-donating acids lose a proton and the proton accepts a share in an electron pair.

Lewis acid: need two from you.

Lewis base: have pair, will share.

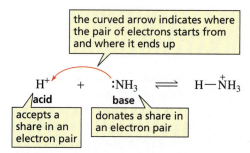

Lewis acids, however, are not limited to compounds that donate protons. According to the Lewis definition, compounds such as aluminum trichloride (AlCl$_3$) and borane (BH$_3$) are acids because they have unfilled valence orbitals and thus can accept a share in an electron pair. These compounds react with a compound that has a lone pair, just as a proton reacts with ammonia. Thus, the Lewis definition of an acid includes all proton-donating compounds and some additional compounds that do not have protons. Throughout this text, the term *acid* is used to mean a proton-donating acid, and the term **Lewis acid** is used to refer to a non-proton-donating acid such as AlCl$_3$ or BH$_3$.

All bases are **Lewis bases** because they all have a pair of electrons that they can share, either with an atom such as aluminum or boron or with a proton.

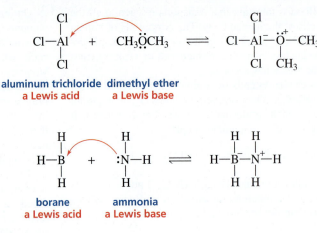

aluminum trichloride   dimethyl ether
a Lewis acid   a Lewis base

borane   ammonia
a Lewis acid   a Lewis base

PROBLEM 19

Write the products of the following reactions using arrows to show where the pair of electrons starts and where it ends up:

a. $ZnCl_2$ + $CH_3\ddot{O}H$ $\rightleftharpoons$

b. $FeBr_3$ + $:\ddot{B}r:^-$ $\rightleftharpoons$

c. $AlCl_3$ + $:\ddot{C}l:^-$ $\rightleftharpoons$

PROBLEM 20

Write the products formed from the reaction of each of the following species with HO^-:

a. CH_3OH
b. $^+NH_4$
c. $CH_3\overset{+}{N}H_3$
d. BF_3
e. $^+CH_3$
f. $FeBr_3$
g. $AlCl_3$
h. CH_3COOH

SUMMARY

An **acid** is a species that donates a proton; a **base** is a species that accepts a proton. A **Lewis acid** is a species that accepts a share in an electron pair; a **Lewis base** is a species that donates a share in an electron pair.

Acidity is a measure of the tendency of a compound to give up a proton. **Basicity** is a measure of a compound's affinity for a proton. The stronger the acid, the weaker is its conjugate base. The strength of an acid is given by the **acid dissociation constant** (K_a). Approximate pK_a values are as follows: protonated alcohols, protonated carboxylic acids, protonated water < 0; carboxylic acids ~ 5; protonated amines ~ 10; alcohols and water ~ 15. The **pH** of a solution indicates the concentration of positively charged hydrogen ions in the solution. In **acid–base reactions**, the equilibrium favors reaction of the stronger acid and formation of the weaker acid.

The strength of an acid is determined by the stability of its conjugate base: the more stable the base, the stronger is its conjugate acid. When atoms are similar in size, the strongest acid will have its hydrogen attached to the most electronegative atom. When atoms are very different in size, the strongest acid will have its hydrogen attached to the largest atom.

A compound exists primarily in its acidic form in solutions more acidic than its pK_a value and primarily in its basic form in solutions more basic than its pK_a value. A **buffer solution** contains both a weak acid and its conjugate base.

PROBLEMS

21. For each of the following compounds, draw the form in which it will predominate at pH = 3, pH = 6, pH = 10, and pH = 14:

a. CH_3COOH
  pK_a = 4.8

b. $CH_3CH_2\overset{+}{N}H_3$
  pK_a = 11.0

c. CF_3CH_2OH
  pK_a = 12.4

22. Write the products of the following acid–base reactions, and indicate whether reactants or products are favored at equilibrium (use the pK_a values that are given in Section 2.3):

a. $CH_3\overset{O}{\overset{\|}{C}}OH + CH_3O^- \rightleftharpoons$

b. $CH_3CH_2OH + {}^-NH_2 \rightleftharpoons$

c. $CH_3\overset{O}{\overset{\|}{C}}OH + CH_3NH_2 \rightleftharpoons$

d. $CH_3CH_2OH + HCl \rightleftharpoons$

23. **a.** Which of the following is the strongest acid?
  b. Which is the weakest acid?
  c. Which acid has the strongest conjugate base?

  1. nitrous acid (HNO_2), K_a = 4.0 × 10^{-4}
  2. nitric acid (HNO_3), K_a = 22
  3. bicarbonate (HCO_3^-), K_a = 6.3 × 10^{-11}

  4. hydrogen cyanide (HCN), K_a = 7.9 × 10^{-10}
  5. formic acid (HCOOH), K_a = 2.0 × 10^{-4}

24. Which is the stronger base?

  a. HS^- or HO^-
  b. CH_3O^- or $CH_3\overset{-}{N}H$
  c. CH_3OH or CH_3O^-
  d. Cl^- or Br^-

25. Locate the three nitrogen atoms in the electrostatic potential map of histamine, the compound that causes the symptoms associated with the common cold and allergic responses. Which of the two nitrogen atoms in the ring is the most basic?

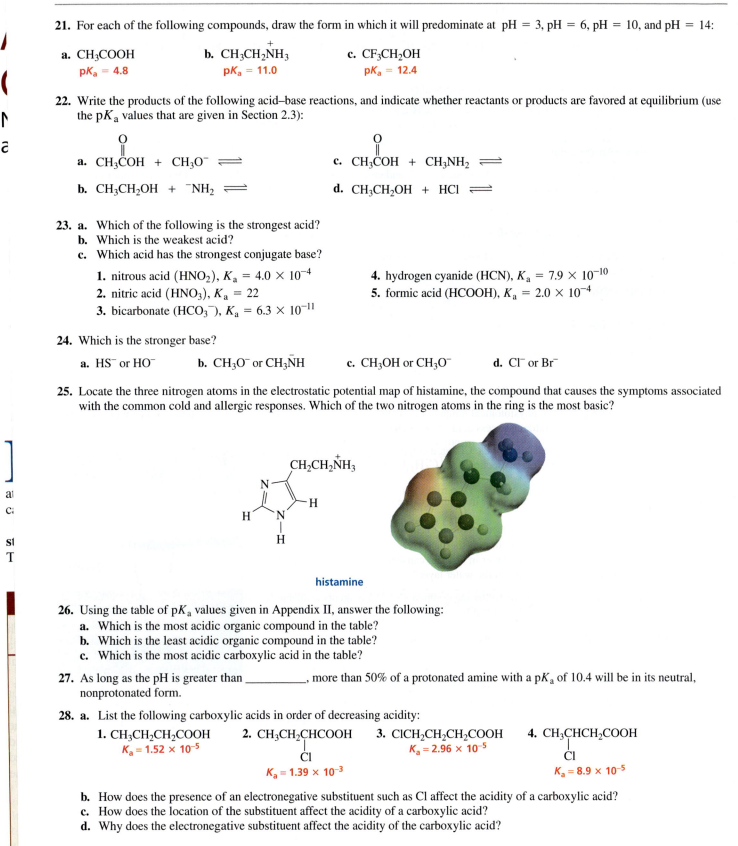

histamine

26. Using the table of pK_a values given in Appendix II, answer the following:
  a. Which is the most acidic organic compound in the table?
  b. Which is the least acidic organic compound in the table?
  c. Which is the most acidic carboxylic acid in the table?

27. As long as the pH is greater than _____, more than 50% of a protonated amine with a pK_a of 10.4 will be in its neutral, nonprotonated form.

28. **a.** List the following carboxylic acids in order of decreasing acidity:

  1. $CH_3CH_2CH_2COOH$
    K_a = 1.52 × 10^{-5}

  2. $CH_3CH_2\underset{\underset{Cl}{|}}{C}HCOOH$
    K_a = 1.39 × 10^{-3}

  3. $ClCH_2CH_2CH_2COOH$
    K_a = 2.96 × 10^{-5}

  4. $CH_3\underset{\underset{Cl}{|}}{C}HCH_2COOH$
    K_a = 8.9 × 10^{-5}

  b. How does the presence of an electronegative substituent such as Cl affect the acidity of a carboxylic acid?
  c. How does the location of the substituent affect the acidity of a carboxylic acid?
  d. Why does the electronegative substituent affect the acidity of the carboxylic acid?

to see if we can predict any difference in their stabilities. The conformer shown on the left has one methyl group in an equatorial position and one methyl group in an axial position. The conformer shown on the right also has one methyl group in an equatorial position and one methyl group in an axial position. Therefore, both chair conformers are equally stable.

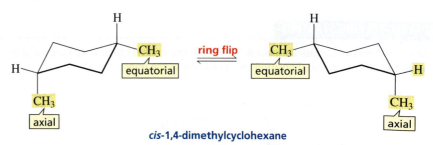

***cis*-1,4-dimethylcyclohexane**

In contrast, the two chair conformers of *trans*-1,4-dimethylcyclohexane have different stabilities because one has both methyl substituents in equatorial positions and the other has both methyl groups in axial positions. The conformer with both methyl groups in equatorial positions is more stable.

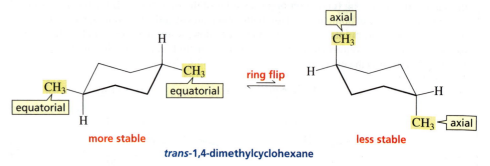

***trans*-1,4-dimethylcyclohexane**

Now let's look at the geometric isomers of 1-*tert*-butyl-3-methylcyclohexane. Both substituents of the cis isomer are in equatorial positions in one conformer and in axial positions in the other conformer. The conformer with both substituents in equatorial positions is more stable.

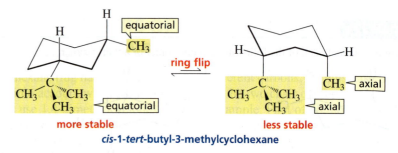

***cis*-1-*tert*-butyl-3-methylcyclohexane**

Both conformers of the trans isomer have one substituent in an equatorial position and the other in an axial position. Because the *tert*-butyl group is larger than the methyl group, the conformer with the *tert*-butyl group in the equatorial position, where there is more room for a substituent, is more stable.

3-D Molecules:
trans-1-*tert*-butyl-3-methyl-
cyclohexane

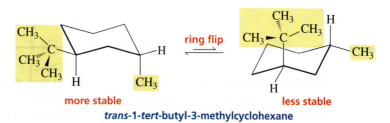

***trans*-1-*tert*-butyl-3-methylcyclohexane**

PROBLEM 33 ***SOLVED***

a. Draw the more stable chair conformer of *cis*-1-ethyl-2-methylcyclohexane.
b. Draw the more stable conformer of *trans*-1-ethyl-2-methylcyclohexane.
c. Which is more stable, *cis*-1-ethyl-2-methylcyclohexane or *trans*-1-ethyl-2-methylcyclohexane?

Solution to 33a If the two substituents of a 1,2-disubstituted cyclohexane are to be on the same side of the ring, one must be in an equatorial position and the other must be in an axial position. The more stable chair conformer is the one in which the larger of the two substituents (the ethyl group) is in the equatorial position.

3.13 FUSED CYCLOHEXANE RINGS

When two cyclohexane rings are fused together, the second ring can be considered to be a pair of substituents bonded to the first ring. As with any disubstituted cyclohexane, the two substituents can be either cis or trans. The trans isomer (one substituent bond points upward and the other downward) has both substituents in the equatorial position. The cis isomer has one substituent in the equatorial position and one in the axial position. **Trans-fused** rings, therefore, are more stable than **cis-fused** rings.

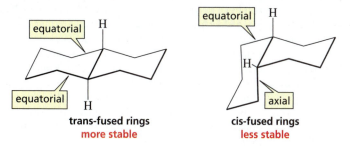

trans-fused rings more stable **cis-fused rings** less stable

Tutorials: Steroids

Hormones are chemical messengers—organic compounds synthesized in glands and delivered by the bloodstream to target tissues in order to stimulate or inhibit some process. Many hormones are **steroids**. The four rings in steroids are designated A, B, C, and D. The B, C, and D rings are all trans fused, and in most naturally occurring steroids, the A and B rings are also trans fused.

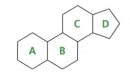

the steroid ring system

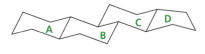

all the rings are trans fused

The most abundant member of the steroid family in animals is **cholesterol**, the precursor of all other steroids. Cholesterol is an important component of cell membranes. Its ring structure makes it more rigid than other membrane components (Section 20.5).

cholesterol

BIOGRAPHY

***Michael S. Brown** and **Joseph Leonard Goldstein** shared the 1985 Nobel Prize in physiology or medicine for their work on the regulation of cholesterol metabolism and the treatment of disease caused by elevated cholesterol levels in the blood. Brown was born in New York in 1941, Goldstein in South Carolina in 1940. They are both professors of medicine at the University of Texas Southwestern Medical Center.*

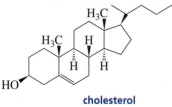

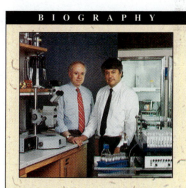

CHOLESTEROL AND HEART DISEASE

Cholesterol is probably the best-known steroid because of the widely publicized correlation between cholesterol levels in the blood and heart disease. Cholesterol is synthesized in the liver and is also found in almost all body tissues. Cholesterol is found in many foods, but we do not require it in our diet because the body can synthesize all we need. A diet high in cholesterol can lead to high levels of cholesterol in the bloodstream, and the excess can accumulate on the walls of arteries, restricting the flow of blood. This disease of the circulatory system is known as *atherosclerosis* and is a primary cause of heart disease. Cholesterol travels through the bloodstream packaged in particles that are classified according to their density. Low-density lipoprotein (LDL)

particles transport cholesterol from the liver to other tissues. Receptors on the surfaces of cells bind LDL particles, allowing them to be brought into the cell so that it can use the cholesterol. High-density lipoprotein (HDL) is a cholesterol scavenger, removing cholesterol from the surfaces of membranes and delivering it back to the liver, where it is converted into bile acids. LDL is the so-called bad cholesterol, whereas HDL is the "good" cholesterol. The more cholesterol we eat, the less the body synthesizes. But this does not mean that dietary cholesterol has no effect on the total amount of cholesterol in the bloodstream, because dietary cholesterol also inhibits the synthesis of the LDL receptors. So the more cholesterol we eat, the less the body synthesizes, but also, the less the body can get rid of by bringing it into target cells.

CLINICAL TREATMENT OF HIGH CHOLESTEROL

Statins are the newest class of cholesterol-reducing drugs. Statins reduce serum cholesterol levels by inhibiting the enzyme that catalyzes the formation of a compound needed for the synthesis of cholesterol. As a consequence of diminished cholesterol synthesis in the liver, the liver forms more LDL receptors—the receptors that help clear LDL from the bloodstream. Studies show that for every 10% that cholesterol is reduced, deaths from coronary heart disease are reduced by 15% and total death risk is reduced by 11%.

Lovastatin and simvastatin are natural statins used clinically under the trade names Mevacor and Zocor. Atorvastatin (Lipitor), a synthetic statin, is now the most popular statin. It has greater potency and a longer half-life than natural statins have, because its metabolites are as active as the parent drug in reducing cholesterol levels. Therefore, smaller doses of the drug may be administered. In addition, Lipitor is less polar than lovastatin and simvastatin, so it has a greater tendency to remain in the endoplasmic reticulum of the liver cells, where it is needed. Lipitor was the second most widely prescribed drug in the United States in 2004 and the first in 2006 (Section 21.0).

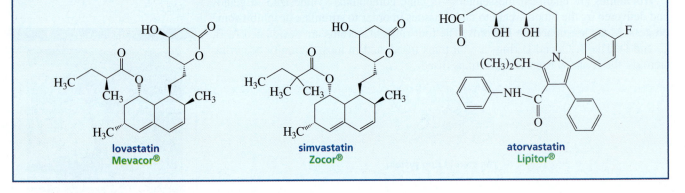

lovastatin
Mevacor®

simvastatin
Zocor®

atorvastatin
Lipitor®

SUMMARY

Alkanes are **hydrocarbons** that contain only single bonds. Their general molecular formula is C_nH_{2n+2}. **Constitutional isomers** have the same molecular formula, but their atoms are linked differently. Alkanes are named by determining the number of carbons in their **parent hydrocarbon**—the longest continuous chain. **Substituents** are listed in alphabetical order, with a number to designate their position on the chain.

Systematic names can contain numbers; **common names** never do. A compound can have more than one name, but a name must specify only one compound.

Whether alkyl halides or alcohols are **primary**, **secondary**, or **tertiary** depends on whether the X (halogen) or OH group is bonded to a primary, secondary, or tertiary carbon. A **primary carbon** is bonded to one carbon, a **secondary carbon** is bonded to two carbons, and a **tertiary carbon** is bonded to three carbons. Whether amines are **primary**, **secondary**, or **tertiary** depends on the number of alkyl groups bonded to the nitrogen.

The oxygen of an alcohol or an ether has the same geometry it has in water; the nitrogen of an amine has the same geometry it has in ammonia. The greater the attractive forces

between molecules—**van der Waals forces, dipole–dipole interactions, hydrogen bonds**—the higher is the **boiling point** of the compound. A **hydrogen bond** is an interaction between a hydrogen bonded to an O, N, or F and a lone pair of an O, N, or F in another molecule. The boiling point of straight-chain compounds increases with increasing molecular weight. Branching lowers the boiling point.

Polar compounds dissolve in **polar solvents**, and **nonpolar compounds** dissolve in **nonpolar solvents**. The interaction between a solvent and a molecule or an ion dissolved in that solvent is called **solvation**. The oxygen of an alcohol or an ether can drag three or four carbons into solution in water.

Rotation about a C—C bond results in two extreme **conformations**, staggered and eclipsed, that rapidly interconvert. A **staggered conformation** is more stable than an **eclipsed conformation**.

Five- and six-membered rings are more stable than three- and four-membered rings because of the **angle strain** that results when bond angles deviate from the ideal bond angle of 109.5°. In a process called **ring flip**, cyclohexane rapidly interconverts between two stable chair conformers. **Bonds** that are **axial** in one chair conformer are **equatorial** in the other and vice versa. The chair conformer with a substituent in the equatorial position is more stable, because there is more room, and hence less **steric strain**, in an equatorial position. In the case of disubstituted cyclohexanes, the more stable conformer will have its larger substituent in the equatorial position. A **cis isomer** has its two substituents on the same side of the ring; a **trans isomer** has its substituents on opposite sides of the ring. Cis and trans isomers are called **cis–trans isomers** or **geometric isomers**. Cis and trans isomers are different compounds; each isomer has two chair conformers.

PROBLEMS

34. Write a structural formula for each of the following:
  a. *sec*-butyl *tert*-butyl ether
  b. isoheptyl alcohol
  c. *sec*-butylamine
  d. 4-*tert*-butylheptane
  e. 1,1-dimethylcyclohexane
  f. 4,5-diisopropylnonane
  g. triethylamine
  h. cyclopentylcyclohexane
  i. 3,4-dimethyloctane
  j. 5,5-dibromo-2-methyloctane

35. Give the systematic name for each of the following:

  a. CH₃CHCH₂CH₂CHCH₂CH₂CH₃
            | |
            Br   CH₃

  (displayed as)
  $$\text{a. } CH_3\underset{|}{CH}CH_2CH_2\underset{|}{CH}CH_2CH_2CH_3 \text{ with Br on C2 and } CH_3 \text{ on C5}$$

  b. $(CH_3)_3CCH_2CH_2CH_2CH(CH_3)_2$

  c. CH₃CHCH₂CHCHCH₃ with CH₃ on middle carbon and CH₃, CH₃ below

  $$\text{c. } CH_3\underset{|}{CH}CH_2\underset{|}{CH}\underset{|}{CH}CH_3$$

  d. $(CH_3CH_2)_4C$

  e. cyclohexane bearing CH_3CHCH_3

  f. cyclohexane bearing $\underset{|}{\overset{CH_3}{N}}CH_3$

36. a. How many primary carbons does the following compound have?

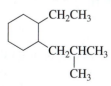

  b. How many secondary carbons does it have?
  c. How many tertiary carbons does it have?

37. Draw the structure and give the systematic name of a compound with a molecular formula C_7H_{14} that has
  a. only one tertiary carbon
  b. only two secondary carbons

38. Which of the following conformations of isobutyl chloride is the most stable?

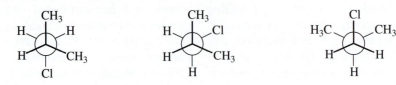

39. Name the following amines and tell whether they are primary, secondary, or tertiary:

a. $CH_3CH_2CH_2NCH_2CH_3$
　　　　　　　|
　　　　　　CH_2CH_3

b. $CH_3CHCH_2NHCHCH_2CH_3$
　　　|　　　　　|
　　CH_3　　　CH_3

c. $CH_3CH_2CH_2NHCH_2CH_2CHCH_3$
　　　　　　　　　　　　　|
　　　　　　　　　　　　CH_3

d. NH_2

40. Draw the structural formula of an alkane that has
　a. six carbons, all secondary
　b. eight carbons and only primary hydrogens
　c. seven carbons with two isopropyl groups

41. Name each of the following:
　a. $CH_3CH_2CH_2OCH_2CH_3$
　b. $CH_3CHCH_2CH_2CH_2OH$
　　　　　|
　　　　CH_3

　c. $CH_3CH_2CHCH_3$
　　　　　　|
　　　　　NH_2

　d. $CH_3CH_2CHCH_3$
　　　　　　|
　　　　　Cl

　e. $CH_3CHCH_2CH_2CH_3$
　　　　　|
　　　　CH_3

　f. CH_3CBr
　　　　|
　　　CH_2CH_3
　　　（CH_3 above）

　g. 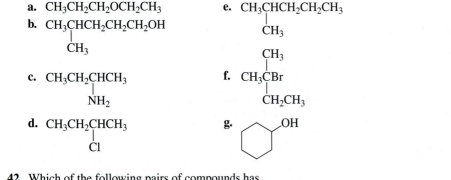 OH

　h. (cyclopentane) Br

　i. CH_3CHNH_2
　　　　　|
　　　　CH_3

　j. $CH_3CH_2CH(CH_3)NHCH_2CH_3$

42. Which of the following pairs of compounds has
　a. the higher boiling point: 1-bromopentane or 1-bromohexane?
　b. the higher boiling point: pentyl chloride or isopentyl chloride?
　c. the greater solubility in water: butyl alcohol or pentyl alcohol?
　d. the higher boiling point: hexyl alcohol or methyl pentyl ether?
　e. the higher melting point: hexane or isohexane?
　f. the higher boiling point: 1-chloropentane or pentyl alcohol?

43. Draw 1,2,3,4,5,6-hexamethylcyclohexane with
　a. all the methyl groups in axial positions.
　b. all the methyl groups in equatorial positions.

44. Ansaid and Motrin belong to the group of drugs known as nonsteroidal anti-inflammatory drugs (NSAIDs). Both are only slightly soluble in water, but one is a little more soluble than the other. Which of the drugs has the greater solubility in water?

45. Al Kane was given the structural formulas of several compounds and was asked to give them systematic names. How many did Al name correctly? Correct those that are misnamed.
　a. 3-isopropyloctane
　b. 2,2-dimethyl-4-ethylheptane
　c. isopentyl bromide
　d. 3,3-dichlorooctane
　e. 5-ethyl-2-methylhexane
　f. 2-methyl-2-isopropylheptane

46. Which of the following is the least stable?

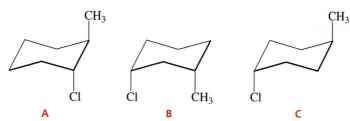

47. Give systematic names for all the alkanes with molecular formula C_7H_{16} that do not have any secondary hydrogens.

48. Draw skeletal structures for the following:
   a. 5-ethyl-2-methyloctane
   b. 1,3-dimethylcyclohexane
   c. propylcyclopentane
   d. 2,3,3,4-tetramethylheptane

49. Which of the following pairs of compounds has
   a. the higher boiling point: 1-bromopentane or 1-chloropentane?
   b. the higher boiling point: diethyl ether or butyl alcohol?
   c. the greater density: heptane or octane?
   d. the higher boiling point: isopentyl alcohol or isopentylamine?
   e. the higher boiling point: hexylamine or dipropylamine?

50. Why are alcohols of lower molecular weight more soluble in water than those of higher molecular weight?

51. For rotation about the C-3—C-4 bond of 2-methylhexane:
   a. Draw the Newman projection of the most stable conformation.
   b. Draw the Newman projection of the least stable conformation.
   c. About which other carbon–carbon bonds may rotation occur?
   d. How many of the carbon–carbon bonds in the compound have staggered conformations that are all equally stable?

52. Which of the following structures represents a cis isomer?

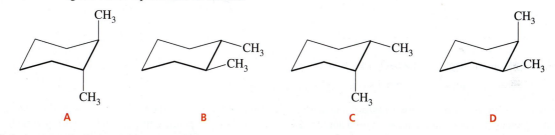

53. Draw all the isomers that have molecular formula $C_5H_{11}Br$. (*Hint:* There are eight of them.)
   a. Give the systematic name for each of the isomers.
   b. How many of the isomers are primary alkyl halides?
   c. How many of the isomers are secondary alkyl halides?
   d. How many of the isomers are tertiary alkyl halides?

54. Give the systematic name for each of the following:

a. 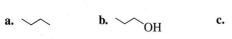 **b.** OH **c.** 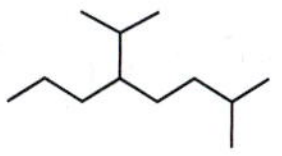 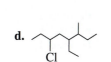 **d.**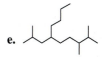

55. Draw the two chair conformers for each of the following, and indicate which conformer is more stable:
   a. *cis*-1-ethyl-3-methylcyclohexane
   b. *trans*-1-ethyl-2-isopropylcyclohexane
   c. *trans*-1-ethyl-2-methylcyclohexane
   d. *trans*-1-ethyl-3-methylcyclohexane
   e. *cis*-1-ethyl-3-isopropylcyclohexane
   f. *cis*-1-ethyl-4-isopropylcyclohexane

56. Explain why
  a. H_2O has a higher boiling point than CH_3OH (65 °C).
  b. H_2O has a higher boiling point than NH_3 (−33 °C).
  c. H_2O has a higher boiling point than HF (20 °C).

57. How many ethers have molecular formula $C_5H_{12}O$? Draw their structures and name them.

58. Draw the most stable conformer of the following molecule:

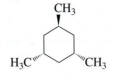

59. Give the systematic name for each of the following:

  a. 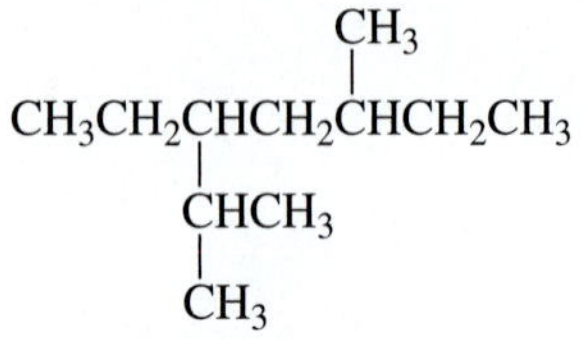 $CH_3CH_2CHCH_2CHCH_2CH_3$
  with CH_3 above the third carbon and $CHCH_3$, CH_3 below the fifth carbon

  b. 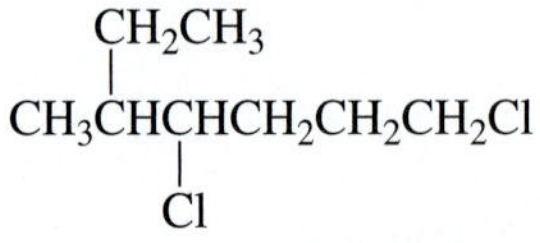 $CH_3CHCHCH_2CH_2CH_2Cl$
  with CH_2CH_3 above the second carbon and Cl below the third carbon

60. Which of the following can be used to verify that carbon is tetrahedral?
  a. Methyl bromide does not have constitutional isomers.
  b. Tetrachloromethane does not have a dipole.
  c. Dibromomethane does not have constitutional isomers.

61. The most stable form of glucose (blood sugar) is a six-membered ring chair conformer with its five substituents all in equatorial positions. Draw the most stable form of glucose by putting the OH groups on the appropriate bonds in the chair conformer.

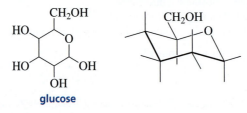

glucose

62. Draw the nine isomeric heptanes and name each isomer.

63. Draw the most stable conformer of 1,2,3,4,5,6-hexachlorocyclohexane.

64. a. Draw all the staggered and eclipsed conformations that result from rotation about the C-2—C-3 bond of pentane.
  b. Draw a potential-energy diagram for rotation about the C-2—C-3 bond of pentane through 360°, starting with the least stable conformation.

65. Using Newman projections, draw the most stable conformation for the following:
  a. 3-methylpentane, considering rotation about the C-2—C-3 bond
  b. 3-methylhexane, considering rotation about the C-3—C-4 bond

66. For each of the following disubstituted cyclohexanes, indicate whether the substituents in its two chair conformers would be both equatorial in one chair conformer and both axial in the other *or* one equatorial and one axial in each of the chair conformers:

  a. *cis*-1,2-   **d.** *trans*-1,3-
  b. *trans*-1,2-   **e.** *cis*-1,4-
  c. *cis*-1,3-   **f.** *trans*-1,4-

67. Which will have a higher percentage of the diequatorial-substituted conformer, compared with the diaxial-substituted conformer: *trans*-1,4-dimethylcyclohexane or *cis*-1-*tert*-butyl-3-methylcyclohexane?

CHAPTER 4

Alkenes

Structure, Nomenclature, Stability, and an Introduction to Reactivity

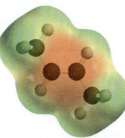

E isomer of 2-butene

Z isomer of 2-butene

In Chapter 3, we saw that alkanes are hydrocarbons that contain only carbon–carbon *single* bonds. Hydrocarbons that contain a carbon–carbon *double* bond are called **alkenes**. Alkenes play many important roles in biology. Ethene, for example, is a plant hormone—a compound that controls growth and other changes in the plant's tissues. Among other things, ethene affects seed germination, flower maturation, and fruit ripening. Many of the flavors and fragrances produced by certain plants also belong to the alkene family.

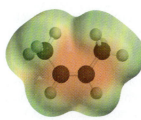

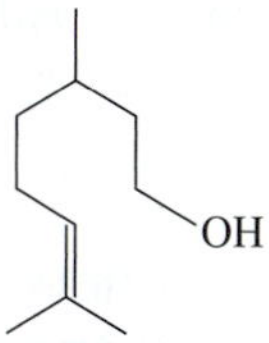

citronellol
in rose and
geranium oils

limonene
in lemon and
orange oils

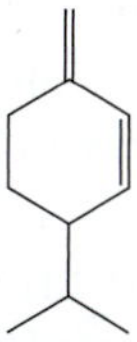

β-phellandrene
oil of eucalyptus

Ethene is the hormone that causes
tomatoes to ripen.

We will begin our study of alkenes by looking at how they are named, their structures, and their relative stabilities. Then we will examine a reaction of an alkene, paying close attention to the steps by which it occurs and the energy changes that accompany them. You will see that some of the discussion in this chapter revolves around concepts with which you are familiar, while some of the information is new and will broaden the foundation of knowledge you will be building on in subsequent chapters.

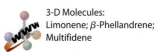

3-D Molecules:
Limonene; β-Phellandrene;
Multifidene

PHEROMONES

Insects communicate by releasing **pheromones**—chemical substances that other insects of the same species detect with their antennae. Many of the sex, alarm, and trail pheromones are alkenes. Interfering with an insect's ability to send or receive chemical signals is an environmentally safe way to control insect populations. For example, traps containing synthetic sex attractants have been used to capture such crop-destroying insects as the gypsy moth and the boll weevil.

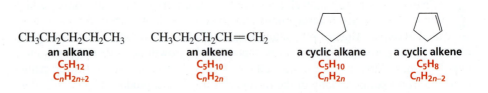

muscalure
sex attractant of the housefly

multifidene
**sex attractant of
brown algae**

4.1 MOLECULAR FORMULAS

We have seen that the general molecular formula for a noncyclic alkane is C_nH_{2n+2} (Section 3.0). We have also seen that the general molecular formula for a cyclic alkane is C_nH_{2n} because the cyclic structure reduces the number of hydrogens by two (Section 3.3).

The general molecular formula for a *noncyclic alkene* is also C_nH_{2n} because, as a result of the double bond, an alkene has two fewer hydrogens than an alkane with the same number of carbons. Thus, the general molecular formula for a *cyclic alkene* must be C_nH_{2n-2}. We can, therefore, make the following statement: *the general molecular formula for a hydrocarbon is C_nH_{2n+2} minus two hydrogens for every π bond or ring in the molecule.*

> The general molecular formula for a hydrocarbon is C_nH_{2n+2}, minus two hydrogens for every π bond or ring present in the molecule.

$CH_3CH_2CH_2CH_2CH_3$
an alkane
C_5H_{12}
C_nH_{2n+2}

$CH_3CH_2CH_2CH{=}CH_2$
an alkene
C_5H_{10}
C_nH_{2n}

a cyclic alkane
C_5H_{10}
C_nH_{2n}

a cyclic alkene
C_5H_8
C_nH_{2n-2}

Because alkanes contain the maximum number of C—H bonds possible—that is, they are saturated with hydrogen—they are called **saturated hydrocarbons**. In contrast, alkenes are called **unsaturated hydrocarbons** because they have fewer than the maximum number of hydrogens.

$CH_3CH_2CH_2CH_3$
a saturated hydrocarbon

$CH_3CH{=}CHCH_3$
an unsaturated hydrocarbon

PROBLEM 1◆ SOLVED

Determine the molecular formula for each of the following:
a. a 5-carbon hydrocarbon with one π bond and one ring.
b. a 4-carbon hydrocarbon with two π bonds and no rings.
c. a 10-carbon hydrocarbon with one π bond and two rings.

Solution to 1a For a 5-carbon hydrocarbon with no π bonds and no rings, $C_nN_{2n+2} = C_5H_{12}$. A 5-carbon hydrocarbon with one π bond and one ring has four fewer hydrogens, because two hydrogens are subtracted for every π bond or ring present in the molecule. The molecular formula, therefore, is C_5H_8.

PROBLEM 2 ♦ **SOLVED**

Determine the total number of double bonds and/or rings for hydrocarbons with the following molecular formulas:

a. $C_{10}H_{16}$ **b.** $C_{20}H_{34}$ **c.** C_4H_8

Solution to 2a   For a 10-carbon hydrocarbon with no π bonds and no rings, $C_nN_{2n+2} = C_{10}H_{22}$. A 10-carbon compound with molecular formula $C_{10}H_{16}$ has six fewer hydrogens. Therefore, it has a total of three double bonds and/or rings.

4.2 THE NOMENCLATURE OF ALKENES

The **functional group** is the center of reactivity in an organic molecule. In an alkene, the double bond is the functional group. The IUPAC system uses a suffix to denote certain functional groups. The systematic name of an alkene, for example, is obtained by replacing the *ane* at the end of the parent hydrocarbon's name with the suffix *ene*. Thus, a two-carbon alkene is called ethene, and a three-carbon alkene is called propene. Ethene also is frequently called by its common name: ethylene.

	$H_2C{=}CH_2$	$CH_3CH{=}CH_2$	cyclopentene	cyclohexene
systematic name:	ethene	propene		
common name:	ethylene	propylene		

The following rules are used to name a compound with a functional group suffix:

1. The longest continuous chain containing the functional group (in this case, the carbon–carbon double bond) is numbered in a direction that gives the functional group suffix the lowest possible number. The position of the double bond is indicated by a number immediately preceding the name of the alkene. For example, 1-butene signifies that the double bond is between the first and second carbons of butene; 2-hexene signifies that the double bond is between the second and third carbons of hexene. (The four alkene names shown above do not need a number because there is no ambiguity.)

> Number the longest continuous chain containing the functional group in the direction that gives the functional group suffix the lowest possible number.

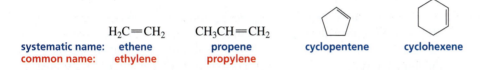

Notice that 1-butene does not have a common name. You might be tempted to call it "butylene," which is analogous to "propylene" for propene, but butylene is not an appropriate name. A name must be unambiguous, and "butylene" could signify either 1-butene or 2-butene.

2. For a compound with two double bonds, the suffix is "diene."

$$\overset{1}{CH_2}{=}\overset{2}{CH}{-}\overset{3}{CH_2}{-}\overset{4}{CH}{=}\overset{5}{CH_2}$$
1,4-pentadiene

$$\overset{1}{CH_3}\overset{2}{CH}{=}\overset{3}{CH}{-}\overset{4}{CH}{=}\overset{5}{CH}\overset{6}{CH_2}\overset{7}{CH_3}$$
2,4-heptadiene

$$\overset{5}{CH_3}\overset{4}{CH}{=}\overset{3}{CH}{-}\overset{2}{CH}{=}\overset{1}{CH_2}$$
1,3-pentadiene

3. The name of a substituent is placed in front of the name of the longest continuous chain that contains the functional group, together with a number to designate the carbon to which the substituent is attached. Notice that *when a compound's name contains both a functional group suffix and a substituent, the functional group suffix gets the lowest possible number.*

> When there is only a substituent, the substituent gets the lowest possible number.
>
> When there is only a functional group suffix, the functional group suffix gets the lowest possible number.
>
> When there is both a functional group suffix and a substituent, the functional group suffix gets the lowest possible number.

$$\overset{}{\underset{1\quad2\quad\;3\;\;4\;|\;\;5}{CH_3CH{=}CHCHCH_3}}\overset{CH_3}{}$$
4-methyl-2-pentene

$$\overset{\overset{2}{CH_2}\overset{1}{CH_3}}{\underset{3\quad4\;\;5\quad6\;\;7}{CH_3C{=}CHCH_2CH_2CH_3}}$$
3-methyl-3-heptene

4. If a chain has more than one substituent, the substituents are listed in alphabetical order, using the same rules for alphabetizing discussed in Section 3.2. Then the appropriate number is assigned to each substituent.

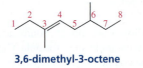

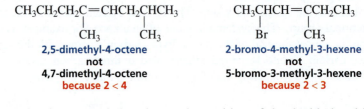

3,6-dimethyl-3-octene   **5-bromo-4-chloro-1-heptene**

Substituents are cited in alphabetical order.

5. If counting in either direction results in the same number for the alkene functional group suffix, the correct name is the one containing the lowest substituent number.

        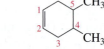

A substituent receives the lowest possible number only if there is no functional group suffix or if the same number for the functional group suffix is obtained in both directions.

$CH_3CH_2CH_2C{=}CHCH_2CHCH_3$
        $|$ $|$
        CH_3 CH_3
2,5-dimethyl-4-octene
not
4,7-dimethyl-4-octene
because 2 < 4

$CH_3CHCH{=}CCH_2CH_3$
    $|$ $|$
    Br CH_3
2-bromo-4-methyl-3-hexene
not
5-bromo-3-methyl-3-hexene
because 2 < 3

6. A number is not needed to denote the position of the double bond in a cyclic alkene because the ring is always numbered so that the double bond is between carbons 1 and 2. To assign numbers to any substituents, count around the ring in the direction (clockwise or counterclockwise) that puts the lowest number into the name.

3-ethylcyclopentene   **4,5-dimethylcyclohexene**   **4-ethyl-3-methylcyclohexene**

7. Numbers are needed to denote the positions of the double bonds if the ring has two double bonds.

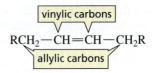

1,3-cyclohexadiene   **1,4-cyclohexadiene**   **2-methyl-1,3-cyclopentadiene**

Remember that the name of a substituent is placed *before* the name of the parent hydrocarbon, and the functional group suffix is placed *after* the name of the parent hydrocarbon.

substituent—parent hydrocarbon—functional group suffix

The sp^2 carbons of an alkene are called **vinylic carbons**. An sp^3 carbon that is adjacent to a vinylic carbon is called an **allylic carbon**.

vinylic carbons

$RCH_2{-}CH{=}CH{-}CH_2R$

allylic carbons

Two groups containing a carbon–carbon double bond are used in common names—the **vinyl group** and the **allyl group**. The vinyl group is the smallest possible group containing a vinylic carbon; the allyl group is the smallest possible group containing an allylic carbon. When "vinyl" or "allyl" is used in a name, the substituent must be attached to the vinylic or allylic carbon, respectively.

Tutorial:
Alkene nomenclature

H₂C=CH— $\quad$ H₂C=CHCH₂—
the vinyl group $\quad$ **the allyl group**

H₂C=CHCl $\quad$ H₂C=CHCH₂Br

systematic name: **chloroethene** $\qquad$ **3-bromopropene**
common name: **vinyl chloride** $\qquad$ **allyl bromide**

PROBLEM 3

Draw the structure for each of the following:

a. 3,3-dimethylcyclopentene $\qquad$ **c.** ethyl vinyl ether
b. 6-bromo-2,3-dimethyl-2-hexene $\qquad$ **d.** allyl alcohol

PROBLEM 4 ◆

Give the systematic name for each of the following:

a. 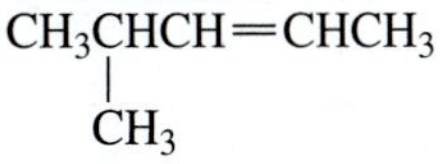 CH₃CHCH=CHCH₃
$\quad\quad$ |
$\quad\quad$ CH₃

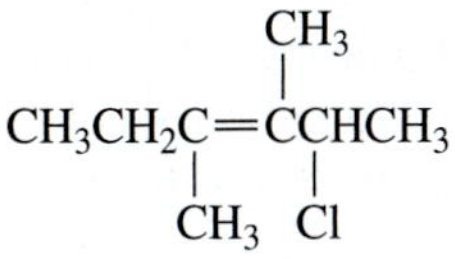

c. 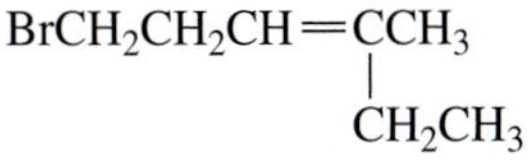 BrCH₂CH₂CH=CCH₃
$\qquad\qquad\qquad$ |
$\qquad\qquad\qquad$ CH₂CH₃

b. CH₃CH₂C=CCHCH₃
$\qquad\quad$ | |
$\qquad\quad$ CH₃ Cl

d. 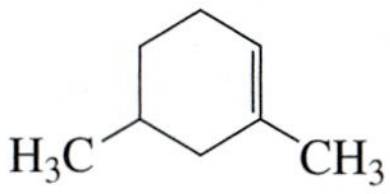 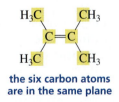 H₃C $\qquad$ CH₃

4.3 THE STRUCTURE OF ALKENES

The structure of the smallest alkene (ethene) was described in Section 1.8. Other alkenes have similar structures. Each double-bonded carbon of an alkene has three sp^2 orbitals that lie in a plane with angles of 120°. Each of these sp^2 orbitals overlaps an orbital of another atom to form a σ bond. Thus, one of the carbon–carbon bonds in a double bond is a σ bond, formed by the overlap of an sp^2 orbital of one carbon with an sp^2 orbital of the other carbon. The second carbon–carbon bond in the double bond (the π bond) is formed by side-to-side overlap of the remaining p orbital of one of the sp^2 carbons with the remaining p orbital of the other sp^2 carbon. Because three points determine a plane, each sp^2 carbon and the two atoms singly bonded to it lie in a plane. In order to achieve maximum orbital–orbital overlap, the two p orbitals must be parallel to each other. Therefore, all six atoms of the double-bond system are in the same plane.

H₃C $\qquad$ CH₃
$\quad$ C=C
H₃C $\qquad$ CH₃

**the six carbon atoms
are in the same plane**

It is important to remember that *the π bond represents the cloud of electrons that is above and below the plane defined by the two sp^2 carbons and the four atoms bonded to them.*

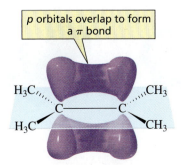

p orbitals overlap to form
a π bond

H₃C,,,,,, $\qquad\qquad$,,,,,CH₃
$\qquad$ C————C
H₃C $\qquad\qquad$ CH₃

3-D Molecules:
2,3-Dimethyl-2-butene

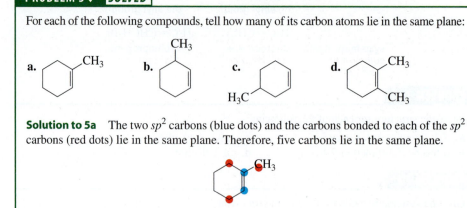

For each of the following compounds, tell how many of its carbon atoms lie in the same plane:

a. **b.** **c.** **d.**

Solution to 5a The two sp^2 carbons (blue dots) and the carbons bonded to each of the sp^2 carbons (red dots) lie in the same plane. Therefore, five carbons lie in the same plane.

4.4 ALKENES CAN HAVE CIS–TRANS ISOMERS

We have just seen that the two p orbitals forming the π bond must be parallel to achieve maximum overlap. Therefore, rotation about a double bond does not readily occur. If rotation were to occur, the two p orbitals would cease to overlap, and the π bond would break (Figure 4.1).

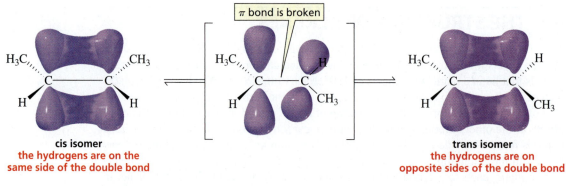

π bond is broken

cis isomer
the hydrogens are on the same side of the double bond

trans isomer
the hydrogens are on opposite sides of the double bond

▲ **Figure 4.1**
Rotation about the carbon–carbon double bond would break the π bond.

Because of the high-energy barrier to rotation about a double bond, an alkene such as 2-butene can exist in two distinct forms: the hydrogens bonded to the sp^2 carbons can be on the same side of the double bond or on opposite sides of the double bond.

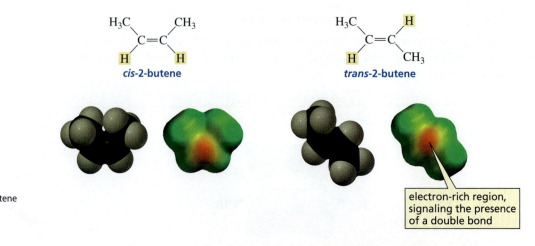

cis-2-butene

trans-2-butene

electron-rich region, signaling the presence of a double bond

3-D Molecules:
cis-2-Butene; *trans*-2-Butene

The isomer with the hydrogens on the same side of the double bond is called the **cis isomer**, and the isomer with the hydrogens on opposite sides of the double bond is called the **trans isomer**. A pair of isomers such as *cis*-2-butene and *trans*-2-butene is called **cis–trans isomers** or **geometric isomers**. This should remind you of the cis–trans isomers of disubstituted cyclohexanes you encountered in Section 3.12— the cis isomer had its substituents on the same side of the ring, and the trans isomer had its substituents on opposite sides of the ring.

If one of the sp^2 carbons of the double bond is attached to two identical substituents, there is only one possible structure for the alkene. In other words, cis and trans isomers are not possible for an alkene that has identical substituents attached to one of the sp^2 carbons.

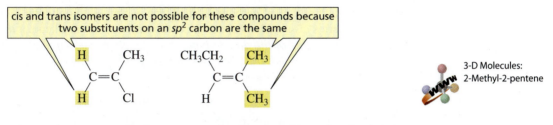

3-D Molecules:
2-Methyl-2-pentene

The cis and trans isomers can be interconverted only when the molecule absorbs sufficient heat or light energy to cause the π bond to break, because once the π bond is broken, rotation can occur easily about the remaining σ bond (Section 3.8).

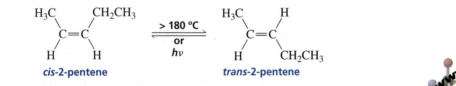

3-D Molecules:
cis-Retinal; *trans*-Retinal

CIS–TRANS INTERCONVERSION IN VISION

When rhodopsin absorbs light, a double bond interconverts between the cis and trans forms. This process plays an important role in vision.

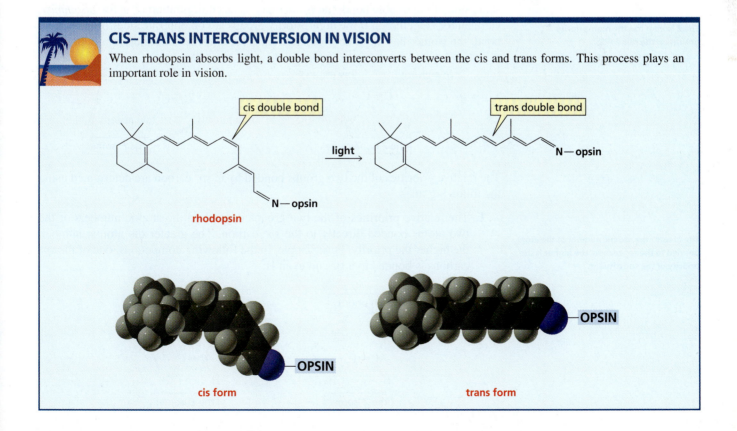

PROBLEM 6 ◆

a. Which of the following compounds can exist as cis–trans isomers?
b. For those compounds, draw and label the cis and trans isomers.

1. $CH_3CH{=}CHCH_2CH_3$ 3. $CH_3CH{=}CHCH_3$

2. $CH_3CH{=}CCH_3$ 4. $CH_3CH_2CH{=}CH_2$
       $|$
      CH_3

4.5 NAMING ALKENES USING THE *E,Z* SYSTEM

As long as each of the sp^2 carbons of an alkene is bonded to only one substituent, we can use the terms *cis* and *trans* to designate the structure of the alkene: *if the hydrogens are on the same side of the double bond, it is the cis isomer; if they are on opposite sides of the double bond, it is the trans isomer.* But how do we designate the isomers of a compound such as 1-bromo-2-chloropropene?

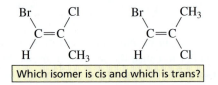

Which isomer is cis and which is trans?

For a compound such as 1-bromo-2-chloropropene, the cis-trans system of nomenclature cannot be used because there are four different groups on the two sp^2 carbons. The *E,Z* system of nomenclature was devised for such compounds.

To name an isomer by the *E,Z* system, we first determine the relative priorities of the two groups bonded to one of the sp^2 carbons and then the relative priorities of the two groups bonded to the other sp^2 carbon. (The rules for assigning relative priorities are explained below.) If the two high-priority groups (one from each carbon) are on the same side of the double bond, the isomer has the Z configuration (Z is for *zusammen*, German for "together"). If the high-priority groups are on opposite sides of the double bond, the isomer has the E configuration (E is for *entgegen*, German for "opposite").

> **The *Z* isomer has the high-priority groups on the same side.**

The relative priorities of the two groups bonded to an sp^2 carbon are determined using the following rules:

> **The greater the atomic number of the atom bonded to the sp^2 carbon, the higher is the priority of the substituent.**

1. The relative priorities of the two groups depend on the atomic numbers of the two atoms bonded directly to the sp^2 carbon. The greater the atomic number, the higher the priority. For example, in the following compounds, one of the sp^2 carbons is bonded to a Br and to an H:

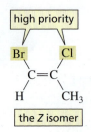

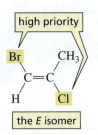

Br has a greater atomic number than H, so **Br** has a higher priority than **H**. The other sp^2 carbon is bonded to a Cl and to a C. Chlorine has the greater atomic number, so **Cl** has a higher priority than **C**. (Notice that you use the atomic number of C, not the mass of the CH_3 group, because the priorities are based on the atomic numbers of atoms, *not* on the masses of groups.) The isomer on the left has the high-priority groups (Br and Cl) on the same side of the double bond, so it is the **Z isomer**. (Zee groups are on Zee Zame Zide.) The isomer on the right has the high-priority groups on opposite sides of the double bond, so it is the **E isomer**.

2. If the two groups bonded to an sp^2 carbon start with the same atom (there is a tie), you then move outward from the point of attachment and consider the atomic numbers of the atoms that are attached to the "tied" atoms. For example, in the following compounds, both atoms bonded the sp^2 carbon on the left are C's (in a CH_2Cl group and a CH_2CH_2Cl group), so there is a tie.

If the atoms attached to an *sp^2* carbon are the same, the atoms attached to the "tied" atoms are compared; the one with the greatest atomic number belongs to the group with the higher priority.

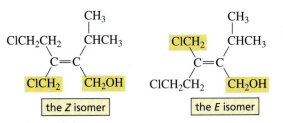

The C of the CH_2Cl group is bonded to **Cl, H, H**, and the C of the CH_2CH_2Cl group is bonded to **C, H, H**. Cl has a greater atomic number than C, so the CH_2Cl group has the higher priority. Both atoms bonded to the other sp^2 carbon are C's (from a CH_2OH group and a $CH(CH_3)_2$ group), so there is a tie on that side as well. The C of the CH_2OH group is bonded to **O, H**, and **H**, and the C of the $CH(CH_3)_2$ group is bonded to **C, C**, and **H**. Of these six atoms, O has the greatest atomic number, so CH_2OH has a higher priority than $CH(CH_3)_2$. (Notice that you do not add the atomic numbers; you consider the single atom with the greatest atomic number.) The *E* and *Z* isomers are as shown above.

3. If an atom is doubly bonded to another atom, the priority system treats it as if it were singly bonded to two of those atoms. If an atom is triply bonded to another atom, the priority system treats it as if it were singly bonded to three of those atoms. For example, one of the sp^2 carbons in the following pair of isomers is bonded to a CH_2CH_2OH group and to a $CH_2C{\equiv}CH$ group:

If an atom is doubly bonded to another atom, treat it as if it were singly bonded to two of those atoms.

If an atom is triply bonded to another atom, treat it as if it were singly bonded to three of those atoms.

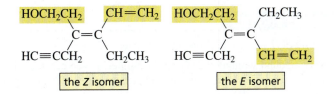

Because the atoms immediately bonded to the sp^2 carbon on the left are both bonded to **C, H, H**, we ignore them and turn our attention to the groups attached to them. One of these is CH_2OH and the other is $C{\equiv}CH$. The triple-bonded C is considered to be bonded to C, C, and C; the other C is bonded to O, H, and H. Of the six atoms, O has the greatest atomic number, so CH_2OH has a higher priority than $C{\equiv}CH$. Both atoms bonded to the other sp^2 carbon are C's, so they are tied. The first carbon of the CH_2CH_3 group is bonded to C, H, and H. The first carbon of the $CH{=}CH_2$ group is bonded to an H and doubly bonded to a C. Therefore, it is considered to be bonded to H, C, and C. One C cancels in each of the two groups, leaving H and H in the CH_2CH_3 group and C and H in the $CH{=}CH_2$ group. C has a greater atomic number than H, so $CH{=}CH_2$ has a higher priority than CH_2CH_3.

Mechanistic Tutorial:
E and *Z* Nomenclature

PROBLEM 7 ◆

Assign relative priorities to each set of substituents:
a. —Br, —I, —OH, —CH$_3$
b. —CH$_2$CH$_2$OH, —OH, —CH$_2$Cl, —CH=CH$_2$

PROBLEM 8

Draw and label the *E* and *Z* isomers for each of the following:

a. CH$_3$CH$_2$CH=CHCH$_3$

b. 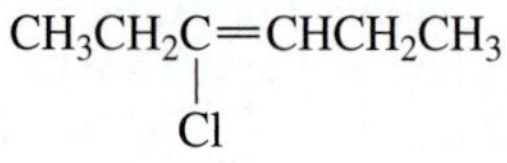 CH$_3$CH$_2$C=CHCH$_2$CH$_3$
            |
            Cl

c. CH$_3$CH$_2$CH$_2$CH$_2$
                  |
CH$_3$CH$_2$C=CCH$_2$Cl
              |
             CH$_3$CHCH$_3$

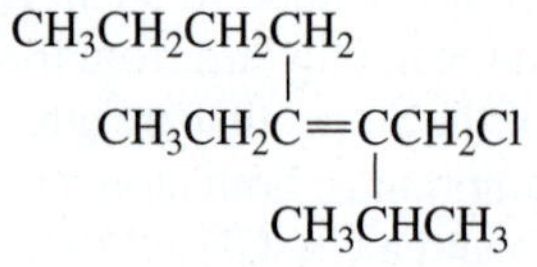

PROBLEM 9

Indicate whether each of the following is an *E* isomer or a *Z* isomer (black = carbon, white = hydrogen, green = chlorine, and red = oxygen):

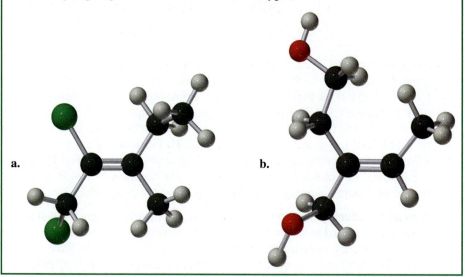

a.

b.

PROBLEM-SOLVING STRATEGY

Draw the structure of (*E*)-1-bromo-2-methyl-2-butene.

First draw the compound without specifying the isomer so you can see what groups are bonded to the *sp*^{2} carbons. Now determine the relative priorities of the two groups on each of the *sp*^{2} carbons.

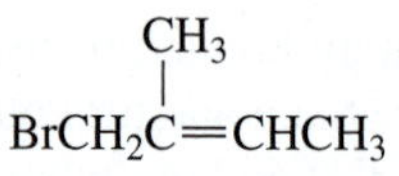

$$\underset{\text{BrCH}_2\text{C}=\text{CHCH}_3}{\overset{\text{CH}_3}{|}}$$

The *sp*^{2} carbon on the right is attached to a CH$_3$ and to an H; CH$_3$ has the higher priority. The other *sp*^{2} carbon is attached to a CH$_3$ and to CH$_2$Br; CH$_2$Br has the higher priority. To get the *E* isomer, draw the compound with the two high-priority groups on opposite sides of the double bond.

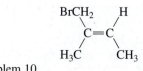

Now continue on to Problem 10.

PROBLEM 10 ♦

Draw the structure of (Z)-3-isopropyl-2-heptene.

4.6 THE RELATIVE STABILITIES OF ALKENES

Alkyl substituents bonded to the sp^2 carbons of an alkene have a stabilizing effect on the alkene.

We can, therefore, make the following statement: *the more alkyl substituents bonded to the sp^2 carbons of an alkene, the greater is its stability.* (Some students find it easier to look at the number of hydrogens bonded to the sp^2 carbons. In terms of hydrogens, the statement is: *the fewer hydrogens bonded to the sp^2 carbons of an alkene, the greater is its stability.*)

The fewer hydrogens bonded to the sp^2 carbons of an alkene, the more stable it is.

relative stabilities of alkyl-substituted alkenes

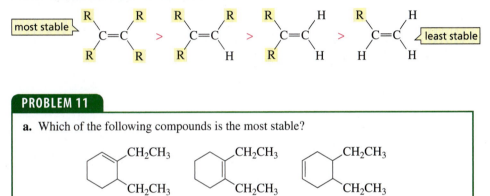

PROBLEM 11

a. Which of the following compounds is the most stable?

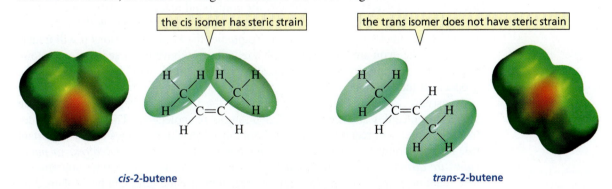

b. Which is the least stable?

Both *cis*-2-butene and *trans*-2-butene have two alkyl groups bonded to their sp^2 carbons. The trans isomer, in which the large substituents are farther apart, is more stable than the cis isomer, in which the large substituents are closer together.

the cis isomer has steric strain   the trans isomer does not have steric strain

cis-**2-butene**   *trans*-**2-butene**

When the large substituents are on the same side of the molecule, their electron clouds can interfere with each other, causing steric strain in the molecule and making it less stable (Section 3.8). When the large substituents are on opposite sides of the molecule, their electron clouds cannot interact so the molecule has less steric strain and is, therefore, more stable.

PROBLEM 12 ♦

Rank the following compounds in order of decreasing stability:

    trans-3-hexene, *cis*-3-hexene, *cis*-2,5-dimethyl-3-hexene, *cis*-3,4-dimethyl-3-hexene

4.7 HOW ALKENES REACT • CURVED ARROWS SHOW THE BONDS THAT BREAK AND THE BONDS THAT FORM

There are many millions of organic compounds. If you had to memorize how each of them reacts, studying organic chemistry would not be a very pleasant experience. Fortunately, organic compounds can be divided into families, and all the members of a family react in similar ways. What makes learning organic chemistry even easier is that there are only a few rules that govern the reactivity of each family.

The family that an organic compound belongs to is determined by its functional group. The **functional group** is the center of reactivity of a molecule. You will find a table of common functional groups inside the back cover of this book. You are already familiar with the functional group of an alkene: the carbon–carbon double bond. All compounds with a carbon–carbon double bond react in similar ways, whether the compound is a small molecule like ethene or a large molecule like cholesterol.

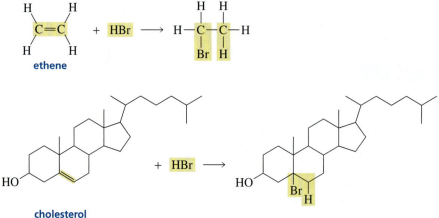

It is not sufficient to know that a compound with a carbon–carbon double bond reacts with HBr to form a product in which the H and Br atoms have taken the place of the π bond; we need to understand *why* the compound reacts with HBr. In each chapter that discusses the reactivity of a particular functional group, we will see how the nature of the functional group allows us to predict the kind of reactions it will undergo. Then, when you are confronted with a reaction you have never seen before, knowing how the structure of the molecule affects its reactivity will help you predict the products of the reaction.

In essence, organic chemistry is all about the interaction between electron-rich atoms or molecules and electron-deficient atoms or molecules. These are the forces that make chemical reactions happen. From this observation follows a very important rule for predicting the reactivity of organic compounds: *electron-rich atoms or molecules are attracted to electron-deficient atoms or molecules.* Each time you study a new functional group, remember that the reactions it undergoes can be explained by this very simple rule.

> **Electron-rich atoms or molecules are attracted to electron-deficient atoms or molecules.**

Therefore, to understand how a functional group reacts, you must first learn to recognize electron-deficient and electron-rich atoms and molecules. An electron-deficient atom or molecule is called an **electrophile**. Literally, "electrophile" means "electron loving" (*phile* is the Greek suffix for "loving"). An electrophile looks for electrons.

$$H^+ \qquad CH_3\overset{+}{C}H_2$$

> these are electrophiles because they can accept a pair of electrons

An electron-rich atom or molecule is called a **nucleophile**. A nucleophile has a pair of electrons it can share. Because a nucleophile has electrons to share and an electrophile is seeking electrons, it should not be surprising that they attract each other. Thus, the preceding rule can be restated as *a nucleophile reacts with an electrophile.*

A nucleophile reacts with an electrophile.

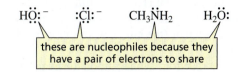

these are nucleophiles because they have a pair of electrons to share

PROBLEM 13 ♦

Identify the electrophile and the nucleophile in each of the following acid–base reactions:

a. $AlCl_3$ + $:NH_3$ ⇌ $Cl_3\bar{Al}-\overset{+}{N}H_3$

b. $H-\overset{..}{\underset{..}{Br}}:$ + $H\overset{..}{\underset{..}{O}}:^-$ ⇌ $:\overset{..}{\underset{..}{Br}}:^-$ + $H_2\overset{..}{O}:$

We have seen that the π bond of an alkene consists of a cloud of electrons above and below the plane defined by the sp^2 carbons and the four atoms bonded to them. As a result of this cloud of electrons, an alkene is an electron-rich molecule—it is a nucleophile. (Notice the relatively electron-rich pale orange area in the electrostatic potential maps for *cis*- and *trans*-2-butene in Section 4.4.) We have also seen that a π bond is weaker than a σ bond (Section 1.14). The π bond, therefore, is the bond that is most easily broken when an alkene undergoes a reaction. For these reasons, we can predict that an alkene will react with an electrophile, and, in the process, the π bond will break. So if a reagent such as hydrogen bromide is added to an alkene, the alkene (a nucleophile) will react with the partially positively charged hydrogen (an electrophile) of hydrogen bromide and a carbocation will be formed. In the second step of the reaction, the positively charged carbocation (an electrophile) will react with the negatively charged bromide ion (a nucleophile) to form an alkyl halide.

$$CH_3CH=CHCH_3 \ + \ \overset{\delta+}{H}-\overset{\delta-}{Br} \ \longrightarrow \ \underset{\underset{\displaystyle H}{|}}{\overset{+}{CH_3CH}}-CHCH_3 \ + \ Br^- \ \longrightarrow \ \underset{\underset{\displaystyle Br \quad H}{|\quad\ |}}{CH_3CH}-CHCH_3$$

a carbocation   2-bromobutane
                                           an alkyl halide

This step-by-step description of the process by which reactants (alkene + HBr) are changed into products (alkyl halide) is called the **mechanism of the reaction**. To help us understand a mechanism, curved arrows are drawn to show how the electrons move as new covalent bonds are formed and existing covalent bonds are broken. These are called "curved" arrows to distinguish them from the "straight" arrow used to link reactants with products in a chemical reaction. Each curved arrow represents the simultaneous movement of two electrons *from an electron-rich center* (at the tail of the arrow) and *toward an electron-deficient center* (at the point of the arrow). In this way, the curved arrows show which bonds are broken and which bonds are formed.

Curved arrows are used to show the bonds that break and the bonds that form; they are drawn from an electron-rich center to an electron-deficient center.

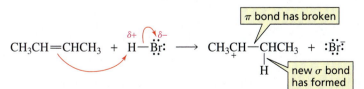

π bond has broken

new σ bond has formed

For the reaction of 2-butene with HBr, an arrow is drawn to show that the two electrons of the π bond of the alkene are attracted to the partially positively charged hydrogen of HBr. The hydrogen is not immediately free to accept this pair of electrons because it is already bonded to a bromine, and hydrogen can be bonded to only one

Mechanistic Tutorial:
Addition of HBr to an alkene

atom at a time (Section 1.4). However, as the π electrons of the alkene move toward the hydrogen, the H—Br bond breaks, with bromine keeping the bonding electrons. Notice that the π electrons are pulled away from one carbon but remain attached to the other. Thus, the two electrons that formerly formed the π bond now form a σ bond between carbon and the hydrogen from HBr. The product of this first step in the reaction is a carbocation because the sp^2 carbon that did not form the new bond with hydrogen has lost a share in an electron pair (the electrons of the π bond) and is, therefore, positively charged.

In the second step of the reaction, a lone pair on the negatively charged bromide ion forms a bond with the positively charged carbon of the carbocation. Notice that in both steps of the reaction *an electrophile reacts with a nucleophile.*

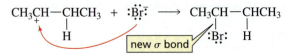

Solely from the knowledge that an electrophile reacts with a nucleophile and a π bond is the weakest bond in an alkene, we have been able to predict that the product of the reaction of 2-butene and HBr is 2-bromobutane. The overall reaction involves the *addition* of 1 mole of HBr to 1 mole of the alkene. The reaction, therefore, is called an **addition reaction**. Because the first step of the reaction is the addition of an electrophile (H^+) to the alkene, the reaction is more precisely called an **electrophilic addition reaction**. *Electrophilic addition reactions are the characteristic reactions of alkenes.*

At this point, you may think that it would be easier just to memorize the fact that 2-bromobutane is the product of the reaction, without trying to understand the mechanism that explains why 2-bromobutane is the product. Keep in mind, however, that you will soon be encountering a great number of reactions, and you will not be able to memorize them all. If you strive to understand the mechanism of each reaction, however, the unifying principles of organic chemistry will soon be clear to you, making mastery of the material much easier and a lot more fun.

It will be helpful to do the exercise on drawing curved arrows in the Study Guide/Solution Manual (Special Topic III).

PROBLEM 14 ◆

Which of the following are electrophiles, and which are nucleophiles?

$$H^- \qquad CH_3O^- \qquad CH_3C\!\equiv\!CH \qquad CH_3\overset{+}{C}HCH_3 \qquad NH_3$$

A FEW WORDS ABOUT CURVED ARROWS

1. Draw a curved arrow so that it points in the direction of electron flow and never away from the flow. This means that an arrow will always be drawn away from a negative charge and/or toward a positive charge. An arrow is used to show both the bond that forms and the bond that breaks.

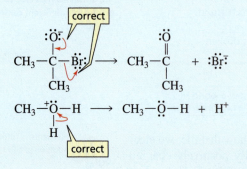

2. Curved arrows are meant to indicate the movement of electrons. Never use a curved arrow to indicate the movement of an atom. For example, do not use an arrow as a lasso to remove the proton, as shown here:

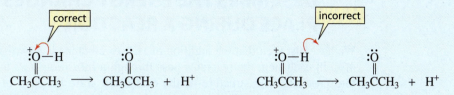

3. A head of a curved arrow always points at an atom or at a bond. Never draw the arrow head pointing out into space.

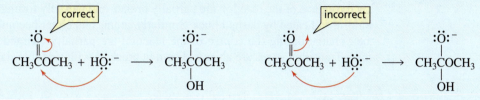

4. The arrow always starts at the electron source. In the following example, the arrow starts at the electron-rich π bond, not at a carbon atom:

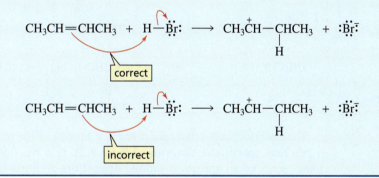

PROBLEM 15

Use curved arrows to show the movement of electrons in each of the following reaction steps. (*Hint:* Look at the starting material and look at the products, then draw the arrows.)

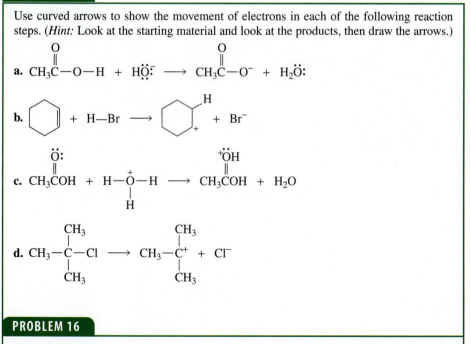

PROBLEM 16

For reactions a–c in Problem 15, indicate which reactant is the nucleophile and which is the electrophile.

4.8 A REACTION COORDINATE DIAGRAM DESCRIBES THE ENERGY CHANGES THAT TAKE PLACE DURING A REACTION

We have just seen that the addition of HBr to 2-butene is a two-step process (Section 4.7). In each step, the reactants pass through a *transition state* as they are converted into products. The structure of the **transition state** for each of the steps is shown below in brackets; it lies somewhere between the structure of the reactants and the structure of the products. Notice that the bonds that break and the bonds that form during the course of the reaction are partially broken and partially formed in the transition state, as indicated by dashed lines. Similarly, atoms that either become charged or lose their charge during the course of the reaction are partially charged in the transition state. Transition states are always shown in brackets with a double-dagger superscript.

$$CH_3CH=CHCH_3 + H-Br \longrightarrow \left[\overset{\delta+}{CH_3CH} \cdots CHCH_3 \atop \underset{\delta-Br}{H} \right]^{\ddagger} \longrightarrow CH_3\underset{+}{C}HCH_2CH_3 + Br^-$$

transition state

$$CH_3\underset{+}{C}HCH_2CH_3 + Br^- \longrightarrow \left[\overset{\delta+}{CH_3CH}CH_2CH_3 \atop \underset{\delta-Br}{} \right]^{\ddagger} \longrightarrow \underset{Br}{CH_3CHCH_2CH_3}$$

transition state

The energy changes that take place in each step of the reaction can be described by a **reaction coordinate diagram** (Figure 4.2). In a reaction coordinate diagram, the total energy of all species is plotted against the progress of the reaction. A reaction progresses from left to right as written in the chemical equation, so the energy of the reactants is plotted on the left-hand side of the *x*-axis and the energy of the products is plotted on the right-hand side.

Figure 4.2a shows that, in the first step of the reaction, the alkene is converted into a carbocation that is less stable than the reactants. Remember that *the more stable the species, the lower is its energy.* Because the product of the first step is less stable than the reactants, we know that this step consumes energy. We see that as the carbocation is formed, the reaction passes through the transition state. Notice that the transition state is a *maximum* energy state on the reaction coordinate diagram.

The carbocation reacts in the second step with the bromide ion to form the final product (Figure 4.2b). Because the product is more stable than the reactants, we know that this step releases energy.

The more stable the species, the lower is its energy.

▶ **Figure 4.2**
The reaction coordinate diagrams for the two steps in the addition of HBr to 2-butene:
(a) the first step;
(b) the second step.

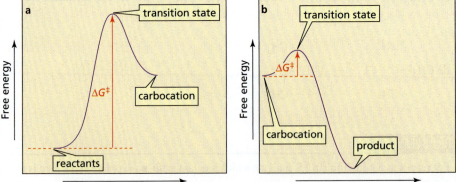

A chemical species that is a product of one step of a reaction and a reactant for the next step is called an **intermediate**. Thus, the carbocation is an intermediate. Although the carbocation is more stable than either of the transition states, it is still too unstable to be isolated. Do not confuse transition states with intermediates: *transition states have partially formed bonds, whereas intermediates have fully formed bonds.*

Because the product of the first step is the reactant of the second step, we can hook the two reaction coordinate diagrams together to obtain the reaction coordinate diagram for the overall reaction (Figure 4.3).

Transition states have partially formed bonds. Intermediates have fully formed bonds.

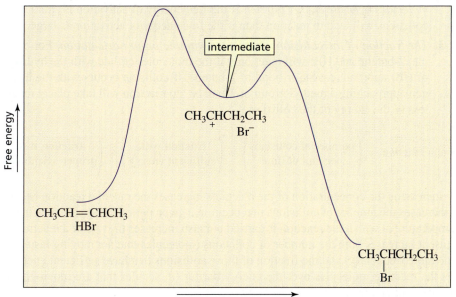

◀ **Figure 4.3**
The reaction coordinate diagram for the addition of HBr to 2-butene.

A reaction is over when the system reaches equilibrium. The relative concentrations of products and reactants at equilibrium depend on their relative stabilities: *the more stable the compound, the greater is its concentration at equilibrium.* Thus, if the products are more stable (have a lower free energy) than the reactants, there will be a higher concentration of products than reactants at equilibrium. On the other hand, if the reactants are more stable than the products, there will be a higher concentration of reactants than products at equilibrium. We see that the free energy of the final products in Figure 4.3 is lower than the free energy of the initial reactants. Therefore, we know that there will be more products than reactants when the reaction has reached equilibrium. A reaction that leads to a higher concentration of products compared with the concentration of reactants is called a **favorable reaction**.

The more stable the compound, the greater is its concentration at equilibrium.

How fast a reaction occurs is indicated by the energy "hill" that must be climbed for the reactants to be converted into products. The higher the energy barrier, the slower is the reaction. The energy barrier is called the **free energy of activation**. The free energy of activation for each step is indicated by $\Delta G^{\ddagger}$ in Figure 4.2. It is the difference between the free energy of the transition state and the free energy of the reactants:

$$\Delta G^{\ddagger} = \text{(free energy of the transition state)} - \text{(free energy of the reactants)}$$

The higher the energy barrier, the slower is the reaction.

We can see from the reaction coordinate diagram that the free energy of activation for the first step of the reaction is greater than the free energy of activation for the second step. In other words, the first step of the reaction is slower than the second step. This is what we would expect, considering that the molecules in the first step of this reaction must collide with sufficient energy to break covalent bonds, whereas no bonds are broken in the second step.

If a reaction has two or more steps, the step that has its transition state *at the highest point on the reaction coordinate* is called the **rate-determining step**. The rate-determining step controls the overall rate of the reaction because the overall rate of a

reaction such as that shown in Figure 4.3 cannot exceed the rate of the rate-determining step. In Figure 4.3, the rate-determining step is the first step—the addition of the electrophile (the proton) to the alkene.

What determines how fast a reaction occurs? The rate of a reaction depends on the following factors:

1. *The number of collisions that take place between the reacting molecules in a given period of time.* The greater the number of collisions, the faster is the reaction.

2. *The fraction of the collisions that occur with sufficient energy to get the reacting molecules over the energy barrier.* If the free energy of activation is small, more collisions will lead to reaction than if the free energy of activation is large.

3. *The fraction of the collisions that occur with the proper orientation.* For example, 2-butene and HBr will react only if the molecules collide with the hydrogen of HBr approaching the π bond of 2-butene. If collision occurs with the hydrogen approaching a methyl group of 2-butene, no reaction will take place, regardless of the energy of the collision.

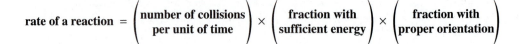

rate of a reaction = $\left(\begin{array}{c}\textbf{number of collisions}\\\textbf{per unit of time}\end{array}\right) \times \left(\begin{array}{c}\textbf{fraction with}\\\textbf{sufficient energy}\end{array}\right) \times \left(\begin{array}{c}\textbf{fraction with}\\\textbf{proper orientation}\end{array}\right)$

Increasing the concentration of the reactants increases the rate of a reaction because it increases the number of collisions that occur in a given period of time. Increasing the temperature at which the reaction is carried out also increases the rate of a reaction because it increases both the number of collisions (molecules that are moving faster collide more frequently) and the fraction of those collisions that have sufficient energy to get the reacting molecules over the energy barrier (molecules that are moving faster collide with greater energy).

PROBLEM 17

Draw a reaction coordinate diagram for
a. a fast reaction with products that are more stable than the reactants.
b. a slow reaction with products that are more stable than the reactants.
c. a slow reaction with products that are less stable than the reactants.

PROBLEM 18 ◆

Given the following reaction coordinate diagram for the reaction of A to give D, answer the following questions:

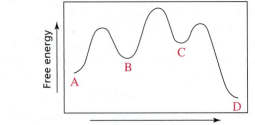

a. How many intermediates are there?
b. Which intermediate is the most stable?
c. How many transition states are there?
d. Which transition state is the most stable?
e. Which is more stable, reactants or products?
f. What is the fastest step in the reaction?
g. What is the reactant of the rate-determining step?
h. Is the overall reaction favorable?

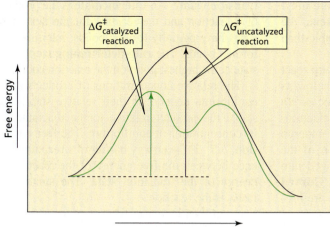

◀ **Figure 4.4**
The reaction coordinate diagrams for an uncatalyzed reaction (black) and for a catalyzed reaction. The catalyzed reaction (green) takes place by an alternative pathway with a lower "energy hill."

The rate of a reaction can also be increased by adding a catalyst to the reaction mixture. A catalyst gives the reactants a new pathway to follow—one with a smaller $\Delta G^{\ddagger}$ (Figure 4.4). Thus, a **catalyst** increases the rate of a reaction by decreasing the energy barrier that has to be overcome in the process of converting the reactants into products. A catalyst must participate in the reaction if it is going to make it go faster, but it is not consumed or changed during the reaction. Because the catalyst is not used up, only a small amount of it is needed to catalyze the reaction (typically 1 to 10% of the number of moles of reactant).

> **A catalyst gives the reagents a new pathway with a lower "energy hill."**

Notice that the stability of the reactants and products is the same in both the catalyzed and uncatalyzed reactions. In other words, the catalyst does not change the relative concentrations of products and reactants when the system reaches equilibrium. Thus, the catalyst does not change the *amount* of product formed; it changes only the *rate* at which it is formed.

PESTICIDES: NATURAL AND SYNTHETIC

Long before chemists learned how to create compounds that would protect plants from predators, plants were doing the job themselves. Plants had every incentive to synthesize pesticides; when you cannot run, you need to find another way to protect yourself. But which pesticides are more harmful, those synthesized by chemists or those synthesized by plants? Unfortunately, we do not know the answer because although federal laws require all human-made pesticides to be tested for any cancer-causing effects, they do not require testing of plant-made pesticides. Besides, risk evaluations of chemicals are usually done on rats, and something that is carcinogenic in a rat may not be carcinogenic in a human. Furthermore, when rats are tested, they are exposed to much greater concentrations of the chemical than would be experienced by a human, and some chemicals are only harmful at high doses. For example, we all need sodium chloride for survival, but high concentrations are poisonous; and although we associate alfalfa sprouts with healthy eating, monkeys fed very large amounts of alfalfa sprouts develop an immune system disorder.

SUMMARY

Alkenes are hydrocarbons that contain a double bond. The double bond is the **functional group** or center of reactivity of the alkene. The **functional group suffix** of an alkene is *ene*. The general molecular formula for a hydrocarbon is C_nH_{2n+2}, minus two hydrogens for every π bond or ring in the molecule. Because alkenes contain fewer than the maximum number of hydrogens, they are called **unsaturated hydrocarbons**.

Rotation about the double bond is restricted, so an alkene can exist as **cis–trans isomers**. The **cis isomer** has its hydrogens on the same side of the double bond; the **trans isomer** has its hydrogens on opposite sides of the double bond. The **Z isomer** has the high-priority groups on the same side of the double bond; the **E isomer** has the high-priority groups on opposite sides of the double bond. The relative priorities depend on the atomic numbers of the

atoms bonded directly to the sp^2 carbon. The more alkyl substituents bonded to the sp^2 carbons of an alkene, the greater is its stability. **Trans alkenes** are more stable than **cis alkenes** because of steric strain.

All compounds with a particular **functional group** react similarly. Due to the cloud of electrons above and below its π bond, an alkene is an electron-rich species (a **nucleophile**). Nucleophiles are attracted to electron-deficient species, called **electrophiles**. Alkenes undergo **electrophilic addition reactions**. The description of the step-by-step process by which reactants are changed into products is called the **mechanism of the reaction. Curved arrows** show which bonds are formed and which are broken in a reaction.

A **reaction coordinate diagram** shows the energy changes that take place in a reaction. The more stable the species, the lower is its energy. As reactants are converted into products, a reaction passes through a maximum energy

transition state. An **intermediate** is the product of one step of a reaction and the reactant for the next step. Transition states have partially formed bonds; intermediates have fully formed bonds. The **rate-determining step** has its transition state at the highest point on the reaction coordinate.

The relative concentrations of products and reactants at equilibrium depend on their relative stabilities. The more stable the product relative to the reactant, the greater is its concentration at equilibrium. The **free energy of activation**, $\Delta G^{\ddagger}$, is the energy barrier of a reaction. It is the difference between the free energy of the reactants and the free energy of the transition state. The smaller the $\Delta G^{\ddagger}$, the faster is the reaction.

A **catalyst** increases the rate of a reaction but is not consumed or changed in the reaction. It changes the *rate* at which a product is formed by providing a pathway with a smaller $\Delta G^{\ddagger}$, but it does not change the *amount* of product formed.

PROBLEMS

19. Give the systematic name for each of the following:

20. Squalene, a hydrocarbon with molecular formula $C_{30}H_{50}$, is obtained from shark liver. (*Squalus* is Latin for "shark.") If squalene is a noncyclic compound, how many π bonds does it have?

21. Draw and label the E and Z isomers for each of the following:

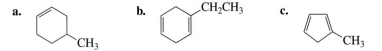

22. For each of the following pairs, indicate which member is more stable:

23. a. Give the structures and the systematic names for all alkenes with molecular formula C_4H_8, ignoring cis–trans isomers. (*Hint:* There are three.)
    b. Which of the compounds have E and Z isomers?

24. Draw the structure for each of the following:
    a. (Z)-1,3,5-tribromo-2-pentene
    b. (Z)-3-methyl-2-heptene
    c. (E)-1,2-dibromo-3-isopropyl-2-hexene
    d. vinyl bromide
    e. 1,2-dimethylcyclopentene
    f. diallylamine

25. Determine the total number of double bonds and/or rings for a hydrocarbon with molecular formula:
    a. $C_{12}H_{20}$ b. $C_{40}H_{56}$

26. Name the following:

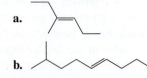 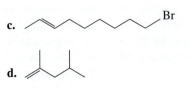 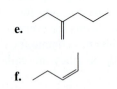 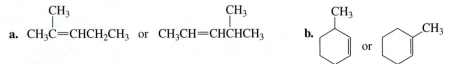

27. Draw curved arrows to show the flow of electrons responsible for the conversion of the reactants into the products:

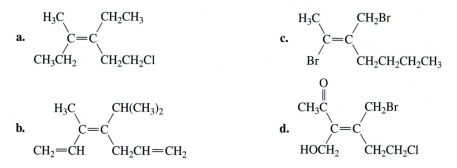

28. Draw three alkenes with molecular formula C_5H_{10} that do not have cis–trans isomers.

29. Tell whether each of the following has the *E* or the *Z* configuration:

a. H_3C ⧵ $C=C$ ⧸ CH_2CH_3 ; CH_3CH_2 ⧸ $C=C$ ⧵ CH_2CH_2Cl

c. H_3C ⧵ $C=C$ ⧸ CH_2Br ; Br ⧸ $C=C$ ⧵ $CH_2CH_2CH_2CH_3$

b. H_3C ⧵ $C=C$ ⧸ $CH(CH_3)_2$; $CH_2=CH$ ⧸ $C=C$ ⧵ $CH_2CH=CH_2$

d. $CH_3\overset{O}{\overset{\|}{C}}$ ⧵ $C=C$ ⧸ CH_2Br ; $HOCH_2$ ⧸ $C=C$ ⧵ CH_2CH_2Cl

30. Which of the following compounds is the most stable? Which is the least stable?

3,4-dimethyl-2-hexene; 2,3-dimethyl-2-hexene; 4,5-dimethyl-2-hexene

31. Assign relative priorities to each set of substituents:
  a. $-CH_2CH_2CH_3$ $\qquad$ $-CH(CH_3)_2$ $\qquad$ $-CH=CH_2$ $\qquad$ $-CH_3$
  b. $-CH_2NH_2$ $\qquad\qquad$ $-NH_2$ $\qquad\qquad$ $-OH$ $\qquad\qquad$ $-CH_2OH$
  c. $-C(=O)CH_3$ $\qquad\quad$ $-CH=CH_2$ $\qquad$ $-Cl$ $\qquad\qquad$ $-C\equiv N$

32. Determine the molecular formula for each of the following:
  a. a 5-carbon hydrocarbon with two π bonds and no rings
  b. an 8-carbon hydrocarbon with three π bonds and one ring

33. Give the systematic name for each of the following:

a. $CH_3CH_2\underset{\underset{Br}{|}}{C}HCH=CHCH_2CH_2\underset{\underset{Br}{|}}{C}HCH_3$

c. (cyclopentene ring with CH_3 and CH_3 substituents)

b. H_3C ⧵ $C=C$ ⧸ CH_2CH_3 ; CH_3CH_2 ⧸ $C=C$ ⧵ $CH_2CH_2\underset{\underset{CH_3}{|}}{C}HCH_3$

d. H_3C ⧵ $C=C$ ⧸ CH_2CH_3 ; H_3C ⧸ $C=C$ ⧵ $CH_2CH_2CH_2CH_3$

34. Draw a reaction coordinate diagram for a two-step reaction in which the products of the first step are less stable than the reactants, the reactants of the second step are less stable than the products of the second step, the final products are less stable than the initial reactants, and the second step is the rate-determining step. Label the reactants, products, intermediates, and transition states.

35. Molly Kule was a lab technician who was asked by her supervisor to add names to the labels on a collection of alkenes that showed only structures on the labels. How many did Molly get right? Correct the incorrect names.
  a. 3-pentene $\qquad\qquad\qquad$ **e.** 5-ethylcyclohexene $\qquad\qquad$ **i.** 2-methylcyclopentene
  b. 2-octene $\qquad\qquad\qquad$ **f.** 5-chloro-3-hexene $\qquad\qquad\quad$ **j.** 2-ethyl-2-butene
  c. 2-vinylpentane $\qquad\qquad$ **g.** 5-bromo-2-pentene
  d. 1-ethyl-1-pentene $\qquad\;$ **h.** (*E*)-2-methyl-1-hexene

36. Determine the number of double bonds and/or π bonds and then draw possible structures for compounds with the following molecular formulas:
  a. C_3H_6 $\qquad$ **b.** C_3H_4 $\qquad$ **c.** C_4H_6

37. Draw a reaction coordinate diagram for the following reaction in which C is the most stable and B is the least stable of the three species and the transition state going from A to B is more stable than the transition state going from B to C:

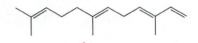

$$A \underset{k_{-1}}{\overset{k_1}{\rightleftharpoons}} B \underset{k_{-2}}{\overset{k_2}{\rightleftharpoons}} C$$

   a. How many intermediates are there?
   b. How many transition states are there?
   c. Which step has the greater rate constant in the forward direction?
   d. Which step has the greater rate constant in the reverse direction?
   e. Of the four steps, which has the greatest rate constant?
   f. Which is the rate-determining step in the forward direction?
   g. Which is the rate-determining step in the reverse direction?

38. The rate constant for a reaction can be increased by _____ the stability of the reactant or by _____ the stability of the transition state.

39. α-Farnesene is a compound found in the waxy coating of apple skins. To complete its systematic name, include the E or Z designation after the number indicating the location of the double bond.

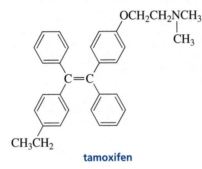

α-**farnesene**
3,7,11-trimethyl-(1,3?,6?,10)-dodecatetraene

40. Give the structures and the systematic names for all alkenes with molecular formula C_6H_{12}, ignoring cis–trans isomers. (*Hint:* There are 13.)

   a. Which of the compounds have E and Z isomers?   **b.** Which of the compounds is the most stable?

41. Tamoxifen slows the growth of some breast tumors by binding to estrogen receptors. Is tamoxifen an E or a Z isomer?

tamoxifen

CHAPTER 5

The Reactions of Alkenes and Alkynes

An Introduction to Multistep Synthesis

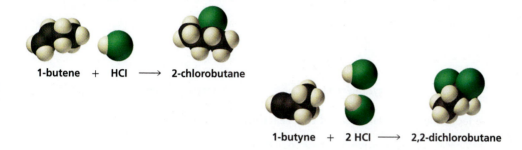

1-butene   +   HCl   ⟶   2-chlorobutane

1-butyne   +   2 HCl   ⟶   2,2-dichlorobutane

W e will start this chapter by looking at the reactions that alkenes undergo. You will see that all the reactions take place by similar mechanisms. As you study each reaction, look for the feature that all alkene reactions have in common: *the relatively loosely held π electrons of the carbon–carbon double bond are attracted to an electrophile. Thus, each reaction starts with the addition of an electrophile to one of the sp² carbons of the alkene and concludes with the addition of a nucleophile to the other sp² carbon.* The end result is that the π bond breaks because it is weaker than the σ bond (Section 1.14), and the *sp²* carbons form new σ bonds with the electrophile and the nucleophile.

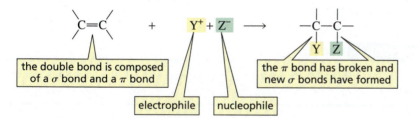

the double bond is composed of a σ bond and a π bond

electrophile

nucleophile

the π bond has broken and new σ bonds have formed

Because both the electrophile and the nucleophile add to the double bond, and the electrophile is the *first* species that adds, this characteristic reaction of alkenes is called an **electrophilic addition reaction.**

This reactivity makes alkenes an important class of organic compounds because they can be used to synthesize a wide variety of other compounds. For example, we will see that alkyl halides, alcohols, ethers, and alkanes all can be synthesized from alkenes by electrophilic addition reactions. The particular product obtained depends only on the *electrophile* and the *nucleophile* used in the addition reaction.

5.1 THE ADDITION OF A HYDROGEN HALIDE TO AN ALKENE

We have seen that an alkene undergoes an **electrophilic addition reaction** with a hydrogen halide (HF, HCl, HBr, or HI); the proton is the electrophile that adds to one of the sp^2 carbons, and the halide ion is the nucleophile that adds to the other sp^2 carbon. Therefore, the product of the reaction is an alkyl halide (Section 4.7).

2,3-dimethyl-2-butene 2-bromo-2,3-dimethylbutane

cyclohexene iodocyclohexane

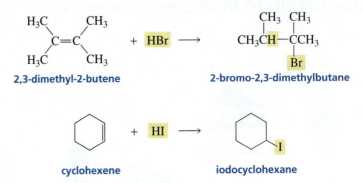

We looked at the **mechanism of this reaction** in Section 4.7.

mechanism for the addition of a hydrogen halide

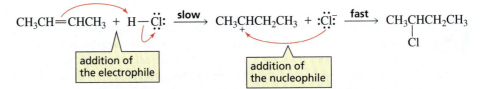

Mechanistic Tutorial: Addition of HBr to an alkene

- The first step of the reaction is a relatively slow addition of the proton (an electrophile) to the alkene (a nucleophile) to form a carbocation intermediate.
- In the second step, the positively charged carbocation intermediate (an electrophile) reacts rapidly with the negatively charged chloride ion (a nucleophile).

PROBLEM 1

Write the mechanism for the reaction of cyclohexene with HI (the second reaction shown above).

Because the alkenes in the two preceding reactions have the same substituents on both sp^2 carbons, it is easy to predict the product of the reaction: the electrophile (H^+) adds to one of the sp^2 carbons, and the nucleophile (X^-) adds to the other sp^2 carbon. It doesn't matter which sp^2 carbon the electrophile attaches to, because the same product will be obtained in either case.

But what happens if the alkene does not have the same substituents on both of the sp^2 carbons? Which sp^2 carbon gets the hydrogen? For example, does the addition of HCl to 2-methylpropene produce *tert*-butyl chloride or isobutyl chloride?

2-methylpropene 2-chloro-2-methylpropane *tert*-butyl chloride 1-chloro-2-methylpropane isobutyl chloride

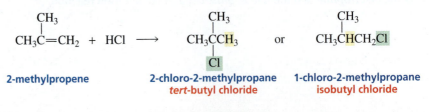

To answer this question, we need to carry out the reaction, isolate the products, and identify them. When we do, we find that the only product of the reaction is *tert*-butyl chloride. Now we need to find out why that compound is the product of the reaction so we can use this knowledge to predict the products of other alkene reactions. To do this, we need to look at the mechanism of the reaction.

The first step of the reaction—the addition of H$^+$ to an *sp^2* carbon to form either the *tert*-butyl cation or the isobutyl cation—is the rate-determining step (Section 4.8). If there is any difference in the rate of formation of these two carbocations, then the one that is formed faster will be the predominant product of the first step. Moreover, because carbocation formation is rate determining, the particular carbocation that is formed in the first step determines the final product of the reaction. That is, if the *tert*-butyl cation is formed, it will react rapidly with Cl$^-$ to form *tert*-butyl chloride. On the other hand, if the isobutyl cation is formed, it will react rapidly with Cl$^-$ to form isobutyl chloride. Since the only product of the reaction is *tert*-butyl chloride, we know that the *tert*-butyl cation is formed much faster than the isobutyl cation.

The *sp^2* carbon that does *not* become attached to the proton is the carbon that is positively charged in the carbocation.

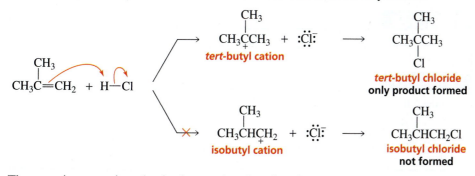

The question now is, why is the *tert*-butyl cation formed faster than the isobutyl cation? To answer this question, we need to take a look at the factors that affect the stability of carbocations and thus the ease with which they are formed.

5.2 CARBOCATION STABILITY DEPENDS ON THE NUMBER OF ALKYL SUBSTITUENTS ATTACHED TO THE POSITIVELY CHARGED CARBON

Carbocations are classified according to the number of alkyl substituents that are bonded to the positively charged carbon: a **primary carbocation** has one alkyl substituent, a **secondary carbocation** has two, and a **tertiary carbocation** has three. The stability of a carbocation increases as the number of alkyl substituents bonded to the positively charged carbon increases. Thus, tertiary carbocations are more stable than secondary carbocations, and secondary carbocations are more stable than primary carbocations.

Carbocation stability: 3° > 2° > 1°

relative stabilities of carbocations

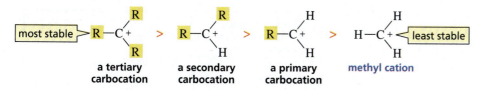

The reason for this pattern of stability is that alkyl groups bonded to the positively charged carbon decrease the concentration of positive charge on the carbon since they can donate electrons through the σ bond better than hydrogens can; decreasing the concentration of positive charge makes the carbocation more stable. Notice that the blue (representing positive charge in these electrostatic potential maps, Section 1.3) is most intense for the least stable methyl cation and is least intense for the most stable *tert*-butyl cation.

The greater the number of alkyl substituents bonded to the positively charged carbon, the more stable the carbocation is.

Alkyl substituents stabilize both alkenes *and* carbocations.

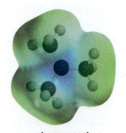

electrostatic
potential map
for the *tert*-butyl cation

electrostatic
potential map
for the isopropyl cation

electrostatic
potential map
for the ethyl cation

electrostatic
potential map
for the methyl cation

PROBLEM 2 ◆

Which is more stable, a methyl cation or an ethyl cation?

PROBLEM 3 ◆

List the following carbocations in order of decreasing stability:

$$\underset{+}{CH_3CH_2\overset{\displaystyle CH_3}{\underset{|}{C}}CH_3} \qquad CH_3CH_2\underset{+}{CHCH_3} \qquad CH_3CH_2CH_2\overset{+}{CH_2}$$

Now we can understand why the *tert*-butyl cation is formed faster than the isobutyl cation when 2-methylpropene reacts with HCl. We know that the *tert*-butyl cation (a tertiary carbocation) is more stable than the isobutyl cation (a primary carbocation). The same factors that stabilize the positively charged carbocation also stabilize the transition state for its formation because the transition state has a partial positive charge (Section 4.8). Therefore, the transition state leading to the *tert*-butyl cation is more stable (that is, lower in energy) than the transition state leading to the isobutyl cation (Figure 5.1).

We have seen that the rate of a reaction is determined by the free energy of activation ($\Delta G^{\ddagger}$), which is the difference between the free energy of the transition state and the free energy of the reactant: the more stable the transition state, the smaller is the free energy of activation, and therefore, the faster is the reaction (Section 4.8). Thus, the *tert*-butyl cation will be formed faster than the isobutyl cation.

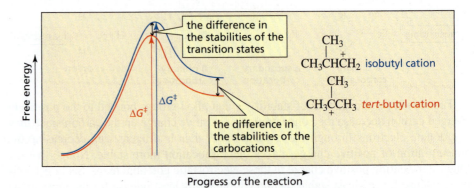

▶ **Figure 5.1**
The reaction coordinate diagram for the addition of H⁺ to 2-methylpropene to form the primary isobutyl cation and the tertiary *tert*-butyl cation.

5.3 ELECTROPHILIC ADDITION REACTIONS ARE REGIOSELECTIVE

We have just seen that the major product of an electrophilic addition reaction is the one obtained by adding the electrophile to the sp^2 carbon that results in the formation of the more stable carbocation. For example, when propene reacts with HCl, the proton can add to the number-1 carbon (C-1) to form a secondary carbocation, or it can add to the number-2 carbon (C-2) to form a primary carbocation. The secondary carbocation is formed more rapidly because it is more stable than the primary carbocation. (Primary carbocations are so unstable that they form only with great difficulty.) The product of the reaction, therefore, is 2-chloropropane.

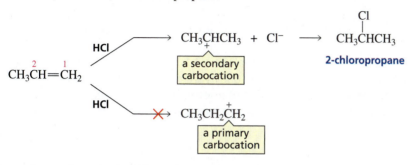

The major product obtained from the addition of HI to 2-methyl-2-butene is 2-iodo-2-methylbutane; only a small amount of 2-iodo-3-methylbutane is obtained. The major product obtained from the addition of HBr to 1-methylcyclohexene is 1-bromo-1-methylcyclohexane. In both cases, the more stable tertiary carbocation is formed more rapidly than the less stable secondary carbocation, so the major product of each reaction is the one that results from forming the tertiary carbocation.

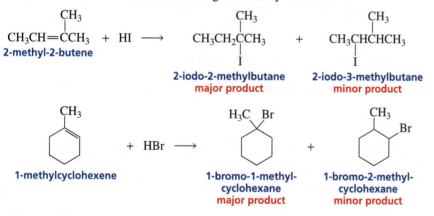

The two different products of each of these reactions are *constitutional isomers*. We saw at the beginning of Chapter 3 that **constitutional isomers** have the same molecular formula, but differ in how their atoms are connected. A reaction (such as either of those just shown) in which two or more constitutional isomers could be obtained as products, but one of them predominates, is called a **regioselective reaction**.

The addition of HBr to 2-pentene is not regioselective. Because the addition of H^+ to either of the sp^2 carbons produces a secondary carbocation, both carbocation intermediates have the same stability, so both will be formed equally easily. Thus, approximately equal amounts of the two alkyl halides will be formed.

> **Regioselectivity is the preferential formation of one constitutional isomer over another.**

$$CH_3CH=CHCH_2CH_3 \ + \ HBr \ \longrightarrow \ \underset{\substack{\textbf{2-bromopentane} \\ \textbf{50\%}}}{CH_3\overset{\overset{\textstyle Br}{|}}{C}HCH_2CH_2CH_3} \ + \ \underset{\substack{\textbf{3-bromopentane} \\ \textbf{50\%}}}{CH_3CH_2\overset{\overset{\textstyle Br}{|}}{C}HCH_2CH_3}$$

2-pentene

Vladimir Vasilevich Markovnikov (1837–1904) *was born in Russia, the son of an army officer. He was a professor of chemistry at Kazan, Odessa, and Moscow Universities.*

The electrophile adds to the *sp*² carbon that is bonded to the greater number of hydrogens.

From the alkene reactions we have seen so far, we can devise a rule that applies to *all* alkene electrophilic addition reactions: *the* **electrophile** *adds to the* sp² *carbon that is bonded to the greater number of hydrogens.* Vladimir Markovnikov was the first to recognize that in the addition of a hydrogen halide to an alkene, the H⁺ adds to the *sp²* carbon that is bonded to the greater number of hydrogens. As a result, this rule is often referred to as **Markovnikov's rule**.

This rule is simply a quick way to determine the relative stabilities of the intermediates that could be formed in the rate-determining step. You will get the same answer, whether you identify the major product of an electrophilic addition reaction by using the rule or whether you identify it by determining relative carbocation stabilities. In the following reaction, for example, H⁺ is the electrophile:

$$CH_3CH_2\overset{2}{C}H=\overset{1}{C}H_2 \ + \ HCl \ \longrightarrow \ CH_3CH_2\overset{\overset{\textstyle Cl}{|}}{C}HCH_3$$

We can say that H⁺ adds preferentially to C-1 because C-1 is bonded to two hydrogens, whereas C-2 is bonded to only one hydrogen. Or we can say that H⁺ adds to C-1 because that results in the formation of a secondary carbocation, which is more stable than the primary carbocation that would have to be formed if H⁺ added to C-2.

PROBLEM 4◆

What would be the major product obtained from the addition of HBr to each of the following compounds?

a. $CH_3CH_2CH=CH_2$

b. $CH_3CH=\overset{\overset{\textstyle CH_3}{|}}{C}CH_3$

c.

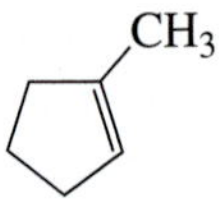

d. $CH_2=\overset{\overset{\textstyle CH_3}{|}}{C}CH_2CH_2CH_3$

e.

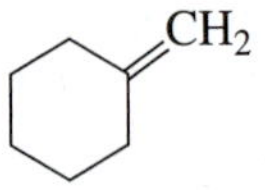

f. $CH_3CH=CHCH_3$

PROBLEM-SOLVING STRATEGY

a. What alkene should be used to synthesize 3-bromohexane?

$$? \ + \ HBr \ \longrightarrow \ CH_3CH_2\overset{\overset{\textstyle }{}}{C}HCH_2CH_2CH_3$$
$$|$$
$$Br$$

3-bromohexane

The best way to answer this kind of question is to begin by listing all the alkenes that could be used. Because you want to synthesize an alkyl halide that has a bromo substituent at the 3-position, the alkene should have an *sp²* carbon at that position. Two alkenes fit the description: 2-hexene and 3-hexene.

$$CH_3CH=CHCH_2CH_2CH_3 \qquad CH_3CH_2CH=CHCH_2CH_3$$
2-hexene **3-hexene**

Because there are two possibilities, we next need to determine whether there is any advantage to using one over the other. The addition of H⁺ to 2-hexene can form two different carbocations, but they are both secondary carbocations. Because they have the same stability, approximately equal amounts of each will be formed. Therefore, half of the product will be the desired 3-bromohexane and half will be 2-bromohexane.

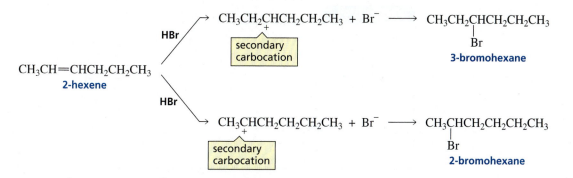

The addition of H$^+$ to either of the sp^2 carbons of 3-hexene, on the other hand, forms the same carbocation because the alkene is symmetrical. Therefore, all of the product will be the desired 3-bromohexane.

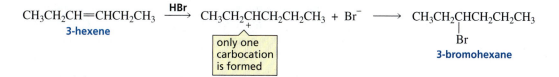

Because all the alkyl halide formed from 3-hexene is 3-bromohexane, but only half the alkyl halide formed from 2-hexene is 3-bromohexane, 3-hexene is the best alkene to use to prepare 3-bromohexane.

b. What alkene should be used to synthesize 2-bromopentane?

?   +   HBr $\longrightarrow$ CH$_3$CHCH$_2$CH$_2$CH$_3$

|

Br

2-bromopentane

Either 1-pentene or 2-pentene could be used because both have an sp^2 carbon at the C-2 position.

CH$_2$=CHCH$_2$CH$_2$CH$_3$ CH$_3$CH=CHCH$_2$CH$_3$

1-pentene   **2-pentene**

When H$^+$ adds to 1-pentene, one of the carbocations that could be formed is secondary and the other is primary. A secondary carbocation is more stable than a primary carbocation, which is so unstable that none will be formed. Thus, 2-bromopentane will be the only product of the reaction.

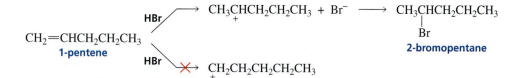

When H$^+$ adds to 2-pentene, on the other hand, each of the two carbocations that can be formed is secondary. Both are equally stable, so they will be formed in approximately equal amounts. Thus, only about half of the product of the reaction will be the desired 2-bromopentane. The other half will be 3-bromopentane.

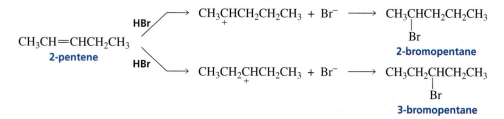

Because all the alkyl halide formed from 1-pentene is 2-bromopentane, but only half the alkyl halide formed from 2-pentene is 2-bromopentane, 1-pentene is the best alkene to use to prepare 2-bromopentane.

Now continue on to Problem 5.

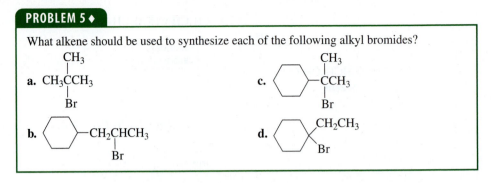

PROBLEM 5 ◆

What alkene should be used to synthesize each of the following alkyl bromides?

5.4 THE ADDITION OF WATER TO AN ALKENE

An alkene does not react with water because there is no electrophile present to start a reaction by adding to the nucleophilic alkene. The O—H bonds of water are too strong—water is too weakly acidic—to allow the hydrogen to act as an electrophile for this reaction.

$$CH_3CH{=}CH_2 \ + \ H_2O \ \longrightarrow \ \text{no reaction}$$

If an acid is added to the solution (the acid most often used is H_2SO_4), then the outcome is much different: a reaction will occur because the acid provides an electrophile (H^+). The product of the reaction is an alcohol. The addition of water to a molecule is called **hydration**, so we can say that an alkene will be *hydrated* in the presence of water and acid.

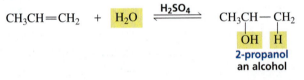

Because H_2SO_4 is a strong acid, it dissociates almost completely in water (Section 2.2). The acid that participates in the reaction, therefore, is most apt to be a hydronium ion.

$$H_2SO_4 \ + \ H_2O \ \rightleftharpoons \ H_3O^+ \ + \ HSO_4^-$$

hydronium ion

The acid is a catalyst—it increases the *rate* at which a product is formed, but it does not affect the *amount* of product formed. Because the catalyst in the hydration of an alkene is an acid, hydration is an **acid-catalyzed reaction**.

Notice that the first two steps of *the mechanism for the acid-catalyzed addition of water to an alkene* are essentially the same as the two steps of *the mechanism for the addition of a hydrogen halide to an alkene*:

mechanism for the acid-catalyzed addition of water

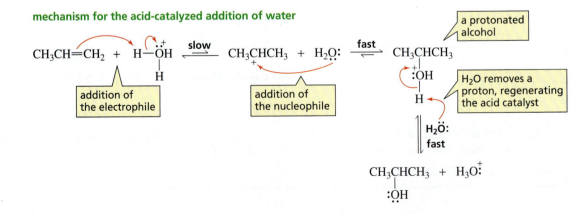

- The electrophile (H⁺) adds to the sp^2 carbon that is bonded to the greater number of hydrogens.
- The nucleophile (H_2O) adds to the carbocation, forming a protonated alcohol.
- The protonated alcohol is a very strong acid (Section 2.2), so it loses a proton. The final product of the addition reaction is an alcohol; the regenerated acid catalyst returns to the reaction mixture.

As we saw in Section 4.7, the addition of the electrophile to the alkene is relatively slow, and the subsequent addition of the nucleophile to the carbocation occurs rapidly. The reaction of the carbocation with a nucleophile is so fast, in fact, that the carbocation combines with whatever nucleophile it collides with: note that there are two nucleophiles in solution—water and the conjugate base of the acid (HSO_4^-) that was used to start the reaction. Because the concentration of water is much greater than the concentration of the conjugate base, the carbocation is much more likely to collide with water. The product of the collision is a protonated alcohol. Notice that the catalyst is not consumed or changed during the reaction.

Movies:
The First Step to Hydration of an Alkene; The Second Step to Hydration of an Alkene; The Third Step to Hydration of an Alkene.

Mechanistic Tutorial:
Addition of water to an alkene

Do not memorize the products of alkene addition reactions. Instead, for each reaction, ask yourself, "What is the electrophile?" and "What nucleophile is present in the greatest concentration?"

PROBLEM 6 ◆

Give the major product obtained from the acid-catalyzed hydration of each of the following alkenes:

a. $CH_3CH_2CH_2CH=CH_2$

c. $CH_3CH_2CH_2CH=CHCH_3$

b.

d.

5.5   THE ADDITION OF AN ALCOHOL TO AN ALKENE

Alcohols react with alkenes in the same way that water does, so this reaction too requires an acid catalyst. The product of the reaction is an ether.

$$CH_3CH=CH_2 \ + \ CH_3OH \ \xrightleftharpoons{H_2SO_4} \ CH_3CH-CH_2$$
$$\underset{\substack{| \quad | \\ OCH_3 \ H}}{}$$

2-methoxypropane
an ether

The mechanism for the acid-catalyzed addition of an alcohol is essentially the same as the mechanism for the acid-catalyzed addition of water. The only difference is that the nucleophile is ROH instead of HOH.

mechanism for the acid-catalyzed addition of an alcohol

$$CH_3CH=CH_2 \ + \ \underset{\underset{H}{|}}{H-\overset{..}{\overset{+}{O}}CH_3} \ \xrightleftharpoons{\textbf{slow}} \ \underset{+}{CH_3CHCH_3} \ + \ CH_3\overset{..}{\underset{..}{O}}H \ \xrightleftharpoons{\textbf{fast}} \ \underset{\substack{| \\ \overset{+}{:}OCH_3 \\ | \\ H}}{CH_3CHCH_3}$$

$$\Bigg\| \ CH_3\overset{..}{\underset{..}{O}}H$$
fast

$$\underset{\substack{| \\ :\underset{..}{O}CH_3}}{CH_3CHCH_3} \ + \ \underset{\substack{| \\ H}}{CH_3\overset{..}{\overset{+}{O}}H}$$

PROBLEM 7

a. Give the major product of each of the following reactions:

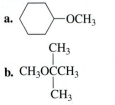

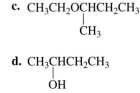

b. What do all the reactions have in common?
c. How do all the reactions differ?

PROBLEM 8

How could the following compounds be prepared, using an alkene as one of the starting materials?

a. ⬡—OCH$_3$

b. CH$_3$OCCH$_3$ (with CH$_3$ above and CH$_3$ below)

c. CH$_3$CH$_2$OCHCH$_2$CH$_3$ (with CH$_3$ below)

d. CH$_3$CHCH$_2$CH$_3$ (with OH below)

PROBLEM 9 ♦

When chemists write reactions, they show reaction conditions, such as the solvent, the temperature, and any required catalyst above or below the arrow.

$$CH_2{=}CHCH_2CH_3 \ + \ H_2O \ \xrightarrow{\text{H}_2\text{SO}_4} \ CH_3CHCH_2CH_3$$
(with OH below)

Sometimes reactions are written by placing only the organic (carbon-containing) reagent on the left-hand side of the arrow; the other reagents are written above or below the arrow.

$$CH_2{=}CHCH_2CH_3 \ \xrightarrow[\text{H}_2\text{O}]{\text{H}_2\text{SO}_4} \ CH_3CHCH_2CH_3$$
(with OH below)

There are two nucleophiles in each of the following reactions. For each reaction, explain why there is a greater concentration of one nucleophile than the other. What will be the major product of each reaction?

a. $CH_3CH{=}CHCH_3 \ + \ H_2O \ \xrightarrow{\text{H}_2\text{SO}_4}$ **b.** $CH_3CH{=}CHCH_3 \ \xrightarrow[\text{CH}_3\text{OH}]{\text{H}_2\text{SO}_4}$

PROBLEM 10

Give the major product obtained from the reaction of HBr with each of the following:

a. 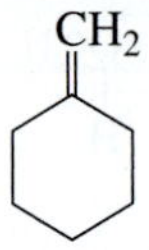 (cyclohexane ring with CH$_2$ double bond)

b. CH$_3$CHCH$_2$CH$=$CH$_2$ (with CH$_3$ below)

c. 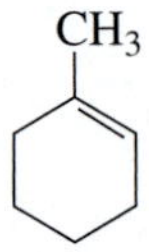 (cyclohexene ring with CH$_3$)

d. 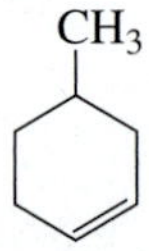 (cyclohexene ring with CH$_3$)

5.6 AN INTRODUCTION TO ALKYNES

An **alkyne** is a hydrocarbon that contains a carbon–carbon triple bond. Because of its triple bond, an alkyne has four fewer hydrogens than an alkane with the same number of carbons. Therefore, while the general molecular formula for an alkene is C_nH_{2n+2}, the general molecular formula for a noncyclic alkyne is C_nH_{2n-2}.

The few drugs in clinical use that contain alkyne functional groups are not naturally occurring compounds; they exist only because chemists have been able to synthesize them. Their brand names are shown below, in green. Brand names are always capitalized; only the company that holds the patent for a product can use the product's brand name for commercial purposes (Section 21.1).

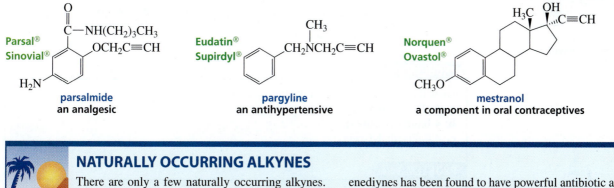

Parsal®
Sinovial®

parsalmide
an analgesic

Eudatin®
Supirdyl®

pargyline
an antihypertensive

Norquen®
Ovastol®

mestranol
a component in oral contraceptives

NATURALLY OCCURRING ALKYNES

There are only a few naturally occurring alkynes. Examples include capillin, which has fungicidal activity, and ichthyothereol, a convulsant used by the indigenous people of the Amazon for poisoned arrowheads. A class of naturally occurring compounds called enediynes has been found to have powerful antibiotic and anticancer properties. These compounds all have a nine- or ten-membered ring that contains two triple bonds separated by a double bond. Some enediynes are currently being tested in clinical trials (Section 21.10).

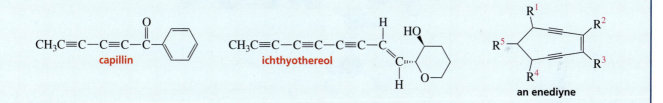

capillin

ichthyothereol

an enediyne

PROBLEM 11◆

What is the general molecular formula for a cyclic alkyne?

PROBLEM 12◆

What is the molecular formula for a cyclic hydrocarbon with 14 carbons and two triple bonds?

5.7 THE NOMENCLATURE OF ALKYNES

The systematic name of an alkyne is obtained by replacing the "ane" ending of the alkane name with "yne." Analogous to the way compounds with other functional groups are named, the longest continuous chain containing the carbon–carbon triple bond is numbered in the direction that gives the alkyne functional group suffix as low a number as possible. If the triple bond is at the end of the chain, the alkyne is classified as a **terminal alkyne**. Alkynes with triple bonds located elsewhere along the chain are called **internal alkynes**. For example, 1-butyne is a terminal alkyne, whereas 2-pentyne is an internal alkyne.

1-hexyne
a terminal alkyne

3-hexyne
an internal alkyne

$$\overset{4\ \ 3\ \ 2\ \ 1}{HC\equiv CH} \qquad \overset{4\ \ 3\ \ 2\ \ 1}{CH_3CH_2C\equiv CH} \qquad \overset{1\ \ 2\ \ 3\ 4\ \ 5}{CH_3C\equiv CCH_2CH_3} \qquad \overset{5\ \ 6}{CH_2CH_3}$$

systematic:	ethyne	1-butyne	2-pentyne
common:	acetylene	a terminal alkyne	an internal alkyne

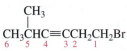

4-methyl-2-hexyne

If counting from either direction leads to the same number for the functional group suffix, the correct systematic name is the one that contains the lowest substituent number. If the compound contains more than one substituent, the substituents are listed in alphabetical order.

$$\underset{6\ \ \ 5\ \ \ 4\ \ \ \ 3\ 2\ \ \ 1}{\overset{CH_3}{CH_3CHC\equiv CCH_2CH_2Br}} \qquad \underset{1\ \ \ 2\ \ \ 3\ \ \ 4\ \ \ \ 5\ 6\ \ \ 7\ \ \ 8}{\overset{Cl\ \ Br}{CH_3CHCHC\equiv CCH_2CH_2CH_3}}$$

1-bromo-5-methyl-3-hexyne   **3-bromo-2-chloro-4-octyne**
not **6-bromo-2-methyl-3-hexyne**   *not* **6-bromo-7-chloro-4-octyne**
because 1 < 2   because 2 < 6

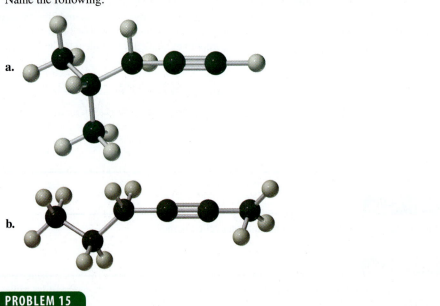

PROBLEM 13 ◆

Draw the structure for each of the following:
a. 1-chloro-3-hexyne
b. 4-bromo-2-pentyne
c. 4,4-dimethyl-1-pentyne

PROBLEM 14 ◆

Name the following:

a.

b.

PROBLEM 15

Draw the structures and give the systematic names for the seven alkynes with molecular formula C_6H_{10}.

PROBLEM 16 ◆

Give the systematic name for each of the following:
a. $BrCH_2CH_2C\equiv CCH_3$

b. $CH_3CH_2\underset{|}{C}HC\equiv CC\underset{|}{H_2}CHCH_3$
               Br   Cl

c. $CH_3CH_2CHC\equiv CCH_2CH_3$
                   CH_3

d. $CH_3CH_2CHC\equiv CH$
            $CH_2CH_2CH_3$

3-D Molecules:
1-Hexyne; 3-Hexyne

5.8 THE STRUCTURE OF ALKYNES

The structure of ethyne was discussed in Section 1.9. We saw that each carbon is *sp* hybridized, so each has two *sp* orbitals and two *p* orbitals. One *sp* orbital overlaps the *s* orbital of a hydrogen, and the other overlaps an *sp* orbital of the other carbon. Because the *sp* orbitals are oriented as far from each other as possible to minimize electron repulsion, ethyne is a linear molecule with bond angles of 180°.

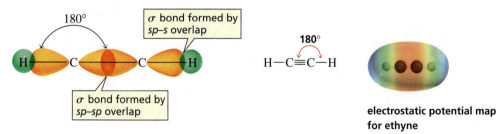

electrostatic potential map
for ethyne

The two remaining *p* orbitals on each carbon are oriented at right angles to one another and to the *sp* orbitals (Figure 5.2). Each of the two *p* orbitals on one carbon overlaps the parallel *p* orbital on the other carbon to form two π bonds. One pair of overlapping *p* orbitals results in a cloud of electrons above and below the σ bond, and the other pair results in a cloud of electrons in front of and behind the σ bond. The electrostatic potential map of ethyne shows that the end result can be thought of as a cylinder of electrons wrapped around the σ bond.

3-D Molecules:
Ethyne

A triple bond is composed of a σ bond and two π bonds.

a.

b.

◀ **Figure 5.2**
(a) Each of the two π bonds of a triple bond is formed by side-to-side overlap of a *p* orbital of one carbon with a parallel *p* orbital of the adjacent carbon. (b) A triple bond consists of a σ bond formed by *sp–sp* overlap (yellow) and two π bonds formed by *p–p* overlap (blue and purple).

PROBLEM 17 ◆

What orbitals are used to form the carbon–carbon σ bond between the highlighted carbons?

a. $CH_3CH{=}CHCH_3$ d. $CH_3C{\equiv}CCH_3$ g. $CH_3CH{=}CHCH_2CH_3$

b. $CH_3CH{=}CHCH_3$ e. $CH_3C{\equiv}CCH_3$ h. $CH_3C{\equiv}CCH_2CH_3$

c. $CH_3CH{=}C{=}CH_2$ f. $CH_2{=}CHCH{=}CH_2$ i. $CH_2{=}CHC{\equiv}CH$

Tutorial:
Orbitals used to form carbon–carbon single bonds

5.9 THE PHYSICAL PROPERTIES
OF UNSATURATED HYDROCARBONS

All hydrocarbons have similar physical properties. In other words, alkenes and alkynes have physical properties similar to those of alkanes (Section 3.7). All are insoluble in water and soluble in nonpolar solvents such as hexane. They are less dense than water and, like any other series of compounds, have boiling points that increase with increasing molecular weight (see Table 3.1 on page 47). Alkynes are more linear than alkenes, causing alkynes to have stronger van der Waals interactions. As a result, an alkyne has a higher boiling point than an alkene containing the same number of carbons (see Appendix I).

5.10 THE ADDITION OF A HYDROGEN HALIDE TO AN ALKYNE

With a cloud of electrons completely surrounding the σ bond, an alkyne is an electron-rich molecule. In other words, it is a nucleophile, and consequently it will react with an electrophile. Thus alkynes, like alkenes, undergo electrophilic addition reactions, and the same reagents that add to alkenes also add to alkynes. Moreover, the mechanism for electrophilic addition to an alkyne is the same as the mechanism for electrophilic addition to an alkene. For example, compare the mechanism for the addition of a hydrogen halide to an alkene shown in Section 5.1 with the mechanism for the addition of a hydrogen halide to an alkyne shown below.

mechanism for the addition of a hydrogen halide

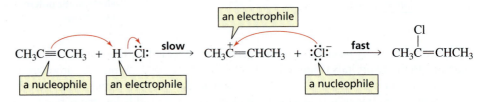

Tutorial:
Addition of HCl to an alkyne

- The relatively weak π bond breaks because the π electrons are attracted to the electrophilic proton.
- The positively charged carbocation intermediate reacts rapidly with the negatively charged chloride ion.

The addition reactions of alkynes, however, have a feature that alkenes do not have: because the product of the addition of an electrophilic reagent to an alkyne is an alkene, a second electrophilic addition reaction can occur if excess hydrogen halide is present. In the second addition reaction, the electrophile (H^+) adds to the sp^2 carbon bonded to the greater number of hydrogens—as predicted by the rule that governs electrophilic addition reactions (Section 5.3).

$$CH_3C{\equiv}CCH_3 \xrightarrow{\text{HCl}} CH_3\overset{\displaystyle Cl}{C}{=}CHCH_3 \xrightarrow{\text{HCl}} CH_3\underset{\displaystyle Cl}{\overset{\displaystyle Cl}{C}}CH_2CH_3$$

a second electrophilic addition reaction occurs

If the alkyne is a *terminal* alkyne, the H^+ will add to the *sp* carbon bonded to the hydrogen, because the *secondary* vinylic cation that results is more stable than the *primary* vinylic cation that would be formed if the H^+ added to the other *sp* carbon. (Recall that alkyl groups stabilize positively charged carbon atoms; see Section 5.2.)

The electrophile adds to the *sp* carbon of a terminal alkyne that is bonded to the hydrogen.

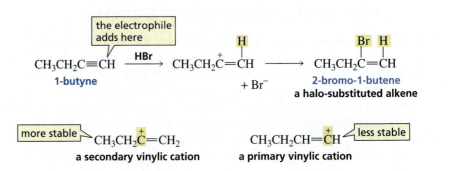

3-D Molecule:
Vinylic cation

A second addition reaction will take place if excess hydrogen halide is present. Once again, the electrophile (H⁺) adds to the sp^2 carbon that is bonded to the greater number of hydrogens.

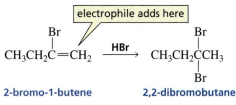

2-bromo-1-butene   2,2-dibromobutane

Addition of a hydrogen halide to an *internal* alkyne forms two products, because the initial addition of the proton can occur with equal ease to either of the *sp* carbons.

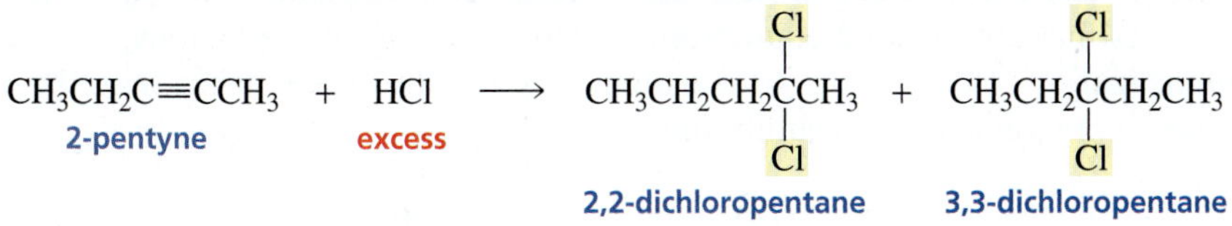

2,2-dichloropentane   3,3-dichloropentane

Note, however, that if the same group is attached to each of the *sp* carbons of the internal alkyne, only one product will be obtained.

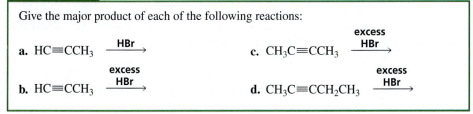

PROBLEM 18 ♦

Give the major product of each of the following reactions:

a. HC≡CCH₃ $\xrightarrow{\text{HBr}}$

b. HC≡CCH₃ $\xrightarrow[\text{HBr}]{\text{excess}}$

c. CH₃C≡CCH₃ $\xrightarrow[\text{HBr}]{\text{excess}}$

d. CH₃C≡CCH₂CH₃ $\xrightarrow[\text{HBr}]{\text{excess}}$

5.11 THE ADDITION OF WATER TO AN ALKYNE

In Section 5.4, we saw that alkenes undergo the acid-catalyzed addition of water. The product of the reaction is an alcohol.

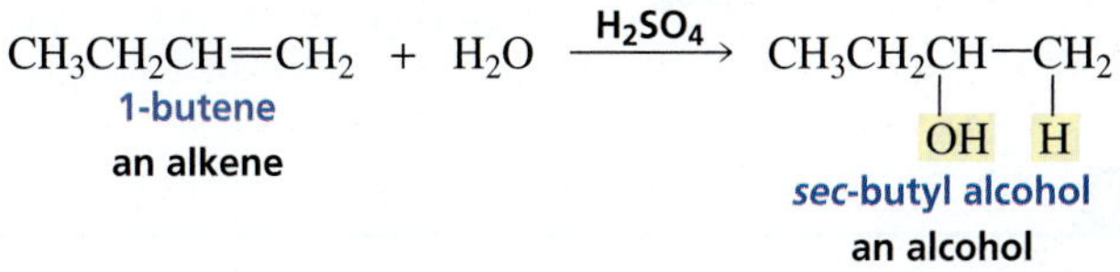

$$CH_3CH_2CH{=}CH_2 + H_2O \xrightarrow{H_2SO_4} CH_3CH_2CH{-}CH_2$$

1-butene
an alkene

sec-butyl alcohol
an alcohol

Alkynes also undergo the acid-catalyzed addition of water. The initial product of the reaction is an *enol*. An **enol** is a compound with a carbon–carbon double bond and an OH group bonded to one of the sp^2 carbons. (The ending "ene" signifies the double bond, and "ol" the OH group; when the two are joined, the second *e* of "ene" is dropped to avoid two consecutive vowels, but the word is pronounced as if the *e* were still there: "ene-ol.")

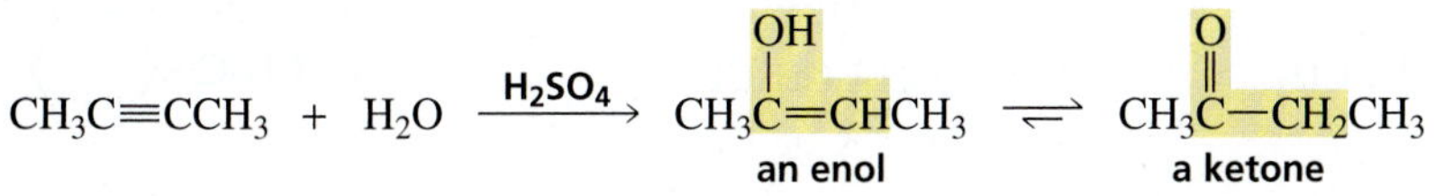

$$CH_3C{\equiv}CCH_3 + H_2O \xrightarrow{H_2SO_4} CH_3C{=}CHCH_3 \rightleftharpoons CH_3C{-}CH_2CH_3$$

an enol   a ketone

Addition of water to an alkyne forms a ketone.

The enol immediately rearranges to a *ketone*, a compound with the general structure shown below. A carbon doubly bonded to an oxygen is called a **carbonyl** ("car-bo-nil") **group**; a **ketone** ("key-tone") is a compound that has two alkyl groups bonded to a carbonyl group.

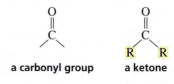

a carbonyl group   a ketone

A ketone and an enol differ only in the location of a double bond and a hydrogen. The ketone and enol are called **keto–enol tautomers**. **Tautomers** ("taw-toe-mers") are isomers that are in rapid equilibrium. Interconversion of the tautomers is called **tautomerization**. Because the keto tautomer is usually more stable than the enol tautomer, it predominates at equilibrium.

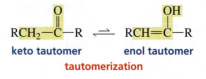

$$\underset{\text{keto tautomer}}{RCH_2{-}\overset{O}{\overset{\|}{C}}{-}R} \rightleftharpoons \underset{\text{enol tautomer}}{RCH{=}\overset{OH}{\overset{|}{C}}{-}R}$$

tautomerization

The addition of water to an internal alkyne that has the same group attached to each of the *sp* carbons forms a single ketone as a product.

$$CH_3CH_2C{\equiv}CCH_2CH_3 + H_2O \xrightarrow{H_2SO_4} CH_3CH_2\overset{O}{\overset{\|}{C}}CH_2CH_2CH_3$$

If the two groups are not identical, two ketones are formed because the initial addition of the proton can occur to either of the *sp* carbons.

$$CH_3C{\equiv}CCH_2CH_3 + H_2O \xrightarrow{H_2SO_4} CH_3\overset{O}{\overset{\|}{C}}CH_2CH_2CH_3 + CH_3CH_2\overset{O}{\overset{\|}{C}}CH_2CH_3$$

Terminal alkynes are less reactive than internal alkynes toward the addition of water. Terminal alkynes will add water if mercuric ion (Hg^{2+}) is added to the acidic mixture. The mercuric ion is a catalyst—it increases the rate of the addition reaction.

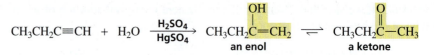

$$CH_3CH_2C{\equiv}CH + H_2O \xrightarrow[HgSO_4]{H_2SO_4} \underset{\text{an enol}}{CH_3CH_2\overset{OH}{\overset{|}{C}}{=}CH_2} \rightleftharpoons \underset{\text{a ketone}}{CH_3CH_2\overset{O}{\overset{\|}{C}}{-}CH_3}$$

PROBLEM 19♦

What ketones would be formed from the acid-catalyzed addition of water to 3-heptyne?

PROBLEM 20♦

Which alkyne would be the best reagent to use for the synthesis of each of the following ketones?

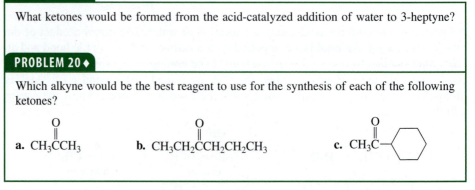

a. $CH_3\overset{O}{\overset{\|}{C}}CH_3$ **b.** $CH_3CH_2\overset{O}{\overset{\|}{C}}CH_2CH_2CH_3$ **c.**

PROBLEM 21 ◆

Draw the enol tautomers for the following ketone:

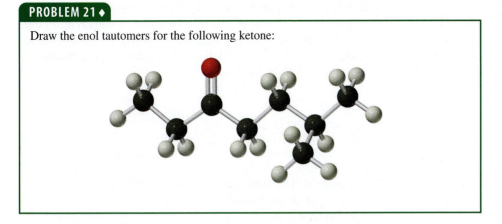

5.12 THE ADDITION OF HYDROGEN TO ALKENES AND ALKYNES

In the presence of a metal catalyst such as platinum or palladium, hydrogen (H_2) adds to the double bond of an alkene to form an alkane. Without the catalyst, the energy barrier to the reaction is enormous because the H—H bond is so strong. The catalyst decreases the energy of activation by breaking the H—H bond (Section 4.8). Platinum and palladium are used in a finely divided state adsorbed on charcoal (Pt/C, Pd/C).

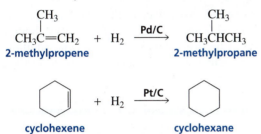

The addition of hydrogen is called **hydrogenation**. Because the preceding reactions require a catalyst, they are **catalytic hydrogenations**. A reaction that increases the number of C—H bonds in a compound is called a **reduction reaction**. Thus, catalytic hydrogenation is a reduction reaction.

The details of the mechanism of catalytic hydrogenation are not completely understood. We know that hydrogen is adsorbed on the surface of the metal and that all the bond-breaking and bond-forming events occur on the surface of the metal. As the alkane product is formed, it diffuses away from the metal surface (Figure 5.3).

A reduction reaction increases the number of C—H bonds.

Movie:
Catalytic hydrogenation
of ethylene

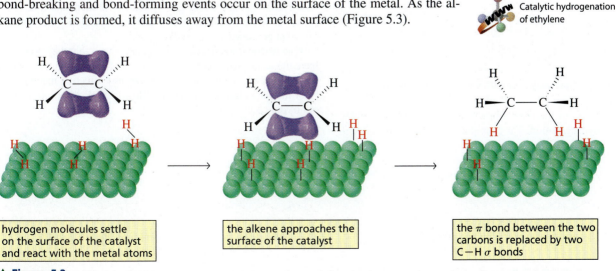

hydrogen molecules settle on the surface of the catalyst and react with the metal atoms

the alkene approaches the surface of the catalyst

the π bond between the two carbons is replaced by two C—H σ bonds

▲ **Figure 5.3**
Catalytic hydrogenation of an alkene.

Hydrogen adds to an alkyne in the presence of a metal catalyst such as palladium or platinum in the same manner that it adds to an alkene. The initial product is an alkene, but it is difficult to stop the reaction at that stage because of hydrogen's strong tendency to add to alkenes in the presence of these efficient metal catalysts. The product of the hydrogenation reaction, therefore, is an alkane.

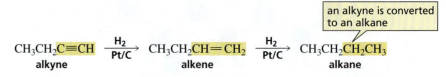

$$CH_3CH_2C{\equiv}CH \xrightarrow[\text{Pt/C}]{H_2} CH_3CH_2CH{=}CH_2 \xrightarrow[\text{Pt/C}]{H_2} CH_3CH_2CH_2CH_3$$

alkyne   alkene   alkane

an alkyne is converted to an alkane

The reaction can be stopped at the alkene stage if a "poisoned" (partially deactivated) metal catalyst is used. The most commonly used partially deactivated metal catalyst is called **Lindlar catalyst**.

Tutorial:
Hydrogenation/Lindlar catalyst

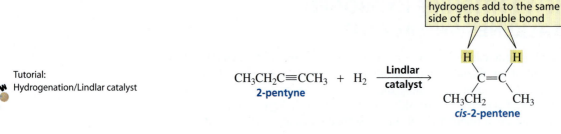

hydrogens add to the same side of the double bond

$$CH_3CH_2C{\equiv}CCH_3 + H_2 \xrightarrow{\text{Lindlar catalyst}}$$

2-pentyne

cis-2-pentene

Because the alkyne sits on the surface of the metal catalyst and the hydrogens are delivered to the triple bond from the surface of the catalyst, both hydrogens are delivered to the same side of the triple bond. Therefore, the addition of hydrogen to an internal alkyne in the presence of Lindlar catalyst forms a **cis alkene**.

TRANS FATS

Fats and oils contain long-chain unbranched carboxylic acids (called fatty acids) with carbon–carbon double bonds in a cis configuration. Fats are solids at room temperature, whereas oils are liquids at room temperature because they contain more double bonds; the double bonds prevent the molecules from packing tightly together (Section 19.1).

ing are prepared by hydrogenating vegetable oils such as soybean oil and safflower oil until they have the desired creamy, solid consistency.

All the double bonds in naturally occurring fats and oils have the cis configuration. The heat used in the hydrogenation process breaks the π bond of these double bonds. Sometimes, instead of being hydrogenated, a double bond reforms; if the sigma bond rotates while the π bond is broken, the double bond can reform in the trans configuration (Section 4.4), forming what is known as a trans fat.

One reason trans fats are a health concern is that they do not have the same shape as natural cis fats but are able to take the place of cis fats in cell membranes, thereby affecting the ability of the membrane to control the flow of molecules into and out of our cells.

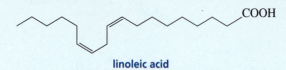

linoleic acid
an 18-carbon fatty acid with two cis double bonds

Some or all of the double bonds in oils can be reduced by catalytic hydrogenation. For example, margarine and shorten-

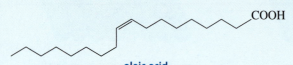

oleic acid
an 18-carbon fatty acid with one cis double bond
before being heated

an 18-carbon fatty acid with one trans double bond
after being heated

PROBLEM-SOLVING STRATEGY

What alkene would you use if you wanted to synthesize methylcyclohexane?

You need to choose an alkene that has the same number of carbons, attached in the same way, as those in the desired product. Several alkenes could be used for this synthesis, because the double bond can be located anywhere in the molecule.

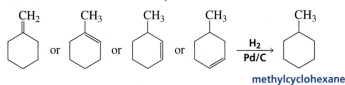

methylcyclohexane

Now continue on to answer the questions in Problem 22.

PROBLEM 22 ♦

What reagents would you use if you wanted to synthesize
a. *cis*-2-butene?
b. 1-hexene?

PROBLEM 23 ♦

How many different alkenes can be hydrogenated to form:
a. butane?
b. pentane?
c. methylcyclopentane?

5.13 A HYDROGEN BONDED TO AN *sp* CARBON IS ACIDIC

Carbon forms nonpolar covalent bonds with hydrogen because carbon and hydrogen, with similar electronegativities, share their bonding electrons almost equally. However, all carbon atoms do not have the same electronegativity. An *sp* carbon is more electronegative than an sp^2 carbon, which is more electronegative than an sp^3 carbon.

> An *sp* carbon is more electronegative than an sp^2 carbon, which is more electronegative than an sp^3 carbon.

relative electronegativities of carbon atoms

$$\boxed{\text{most electronegative}} \quad sp \;>\; sp^2 \;>\; sp^3 \quad \boxed{\text{least electronegative}}$$

Because the most acidic compound is the one with the hydrogen attached to the most electronegative atom (when the atoms are the same size; see Section 2.6), ethyne is a stronger acid than ethene, and ethene is a stronger acid than ethane. (Recall that the stronger the acid, the lower its pK_a.)

HC≡CH	$H_2C=CH_2$	CH_3CH_3
ethyne	ethene	ethane
pK_a = 25	pK_a = 44	pK_a > 60

In order to remove a proton from an acid (in a reaction that strongly favors products), the base that removes the proton must be stronger than the base that is generated as a result of removing the proton (Section 2.5). In other words, you must start with a stronger base than the base that will be formed. Because NH_3 is a weaker acid (pK_a = 36) than a terminal alkyne (pK_a = 25), an amide ion ($^-NH_2$) is a stronger base than the

The stronger the acid, the weaker its conjugate base.

To remove a proton from an acid in a reaction that favors products, the base that removes the proton must be stronger than the base that is formed.

carbanion—called an **acetylide ion**—that is formed when a hydrogen is removed from the *sp* carbon of a terminal alkyne. (Remember, the stronger the acid, the weaker its conjugate base.) Therefore, an amide ion can be used to remove a proton from a terminal alkyne to form an acetylide ion.

$$RC{\equiv}CH \quad + \quad {}^-NH_2 \quad \rightleftharpoons \quad RC{\equiv}C^- \quad + \quad NH_3$$

amide ion   acetylide ion

stronger acid   stronger base   weaker base   weaker acid

An amide ion cannot remove a hydrogen bonded to an sp^2 or an sp^3 carbon. Only a hydrogen bonded to an *sp* carbon is sufficiently acidic to be removed by an amide ion. Consequently, a hydrogen bonded to an *sp* carbon sometimes is referred to as an "acidic" hydrogen. The "acidic" property of terminal alkynes is one way their reactivity differs from that of alkenes. Be careful not to misinterpret what is meant when we say that a hydrogen bonded to an *sp* carbon is "acidic." It is more acidic than most other carbon-bound hydrogens, but it is much less acidic than a hydrogen of a water molecule, and water is only a very weakly acidic compound (pK_a = 15.7).

relative acid strengths

strongest acid											weakest acid
HF	>	H_2O	>	$HC{\equiv}CH$	>	NH_3	>	$H_2C{=}CH_2$	>	CH_3CH_3	
pK_a = 3.2		pK_a = 15.7		pK_a = 25		pK_a = 36		pK_a = 44		pK_a > 60	

PROBLEM 24 ◆

Explain why sodium amide cannot be used to form a carbanion from an alkane in a reaction that favors products.

PROBLEM-SOLVING STRATEGY

a. List the following compounds in order of decreasing acidity:

$$CH_3CH_2\overset{+}{N}H_3 \qquad CH_3CH{=}\overset{+}{N}H_2 \qquad CH_3CH{\equiv}\overset{+}{N}H$$

To compare the acidities of a group of compounds, first look at how the compounds differ. These three compounds differ in the hybridization of the nitrogen to which the acidic hydrogen is attached. Now recall what you know about hybridization and acidity. You know that hybridization of an atom affects its electronegativity (*sp* is more electronegative than sp^2, and sp^2 is more electronegative than sp^3), and you know that the more electronegative the atom to which a hydrogen is attached, the more acidic is the hydrogen. Now you can answer the question.

relative acidities $CH_3C{\equiv}\overset{+}{N}H$ > $CH_3CH{=}\overset{+}{N}H_2$ > $CH_3CH_2\overset{+}{N}H_3$

b. Draw the conjugate bases of the above compounds and list them in order of decreasing basicity. First remove a proton from each acid to obtain the structures of the conjugate bases. The stronger the acid, the weaker is its conjugate base, so using the relative acid strengths obtained in part a, we find that the order of basicity is.

relative basicities $CH_3CH_2NH_2$ > $CH_3CH{=}NH$ > $CH_3C{\equiv}N$

Now continue on to Problem 25.

PROBLEM 25 ♦

List the following species in order of decreasing basicity:

a. $CH_3CH_2CH=\bar{C}H$ $CH_3CH_2C\equiv C^-$ $CH_3CH_2CH_2\bar{C}H_2$

b. $CH_3CH_2O^-$ F^- $CH_3C\equiv C^-$ $^-NH_2$

PROBLEM 26   *SOLVED*

Which carbocation in each of the following pairs is more stable?

a. $CH_3\overset{+}{C}H_2$ or $H_2C=\overset{+}{C}H$ **b.** $H_2C=\overset{+}{C}H$ or $HC\equiv\overset{+}{C}$

Solution to 26a   A double-bonded carbon is more electronegative than a single-bonded carbon. Therefore, a double-bonded carbon with a positive charge would be less stable than a single-bonded carbon with a positive charge. Therefore, the ethyl carbocation is more stable.

5.14 SYNTHESIS USING ACETYLIDE IONS

Reactions that form carbon–carbon bonds are important in the synthesis of organic compounds because, without such reactions, we could not convert molecules with small carbon skeletons into molecules with larger carbon skeletons. Instead, the product of a reaction would always have the same number of carbons as the starting material.

One reaction that forms a carbon–carbon bond is the reaction of an acetylide ion with an alkyl halide. Only primary alkyl halides or methyl halides should be used in this reaction.

$CH_3CH_2C\equiv C^-$ + $CH_3CH_2CH_2Br$ $\longrightarrow$ $CH_3CH_2C\equiv CCH_2CH_2CH_3$ + Br^-
an acetylide ion   **an alkyl halide**   **3-heptyne**

The mechanism of this reaction is well understood. Bromine is more electronegative than carbon, and as a result, the electrons in the C—Br bond are not shared equally by the two atoms. There is a partial positive charge on carbon and a partial negative charge on bromine. The negatively charged acetylide ion (a nucleophile) is attracted to the partially positively charged carbon (an electrophile) of the alkyl halide. As the electrons of the acetylide ion approach the carbon to form the new C—C bond, they push out the bromine and its bonding electrons because carbon can bond to no more than four atoms at a time.

$CH_3CH_2C\equiv\overset{..}{C}^-$ + $CH_3CH_2CH_2\overset{\delta+}{\underset{\delta-}{-}}Br$ $\longrightarrow$ $CH_3CH_2C\equiv CCH_2CH_2CH_3$ + Br^-

3-D Molecules:
1-bromobutane; 3-octyne

We can convert terminal alkynes into internal alkynes of any desired chain length simply by choosing an alkyl halide with the appropriate structure. Just count the number of carbons in the terminal alkyne and the number of carbons in the product to see how many carbons are needed in the alkyl halide.

$CH_3CH_2CH_2C\equiv CH$ $\xrightarrow{\text{NaNH}_2}$ $CH_3CH_2CH_2C\equiv C^-$ $\xrightarrow{\text{CH}_3\text{CH}_2\text{Br}}$ $CH_3CH_2CH_2C\equiv CCH_2CH_3$
1-pentyne   **3-heptyne**

PROBLEM 27 SOLVED

A chemist wants to synthesize 3-heptyne but cannot find any 1-pentyne, the starting material used in the synthesis shown on page 127. How else can 3-heptyne be synthesized?

Solution The *sp* carbons of 3-heptyne are bonded to an ethyl group and to a propyl group. Therefore, to produce 3-heptyne, the acetylide ion of 1-pentyne can react with an ethyl halide (as on page 127), or the acetylide ion of 1-butyne can react with a propyl halide. Since 1-pentyne is not available, the chemist should use 1-butyne and a propyl halide.

$$CH_3CH_2C{\equiv}C^- \ + \ CH_3CH_2CH_2Br \ \longrightarrow \ CH_3CH_2C{\equiv}CCH_2CH_2CH_3 \ + \ Br^-$$
3-heptyne

5.15 AN INTRODUCTION TO MULTISTEP SYNTHESIS

Synthetic chemists consider time, cost, and yield in designing syntheses. In the interest of time, a well-designed synthesis will consist of as few steps (sequential reactions) as possible, and each of those steps will be a reaction that is easy to carry out. If two chemists in a pharmaceutical company were each asked to prepare a new drug, and one synthesized the drug in 3 simple steps while the other used 20 difficult steps, which chemist would not get a raise? The costs of the starting materials are also taken into consideration; the more reactant needed to synthesize one gram of product, the more expensive it is to produce. Moreover, each step in the synthesis should provide the greatest possible yield of the desired product. Sometimes a synthesis involving several steps is preferred because the starting materials are inexpensive, the reactions are easy to carry out, and the yield of each step is high. Such a synthesis is better than one with fewer steps if those steps require expensive starting materials and reactions that are more difficult to run or give lower yields. At this point in your chemical education, however, you are not yet familiar with the costs of different chemicals or the difficulty of carrying out specific reactions. So, for the time being, when you design a synthesis, just focus on finding the route with the fewest steps.

The following examples will give you an idea of the type of thinking required for designing a successful synthesis.

Example 1. Starting with 1-butyne, how could you make the ketone shown below? You can use any organic and inorganic reagents.

$$CH_3CH_2C{\equiv}CH \ \xrightarrow{?} \ CH_3CH_2\overset{\overset{\displaystyle O}{\|}}{C}CH_2CH_2CH_3$$
1-butyne

Many chemists find that the easiest way to design a synthesis is to work backward. Instead of looking at the starting material and deciding how to do the first step of the synthesis, look at the product and decide how to do the last step. The product is a ketone. At this point, the only reaction you know that forms a ketone is the addition of water (in the presence of an acid catalyst) to an alkyne. If the alkyne used in the reaction has identical substituents on both *sp* carbons, only one ketone will be obtained. Thus, 3-hexyne is the best alkyne to use for the synthesis of the desired ketone.

$$CH_3CH_2C{\equiv}CCH_2CH_3 \ \xrightarrow[H_2SO_4]{H_2O} \ CH_3CH_2\overset{\overset{\displaystyle OH}{|}}{C}{=}CHCH_2CH_3 \ \rightleftharpoons \ CH_3CH_2\overset{\overset{\displaystyle O}{\|}}{C}CH_2CH_2CH_3$$
3-hexyne

3-Hexyne can be obtained from the four-carbon starting material by removing the proton from its *sp* carbon, followed by reaction with a two-carbon alkyl halide. (The numbers 1 and 2 in front of the reagents above and below the reaction arrow indicate two sequential reactions; the second reagent is not added until the reaction with the first reagent is completely over.)

$$CH_3CH_2C\equiv CH \quad \xrightarrow[\text{2. CH}_3\text{CH}_2\text{Br}]{\text{1. NaNH}_2} \quad CH_3CH_2C\equiv CCH_2CH_3$$

1-butyne   **3-hexyne**

Thus, the synthetic scheme for the synthesis of the desired ketone is given by

$$CH_3CH_2C\equiv CH \quad \xrightarrow[\text{2. CH}_3\text{CH}_2\text{Br}]{\text{1. NaNH}_2} \quad CH_3CH_2C\equiv CCH_2CH_3 \quad \xrightarrow[\text{H}_2\text{SO}_4]{\text{H}_2\text{O}} \quad CH_3CH_2\overset{\displaystyle O}{\overset{\displaystyle \|}{C}}CH_2CH_2CH_3$$

Example 2. Starting with ethyne, how could you make 2-bromopentane?

$$HC\equiv CH \quad \xrightarrow{\text{?}} \quad CH_3CH_2CH_2\underset{\displaystyle Br}{CH}CH_3$$

ethyne

2-bromopentane

The desired product can be prepared from 1-pentene, which can be prepared from 1-pentyne. 1-Pentyne can be prepared from ethyne and an alkyl halide with three carbons.

$$HC\equiv CH \quad \xrightarrow[\text{2. CH}_3\text{CH}_2\text{CH}_2\text{Br}]{\text{1. NaNH}_2} \quad CH_3CH_2CH_2C\equiv CH \quad \xrightarrow[\substack{\text{Lindlar}\\\text{catalyst}}]{\text{H}_2} \quad CH_3CH_2CH_2CH=CH_2 \quad \xrightarrow{\text{HBr}} \quad CH_3CH_2CH_2\underset{\displaystyle Br}{CH}CH_3$$

Example 3. How could you prepare 3,3-dibromohexane from reagents that contain no more than two carbons?

$$\text{reagents with no more than 2 carbons} \quad \xrightarrow{\text{?}} \quad CH_3CH_2\underset{\displaystyle Br}{\overset{\displaystyle Br}{C}}CH_2CH_2CH_3$$

3,3-dibromohexane

The desired product can be prepared from an alkyne and excess HBr. 3-Hexyne is the alkyne of choice, because it will form one dibromide, whereas 2-hexyne would form two different dibromides (3,3-dibromohexane and 2,2-dibromohexane) because it is not a symmetrical alkyne. 3-Hexyne can be prepared from 1-butyne and ethyl bromide, and 1-butyne can be prepared from ethyne and ethyl bromide.

$$HC\equiv CH \quad \xrightarrow[\text{2. CH}_3\text{CH}_2\text{Br}]{\text{1. NaNH}_2} \quad CH_3CH_2C\equiv CH \quad \xrightarrow[\text{2. CH}_3\text{CH}_2\text{Br}]{\text{1. NaNH}_2} \quad CH_3CH_2C\equiv CCH_2CH_3 \quad \xrightarrow{\text{excess HBr}} \quad CH_3CH_2\underset{\displaystyle Br}{\overset{\displaystyle Br}{C}}CH_2CH_2CH_3$$

PROBLEM 28

Starting with acetylene, how could the following compounds be synthesized?

a. $CH_3CH_2CH_2C\equiv CH$

c. $CH_3CH=CH_2$

e. $\underset{\displaystyle H}{\overset{\displaystyle CH_3}{C}}=\underset{\displaystyle H}{\overset{\displaystyle CH_3}{C}}$

b. $CH_3CH_2CH_2\overset{\displaystyle O}{\overset{\displaystyle \|}{C}}CH_3$

d. $CH_3\underset{\displaystyle Br}{CH}CH_3$

f. $CH_3\underset{\displaystyle Cl}{\overset{\displaystyle Cl}{C}}CH_3$

5.16 SYNTHETIC POLYMERS

A **polymer** is a large molecule made by linking together repeating units of small molecules called **monomers**. The process of linking them together is called **polymerization**.

$$n\text{M} \xrightarrow{\text{polymerization}} \text{—M—M—M—M—M—M—M—M—M—}$$

monomers   polymer

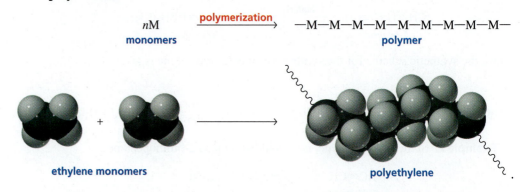

ethylene monomers   polyethylene

Polymers can be divided into two broad groups: **synthetic polymers** and **biopolymers**. Synthetic polymers are synthesized by scientists, whereas biopolymers are synthesized by organisms. Examples of biopolymers are DNA, which is the storage molecule for genetic information—the molecule that determines whether a fertilized egg becomes a human or a honeybee; RNA and proteins, the molecules that induce biochemical transformations; and polysaccharides, which store energy, act as recognition sites on cell surfaces, and also function as structural materials. The structures and properties of these biopolymers are presented in other chapters. Here, we will explore synthetic polymers.

Probably no group of synthetic compounds is more important to modern life than synthetic polymers. Some synthetic polymers resemble natural substances, but most are quite different from materials found in nature. Such diverse products as photographic film, compact discs, food wrap, artificial joints, Super Glue, toys, plastic bottles, weather stripping, automobile body parts, and shoe soles are made of synthetic polymers. More than 2.5×10^{13} kilograms of synthetic polymers are produced in the United States each year, and approximately 30,000 polymer patents are currently in force. We can expect scientists to develop many more new materials in the years to come.

Synthetic polymers can be divided into two major classes, depending on their method of preparation. Here we will look at *chain-growth polymers*. The second major class of polymers, *step-growth polymers*, are discussed in Section 11.14.

Chain-growth polymers are made by **chain reactions**—the addition of monomers to the end of a growing chain. The monomers used most commonly in chain-growth polymerization are ethylene and substituted ethylenes. Polystyrene—used for disposable food containers, insulation, and toothbrush handles, among other things—is an example of a chain-growth polymer. As its name suggests, the monomer used to form polystyrene is a substituted ethylene called styrene. Polystyrene can be pumped full of air to produce the material known as Styrofoam.

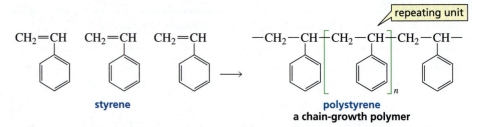

styrene   polystyrene
a chain-growth polymer

Some of the many polymers synthesized by chain-growth polymerization are listed in Table 5.1.

Table 5.1   Some Important Chain-Growth Polymers and Their Uses

Monomer	Repeating unit	Polymer name	Uses
$CH_2{=}CH_2$	$-CH_2-CH_2-$	polyethylene	film, toys, bottles, plastic bags
$CH_2{=}CH$ $\quad\mid$ $\quad Cl$	$-CH_2-CH-$ $\qquad\mid$ $\qquad Cl$	poly(vinyl chloride)	"squeeze" bottles, pipe, siding, flooring
$CH_2{=}CH-CH_3$ $\quad\mid$ $\quad CH_3$	$-CH_2-CH-$ $\qquad\mid$ $\qquad CH_3$	polypropylene	molded caps, margarine tubs, indoor/outdoor carpeting, upholstery
$CH_2{=}CH$ ⬡	$-CH_2-CH-$ ⬡	polystyrene	packaging, toys, clear cups, egg cartons, hot drink cups
$CF_2{=}CF_2$	$-CF_2-CF_2-$	poly(tetrafluoroethylene) Teflon	nonsticking surfaces, liners, cable insulation
$CH_2{=}CH$ $\quad\mid$ $\quad C{\equiv}N$	$-CH_2-CH-$ $\qquad\mid$ $\qquad C{\equiv}N$	poly(acrylonitrile) Orlon, Acrilan	rugs, blankets, yarn, apparel, simulated fur
$CH_2{=}C-CH_3$ $\quad\mid$ $\quad COCH_3$ $\quad\parallel$ $\quad O$	$\qquad CH_3$ $\qquad\mid$ $-CH_2-C-$ $\qquad\mid$ $\qquad COCH_3$ $\qquad\parallel$ $\qquad O$	poly(methyl methacrylate) Plexiglas, Lucite	lighting fixtures, signs, solar panels, skylights
$CH_2{=}CH$ $\quad\mid$ $\quad OCCH_3$ $\quad\parallel$ $\quad O$	$-CH_2-CH-$ $\qquad\mid$ $\qquad OCCH_3$ $\qquad\parallel$ $\qquad O$	poly(vinyl acetate)	latex paints, adhesives

The two most common mechanisms for chain-growth polymerization are cationic polymerization and radical polymerization. Each of these mechanisms has three distinct phases; *initiation steps* that start the polymerization, *propagation steps* that allow the polymer chain to grow, and *termination* steps that stop the growth of the chain.

Chain reactions have initiation, propagation, and termination steps.

Cationic Polymerization

In cationic polymerization, the initiator is an electrophile that adds to the alkene monomer, causing it to become a cation. The initiator most often used in cationic polymerization is a proton, generated from the reaction of BF_3 with water (Section 2.9); because boron does not have a complete octet, it accepts a share in an electron pair from the oxygen of water.

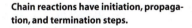

initiation steps

$$F_3B + H_2\ddot{O}{:} \;\rightleftharpoons\; F_3\bar{B}{:}\overset{+}{\underset{\cdot\cdot}{O}}H_2 \;\rightleftharpoons\; F_3\bar{B}{:}\ddot{O}H + H^+$$

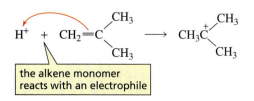

the alkene monomer reacts with an electrophile

- The cation (an electrophile) formed in the initiation steps reacts with a second monomer, forming a new cation that reacts in turn with a third monomer. These are called **propagation steps** because they propagate the chain reaction. The cation is now at the end of the unit that was most recently added to the end of the chain. This is called the **propagating site**.

propagation steps

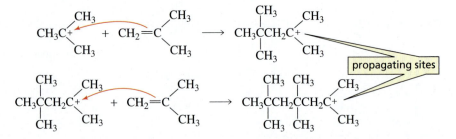

As each subsequent monomer adds to the chain, the new positively charged propagating site is at the end of the last unit added. This process is repeated over and over. Hundreds or even thousands of alkene monomers can be added one at a time to the growing chain. Notice that in both the chain-initiating and chain-propagating steps, the rule governing electrophilic addition reactions is followed: the electrophile adds to the carbon bonded to the greater number of hydrogens (Section 5.3).

- Eventually, the chain reaction stops because the propagating sites are destroyed. A propagating site is destroyed when it reacts with a nucleophile. This is called a **termination step**.

termination step

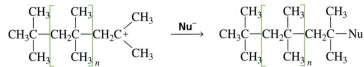

DESIGNING A POLYMER

A polymer used for making contact lenses must be sufficiently hydrophilic (water-loving) to allow lubrication of the eye. Such a polymer, therefore, has many OH groups.

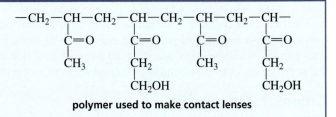

polymer used to make contact lenses

PROBLEM 29

Draw a short segment of the polymer that would be formed from cationic polymerization of methyl vinyl ether with H^+ as the initiator.

Radical Polymerization

In radical polymerization, the initiator is a species that breaks into radicals. Most radical initiators have an $O—O$ bond because such a bond easily breaks in a way that allows each of the atoms that formed the bond to retain one of the bonding electrons. Each of the radicals that is formed seeks an electron to complete its octet. A radical can obtain an electron by adding to the electron-rich π bond of the alkene, thereby forming a new radical. The curved arrows that we have previously seen have arrowheads with two barbs because they represent the movement of two electrons. Notice

that the arrowheads in the mechanism shown below have only one barb because they represent the movement of only one electron.

- Radical polymerization has two initiation steps; one creates radicals, and the other forms the radical that propagates the chain reaction. Because the radical is seeking an electron, a radical is an electrophile and, like other electrophiles, it adds to the sp^2 carbon bonded to the greater number of hydrogens.

initiation steps

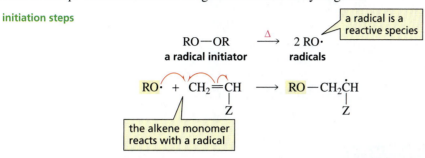

- The radical adds to another alkene monomer, converting it into a radical. This radical reacts with another monomer, adding a new subunit that propagates the chain reaction. Notice that when a radical is used to initiate polymerization, the propagating sites are also radicals.

propagation steps

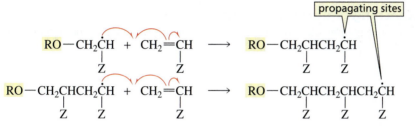

- The chain reaction stops when the propagating sites are destroyed. A propagating site is destroyed when it reacts with a species (XY) that allows it to pair up its electron.

termination step

RECYCLING SYMBOLS

When plastics are recycled, the various types must be separated from one another. To aid in the separation, many states require manufacturers to place a recycling symbol on their products to indicate the type of plastic it is. You are probably familiar with these symbols, which are often embossed on the bottom of plastic containers. The symbols consist of three arrows around one of seven numbers; an abbreviation below the symbol indicates the type of polymer from which the container is made. The lower the number in the middle of the symbol, the greater is the ease with which the material can be recycled: 1 (PET) stands for poly(ethylene terephthalate), 2 (HDPE) for high-density polyethylene, 3 (V) for poly(vinyl chloride), 4 (LDPE) for low-density polyethylene, 5 (PP) for polypropylene, 6 (PS) for polystyrene, and 7 for all other plastics.

PROBLEM 30◆

What monomer would you use to form each of the following polymers?

a. —CH₂CHCH₂CHCH₂CHCH₂CHCH₂CH—
  │   │   │   │   │
  Cl Cl Cl Cl Cl

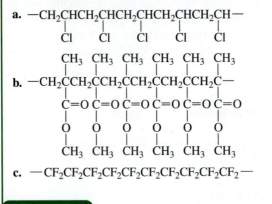

b. —CH₂CCH₂CCH₂CCH₂CCH₂CCH₂C—
with CH₃ groups and C=O, O, CH₃ groups

c. —CF₂CF₂CF₂CF₂CF₂CF₂CF₂CF₂CF₂CF₂—

PROBLEM 31

Show the mechanism for the formation of a segment of poly(vinyl chloride) containing three units of vinyl chloride and initiated by HO·.

Branching of the Polymer Chain

If the propagating site removes a hydrogen atom from the polymer chain, a branch can grow off the chain at that point.

—CH₂CH₂CH₂ĊH₂ + —CH₂CH₂CHCH₂CH₂CH₂—
     H

—CH₂CH₂CH₂ĊH₂ + —CH₂CH₂ĊHCH₂CH₂CH₂— →(CH₂=CH₂)→ —CH₂CH₂ĊHCH₂CH₂CH₂—
     H   ĊH₂
                                                            CH₂

Removing a hydrogen atom from a carbon near the end of a chain leads to short branches, whereas removing a hydrogen atom from a carbon near the middle of a chain results in long branches. Short branches are more likely to be formed than long ones because the ends of the chain are more accessible.

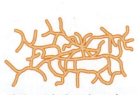

chain with short branches   chain with long branches

Branching greatly affects the physical properties of the polymer. Unbranched chains can pack together more closely than branched chains can. Consequently, linear polyethylene (known as high-density polyethylene) is a relatively hard plastic, used for the production of such things as artificial hip joints, whereas branched polyethylene (low-density polyethylene) is a much more flexible polymer, used for trash bags and dry-cleaning bags.

PROBLEM 32 ◆

Polyethylene can be used for the production of beach chairs as well as beach balls. Which of these items is made from more highly branched polyethylene?

5.17 RADICALS IN BIOLOGICAL SYSTEMS

Radicals are extremely reactive species. Fats and oils react with radicals to form compounds with strong odors that are responsible for the unpleasant taste and smell associated with sour milk and rancid butter. The molecules that form cell membranes can undergo this same reaction (Section 19.4). Radical reactions in biological systems also have been implicated in the aging process.

Clearly, unwanted radicals in biological systems must be destroyed before radical reactions have an opportunity to damage cells. Compounds known as **radical inhibitors** destroy radicals by creating compounds with only paired electrons. Hydroquinone is an example of a radical inhibitor. Two unwanted radicals each obtain an electron by removing a hydrogen atom from hydroquinone; rearrangement of the electrons in the product forms quinone, a compound in which all the electrons are paired.

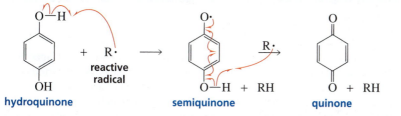

hydroquinone   semiquinone   quinone

Vitamin C and vitamin E are radical inhibitors present in biological systems. Vitamin C (also called ascorbic acid) is a water-soluble compound that traps radicals formed in the aqueous environment of the cell and in blood plasma. Vitamin E is a water-insoluble (therefore, fat-soluble) compound that traps radicals formed in nonpolar membranes. Why one vitamin functions in aqueous environments and the other in nonaqueous environments is apparent from their structures and electrostatic potential maps, which show that vitamin C is a relatively polar compound, whereas vitamin E is nonpolar.

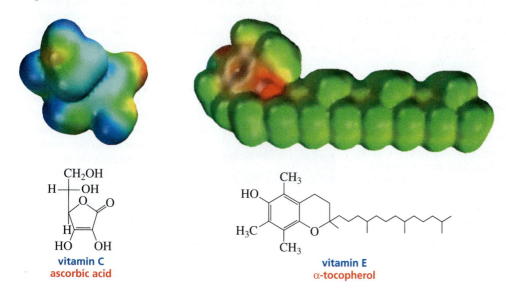

vitamin C
ascorbic acid

vitamin E
α-tocopherol

FOOD PRESERVATIVES

Radical inhibitors that are present in food are known as *preservatives* or *antioxidants*. They preserve food by destroying radicals, thereby preventing undesir- able radical reactions. BHA and BHT are synthetic preservatives that are added to many packaged foods. Vitamin E is a naturally occurring preservative found in vegetable oil.

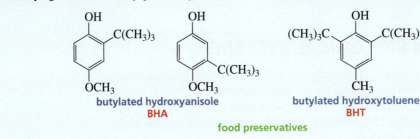

butylated hydroxyanisole
BHA

butylated hydroxytoluene
BHT

food preservatives

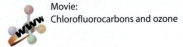

Movie:
Chlorofluorocarbons and ozone

RADICALS AND STRATOSPHERIC OZONE

Ozone (O_3), a major constituent of smog, is a health hazard at ground level. In the stratosphere, however, a layer of ozone shields the Earth from harmful solar radiation. The ozone layer is thinnest at the equator and dens- est toward the poles, with the greatest concentrations lying be- tween 12 and 15 miles above the Earth's surface. Ozone is formed in the atmosphere from the interaction of molecular oxygen with very short wavelength ultraviolet light ($h\nu$).

$$O_2 \xrightarrow{h\nu} O + O$$
$$O + O_2 \longrightarrow O_3$$
ozone

The stratospheric ozone layer acts as a filter for biological- ly harmful ultraviolet radiation that otherwise would reach the surface of the Earth. Among other effects, high-energy short- wavelength ultraviolet light can damage DNA in skin cells, causing mutations that trigger skin cancer. We owe our very existence to this protective ozone layer. According to current theories of evolution, life could not have developed on land in the absence of this ozone layer. Instead, most if not all living things would have had to remain in the ocean, where water screens out the harmful ultraviolet radiation.

Since about 1985, scientists have noted a precipitous drop in stratospheric ozone over Antarctica. This area of ozone de- pletion, dubbed the "ozone hole," is unprecedented in the his- tory of ozone observations. Scientists subsequently noted a similar decrease in ozone over Arctic regions; then, in 1988, they detected a depletion of ozone over the United States for the first time. Three years later, scientists determined that the rate of ozone depletion was two to three times faster than orig- inally anticipated. Many in the scientific community blame re- cently observed increases in cataracts and skin cancer as well as diminished plant growth on the ultraviolet radiation that has penetrated the reduced ozone layer. Some predict that erosion of the protective ozone layer will cause an additional 200,000 deaths from skin cancer over the next 50 years.

Strong circumstantial evidence implicates synthetic chloro- fluorocarbons (CFCs)—alkanes in which all the hydrogens have been replaced by fluorine and chlorine, such as $CFCl_3$ and CF_2Cl_2—as a major cause of ozone depletion. These gases, known commercially as Freons, have been used exten- sively as cooling fluids in refrigerators and air conditioners. They were also once widely used as propellants in aerosol spray cans (deodorant, hair spray, and so on) because of their odorless, nontoxic, and nonflammable properties, and, being chemically inert, they do not react with the contents of the can. Such use now, however, has been banned.

Polar stratospheric clouds increase the rate of ozone destruction. These clouds form over Antarctica during the cold winter months. Ozone depletion in the Arctic is less severe because the temperature generally does not get cold enough for the polar stratospheric clouds to form there.

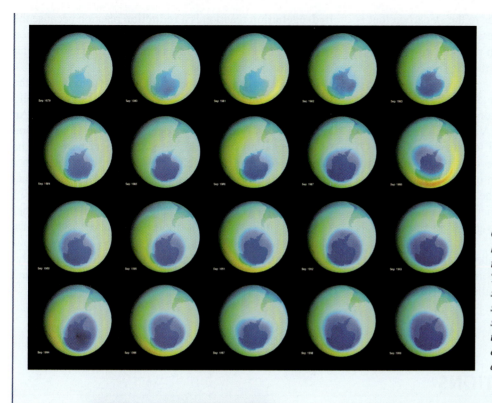

Growth of the Antarctic ozone hole, located mostly over the continent of Antarctica, since 1979. The images were made from data supplied by total ozone-mapping spectrometers (TOMSs). The color scale depicts the total ozone values in Dobson units, with the lowest ozone densities represented by dark blue.

Dobson Units

| 100 | 200 | 300 | 400 | 500 |

Chlorofluorocarbons remain very stable in the atmosphere until they reach the stratosphere. There they encounter wavelengths of ultraviolet light that cause the carbon–chlorine bond to break, generating chlorine radicals.

The chlorine radicals are the ozone-removing agents. They react with ozone to form chlorine monoxide radicals and oxygen (O_2). The chlorine monoxide radicals then react with ozone to form chlorine dioxide, which dissociates to regenerate a chlorine radical. These three steps—two of which each destroy an ozone molecule—are repeated over and over. It has been calculated that each chlorine atom destroys 100,000 ozone molecules!

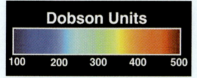

$$Cl\cdot + O_3 \longrightarrow ClO\cdot + O_2$$
$$ClO\cdot + O_3 \longrightarrow \cdot ClO_2 + O_2$$
$$\cdot ClO_2 \longrightarrow Cl\cdot + O_2$$

SUMMARY

Alkenes undergo **electrophilic addition reactions**. These start with the addition of an electrophile to one of the sp^2 carbons and conclude with the addition of a nucleophile to the other sp^2 carbon. In all electrophilic addition reactions, the *electrophile* adds to the sp^2 carbon bonded to the greater number of hydrogens. **Regioselectivity** is the preferential formation of one **constitutional isomer** over another.

Regioselectivity results from the fact that the addition of hydrogen halides and the acid-catalyzed addition of water and alcohols form the most stable **carbocation intermediate**; **tertiary carbocations** are more stable than **secondary**

carbocations, which are more stable than **primary carbocations**. We have now seen that alkyl groups stabilize both alkenes and carbocations.

An **alkyne** is a hydrocarbon that contains a carbon–carbon triple bond. The functional group suffix of an alkyne is "yne." A **terminal alkyne** has the triple bond at the end of the chain; an **internal alkyne** has the triple bond located elsewhere along the chain.

Alkynes, like alkenes, undergo electrophilic addition reactions. The same reagents that add to alkenes add to alkynes. Electrophilic addition to a *terminal* alkyne is

regioselective; the *electrophile* adds to the *sp* carbon that is bonded to the hydrogen. If excess reagent is available, alkynes undergo a second addition reaction with hydrogen halides because the product of the first reaction is an alkene.

When an alkyne undergoes the acid-catalyzed addition of water, the product of the reaction is an enol, which immediately rearranges to a ketone. A **ketone** is a compound that has two alkyl groups bonded to a **carbonyl** ($C=O$) **group**. The ketone and enol are called **keto–enol tautomers**; they differ in the location of a double bond and a hydrogen. Interconversion of the tautomers is called **tautomerization**. The keto tautomer predominates at equilibrium. Terminal alkynes add water if mercuric ion is added to the acidic mixture.

Hydrogen adds to alkenes and alkynes in the presence of a metal catalyst (Pd/C or Pt/C) to form an alkane. The addition of H_2 to a compound is called **hydrogenation**. A

hydrogenation reaction is a **reduction** reaction because the product has more $C-H$ bonds than the reactant. Addition of hydrogen to an internal alkyne in the presence of Lindlar catalyst forms a *cis alkene*.

The electronegativities of carbon atoms decrease in the order: $sp > sp^2 > sp^3$. Ethyne is, therefore, a stronger acid than ethene, and ethene is a stronger acid than ethane. An amide ion can remove a hydrogen bonded to an *sp* carbon of a terminal alkyne because it is a stronger base than the **acetylide ion** that is formed. An acetylide ion reacts with a methyl halide or a primary alkyl halide to form an internal alkyne.

A **polymer** is a giant molecule made by linking together repeating units of small molecules called **monomers**. **Chain-growth polymers** are formed by chain reactions with **initiation**, **propagation**, and **termination steps**. Polymerization of alkenes can be initiated by electrophiles and by radicals.

SUMMARY OF REACTIONS

1. Electrophilic addition reactions of alkenes

    a. Addition of hydrogen halides (Section 5.1)

$$RCH=CH_2 + \boxed{HX} \longrightarrow RCHCH_3$$
$$\qquad\qquad\qquad\qquad\quad |$$
$$\qquad\qquad\qquad\qquad\quad X$$

$$HX = HF, HCl, HBr, HI$$

    b. Acid-catalyzed addition of water and alcohols (Sections 5.4 and 5.5)

$$RCH=CH_2 + \boxed{H_2O} \xrightarrow{H_2SO_4} RCHCH_3$$
$$\qquad\qquad\qquad\qquad\qquad\qquad |$$
$$\qquad\qquad\qquad\qquad\qquad\qquad OH$$

$$RCH=CH_2 + \boxed{CH_3OH} \xrightarrow{H_2SO_4} RCHCH_3$$
$$\qquad\qquad\qquad\qquad\qquad\qquad\qquad |$$
$$\qquad\qquad\qquad\qquad\qquad\qquad\qquad OCH_3$$

2. Electrophilic addition reactions of alkynes

    a. Addition of hydrogen halides (Section 5.10)

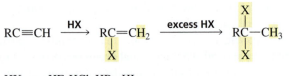

$$HX = HF, HCl, HBr, HI$$

    b. Acid-catalyzed addition of water (Section 5.11); R' means that R and R' do not have to be the same alkyl group.

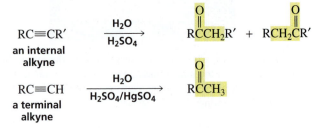

3. Addition of hydrogen to alkenes and alkynes (Section 5.12)

$$RCH=CH_2 + \boxed{H_2} \xrightarrow{\text{Pd/C or Pt/C}} R\boxed{CH_2CH_3}$$

$$RC\equiv CR' + 2\,H_2 \xrightarrow{\text{Pd/C or Pt/C}} R\boxed{CH_2CH_2}R'$$

$$R-C\equiv C-R' + H_2 \xrightarrow[\text{catalyst}]{\text{Lindlar}} \begin{array}{c} \boxed{H}\quad\boxed{H}\\ \diagdown\;\diagup\\ C=C\\ \diagup\;\diagdown\\ R\qquad R'\end{array}$$

4. Removal of a proton from a terminal alkyne, followed by reaction with an alkyl halide (Sections 5.13 and 5.14)

$$RC\equiv CH \xrightarrow{NaNH_2} RC\equiv C^- \xrightarrow{R'CH_2Br} RC\equiv C\boxed{CH_2}R'$$

PROBLEMS

33. Identify the electrophile and the nucleophile in each of the following reaction steps. Then draw curved arrows to illustrate the bond-making and bond-breaking processes.

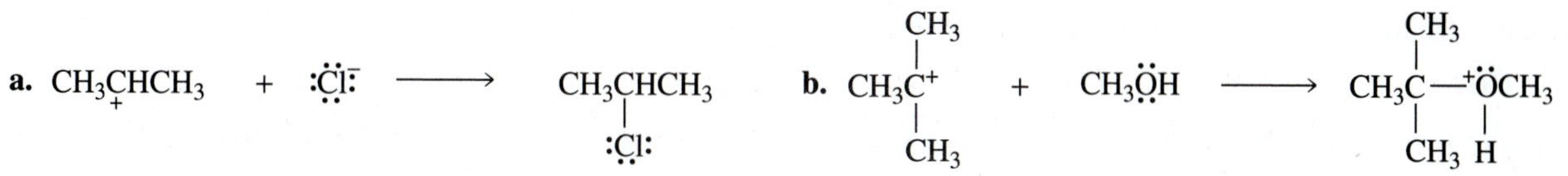

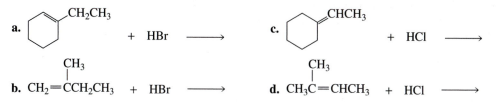

34. What will be the major product of the reaction of 2-methyl-2-butene with each of the following reagents?
a. HBr
b. HI
c. H$_2$, Pd/C
d. H$_2$O + trace H$_2$SO$_4$
e. CH$_3$OH + trace H$_2$SO$_4$
f. CH$_3$CH$_2$OH + trace H$_2$SO$_4$

35. Give the major product of each of the following reactions:

a. (cyclohexene with CH$_2$CH$_3$) + HBr ⟶

c. (cyclohexane ring with =CHCH$_3$) + HCl ⟶

b. CH$_2$=CCH$_2$CH$_3$ (with CH$_3$) + HBr ⟶

d. CH$_3$C=CHCH$_3$ (with CH$_3$) + HCl ⟶

36. Draw curved arrows to show the flow of electrons responsible for the conversion of reactants into products.

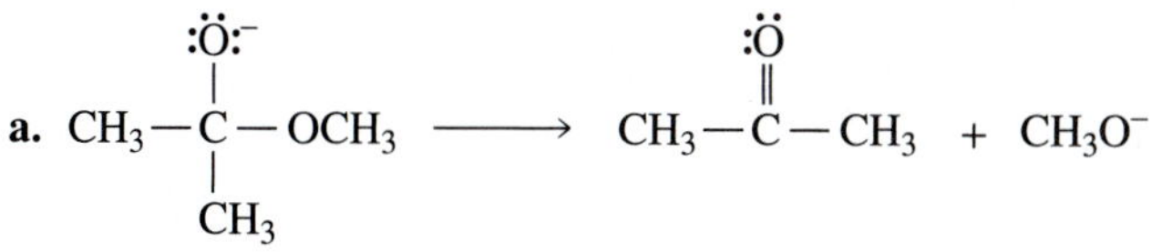

a. CH$_3$—C(—OCH$_3$)(:Ö:⁻)—CH$_3$ ⟶ CH$_3$—C(=Ö)—CH$_3$ + CH$_3$O⁻

b. CH$_3$C≡C—H + :N̈H$_2$ ⟶ CH$_3$C≡C⁻ + N̈H$_3$

c. CH$_3$CH$_2$—Br + CH$_3$Ö:⁻ ⟶ CH$_3$CH$_2$—ÖCH$_3$ + Br⁻

37. Give the reagents that would be required to carry out the following syntheses:

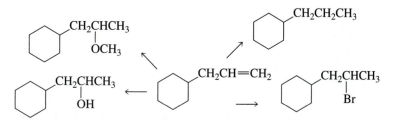

38. Draw all the enol tautomers for each of the ketones in Problem 20.

39. What ketones are formed when the following alkyne undergoes the acid-catalyzed addition of water?

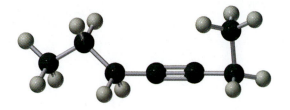

40. Give the major product of each of the following reactions:

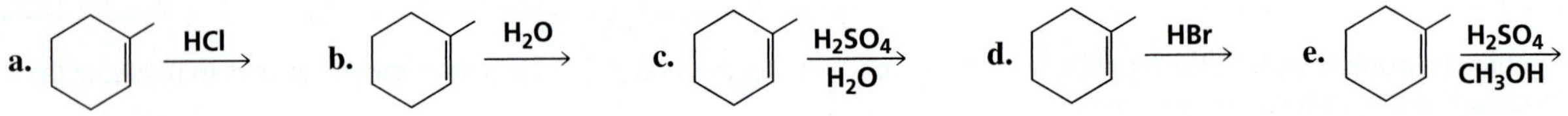

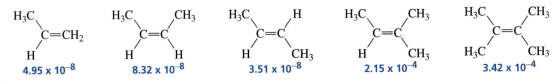

a. $\xrightarrow{\text{HCl}}$ **b.** $\xrightarrow{\text{H}_2\text{O}}$ **c.** $\xrightarrow[\text{H}_2\text{O}]{\text{H}_2\text{SO}_4}$ **d.** $\xrightarrow{\text{HBr}}$ **e.** $\xrightarrow[\text{CH}_3\text{OH}]{\text{H}_2\text{SO}_4}$

41. For each of the following pairs, indicate which member is more stable:

a. $CH_3\overset{+}{\underset{\underset{\displaystyle CH_3}{|}}{C}}CH_3$ or $CH_3\overset{+}{C}HCH_2CH_3$ **b.** $CH_3CH_2\overset{+}{C}H_2$ or $CH_3\overset{+}{C}HCH_3$ **c.** $CH_3\overset{+}{C}H_2$ or $CH_2=\overset{+}{C}H$

42. Using an alkene and any other reagents, how would you prepare the following compounds?

a. (cyclohexane ring)

b. $CH_3CH_2CH_2\underset{\underset{\displaystyle Cl}{|}}{C}HCH_3$

c. (cyclohexyl) $CH_2\underset{\underset{\displaystyle OH}{|}}{C}HCH_3$

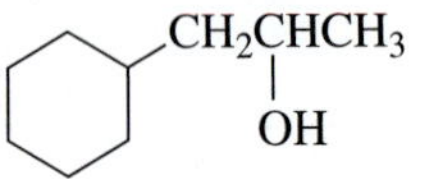

43. Identify the two alkenes that react with HBr to give 1-bromo-1-methylcyclohexane.

44. The second-order rate constant (in units of $M^{-1}s^{-1}$) for acid-catalyzed hydration at 25 °C is given for each of the following alkenes:

| $\underset{4.95 \times 10^{-8}}{\underset{\displaystyle H}{\overset{\displaystyle H_3C}{}}C=CH_2}$ | $\underset{8.32 \times 10^{-8}}{\overset{\displaystyle H_3C \quad CH_3}{\underset{\displaystyle H \quad\;\; H}{C=C}}}$ | $\underset{3.51 \times 10^{-8}}{\overset{\displaystyle H_3C \quad\;\; H}{\underset{\displaystyle H \quad CH_3}{C=C}}}$ | $\underset{2.15 \times 10^{-4}}{\overset{\displaystyle H_3C \quad CH_3}{\underset{\displaystyle H \quad CH_3}{C=C}}}$ | $\underset{3.42 \times 10^{-4}}{\overset{\displaystyle H_3C \quad CH_3}{\underset{\displaystyle H_3C \quad CH_3}{C=C}}}$ |

a. Why does (Z)-2-butene react faster than (E)-2-butene?
b. Why does 2-methyl-2-butene react faster than (Z)-2-butene?
c. Why does 2,3-dimethyl-2-butene react faster than 2-methyl-2-butene?

45. a. Propose a mechanism for the following reaction (remember to use curved arrows when showing a mechanism):

$$CH_3CH_2CH=CH_2 + CH_3OH \xrightarrow{\text{H}_2\text{SO}_4} CH_3CH_2\underset{\underset{\displaystyle OCH_3}{|}}{C}HCH_3$$

b. Which step is the rate-determining step?
c. What is the electrophile in the first step?
d. What is the nucleophile in the first step?
e. What is the electrophile in the second step?
f. What is the nucleophile in the second step?

46. The pK_a of protonated ethyl alcohol is -2.4 and the pK_a of ethyl alcohol is 15.9. Therefore, as long as the pH of the solution is greater than _____ and less than _____, more than 50% of ethyl alcohol will be in its neutral, nonprotonated form. (*Hint:* See Section 2.7.)

47. a. How many alkenes could you treat with H_2, Pt/C in order to prepare methylcyclopentane?
b. Which of the alkenes is the most stable?

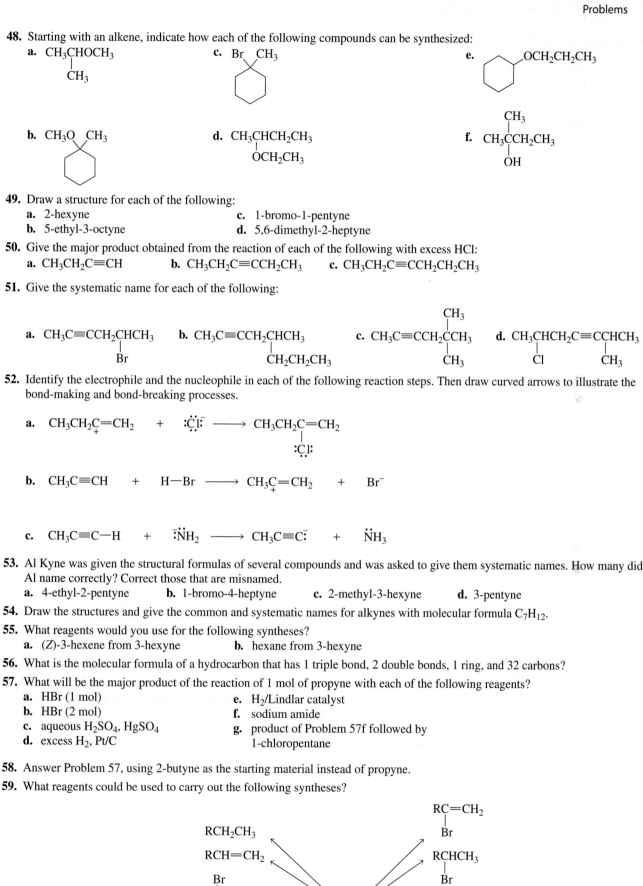

48. Starting with an alkene, indicate how each of the following compounds can be synthesized:

a. CH₃CHOCH₃
    |
    CH₃

c. Br CH₃ (cyclohexane with Br and CH₃)

e. (cyclohexane)–OCH₂CH₂CH₃

b. CH₃O CH₃ (on cyclohexane)

d. CH₃CHCH₂CH₃
    |
    OCH₂CH₃

f.
      CH₃
      |
  CH₃CCH₂CH₃
      |
      OH

49. Draw a structure for each of the following:
a. 2-hexyne
b. 5-ethyl-3-octyne
c. 1-bromo-1-pentyne
d. 5,6-dimethyl-2-heptyne

50. Give the major product obtained from the reaction of each of the following with excess HCl:
a. CH₃CH₂C≡CH
b. CH₃CH₂C≡CCH₂CH₃
c. CH₃CH₂C≡CCH₂CH₂CH₃

51. Give the systematic name for each of the following:

a. CH₃C≡CCH₂CHCH₃
             |
             Br

b. CH₃C≡CCH₂CHCH₃
             |
             CH₂CH₂CH₃

c.
            CH₃
            |
  CH₃C≡CCH₂CCH₃
            |
            CH₃

d. CH₃CHCH₂C≡CCHCH₃
      | |
      Cl   CH₃

52. Identify the electrophile and the nucleophile in each of the following reaction steps. Then draw curved arrows to illustrate the bond-making and bond-breaking processes.

a. CH₃CH₂C⁺=CH₂ + :C̈l:⁻ ⟶ CH₃CH₂C=CH₂
                                    |
                                    :C̈l:

b. CH₃C≡CH + H—Br ⟶ CH₃C⁺=CH₂ + Br⁻

c. CH₃C≡C—H + :N̈H₂⁻ ⟶ CH₃C≡C:⁻ + N̈H₃

53. Al Kyne was given the structural formulas of several compounds and was asked to give them systematic names. How many did Al name correctly? Correct those that are misnamed.
a. 4-ethyl-2-pentyne   **b.** 1-bromo-4-heptyne   **c.** 2-methyl-3-hexyne   **d.** 3-pentyne

54. Draw the structures and give the common and systematic names for alkynes with molecular formula C₇H₁₂.

55. What reagents would you use for the following syntheses?
a. (Z)-3-hexene from 3-hexyne   **b.** hexane from 3-hexyne

56. What is the molecular formula of a hydrocarbon that has 1 triple bond, 2 double bonds, 1 ring, and 32 carbons?

57. What will be the major product of the reaction of 1 mol of propyne with each of the following reagents?
a. HBr (1 mol)
b. HBr (2 mol)
c. aqueous H₂SO₄, HgSO₄
d. excess H₂, Pt/C
e. H₂/Lindlar catalyst
f. sodium amide
g. product of Problem 57f followed by 1-chloropentane

58. Answer Problem 57, using 2-butyne as the starting material instead of propyne.

59. What reagents could be used to carry out the following syntheses?

RCH₂CH₃

RCH=CH₂
 |
 Br

    Br
    |
RCCH₃
    |
    Br

RC≡CH

    RC=CH₂
    |
    Br

RCHCH₃
 |
 Br

    O
    ‖
RCCH₃

60. a. Starting with 5-methyl-2-hexyne, how could you prepare the following compound?

$$CH_3CH_2\underset{\underset{OH}{|}}{C}HCH_2\underset{\underset{CH_3}{|}}{C}HCH_3$$

   b. What other alcohol would also be obtained?

61. How many of the following names are correct? Correct the incorrect names.
   a. 4-heptyne
   b. 2-ethyl-3-hexyne
   c. 4-chloro-2-pentyne
   d. 2,3-dimethyl-5-octyne
   e. 4,4-dimethyl-2-pentyne
   f. 2,5-dimethyl-3-hexyne

62. Which of the following pairs are keto–enol tautomers?

   a. $CH_3\underset{\underset{OH}{|}}{C}HCH_3$ and $CH_3\underset{\overset{O}{\|}}{C}CH_3$
      c. $CH_3CH_2CH_2CH=CHOH$ and $CH_3CH_2CH_2\underset{\overset{O}{\|}}{C}CH_3$

   b. $CH_3CH_2CH_2\underset{\underset{OH}{|}}{C}=CH_2$ and $CH_3CH_2CH_2\underset{\overset{O}{\|}}{C}CH_3$

63. Using ethyne as the starting material, how can the following compounds be prepared?

   a. $CH_3\underset{\overset{O}{\|}}{C}CH_3$
   b. 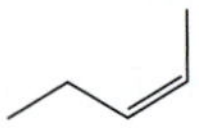 ⌁
   c.  ⌁

64. Draw the keto tautomer for each of the following:

   a. $CH_3CH=\underset{\overset{|}{OH}}{C}CH_3$
      c. ⬡—OH

   b. $CH_3CH_2CH_2\underset{\overset{|}{OH}}{C}=CH_2$
      d. 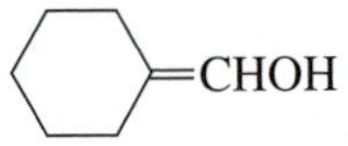 ⬡=CHOH

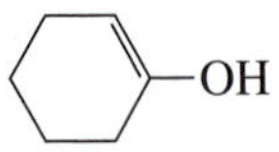

65. Show how each of the following compounds could be prepared using the given starting material, any necessary inorganic reagents, and any necessary organic compound that has no more than four carbon atoms:

   a. $HC≡CH \longrightarrow CH_3CH_2CH_2CH_2\underset{\overset{O}{\|}}{C}CH_3$
      c. $HC≡CH \longrightarrow CH_3CH_2CH_2\underset{\underset{OH}{|}}{C}HCH_3$

   b. $HC≡CH \longrightarrow CH_3CH_2\underset{\underset{Br}{|}}{C}HCH_3$
      d. 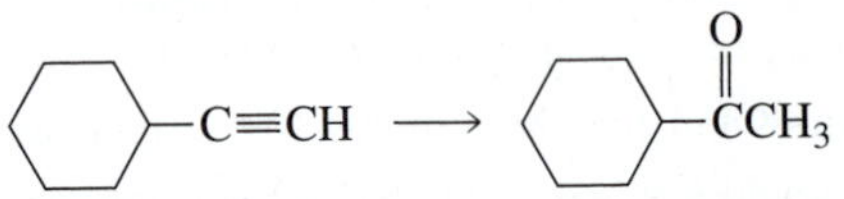 ⬡—C≡CH $\longrightarrow$ ⬡—$\underset{\overset{O}{\|}}{C}CH_3$

66. Any base whose conjugate acid has a pK_a greater than _____ can remove a proton from a terminal alkyne to form an acetylide ion in a reaction that favors products.

67. Dr. Polly Meher was planning to synthesize 3-octyne by adding 1-bromobutane to the product obtained from the reaction of 1-butyne with sodium amide. Unfortunately, however, she had forgotten to order 1-butyne. How else can she prepare 3-octyne?

68. Draw short segments of the polymers obtained from the following monomers:
   a. $CH_2=CHF$
      b. $CH_2=CHCO_2H$

69. Draw the structure of the monomer or monomers used to synthesize the following polymers:

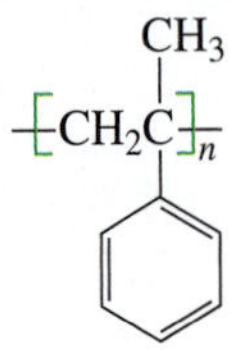

70. Draw short segments of the polymer obtained from 1-pentene, using H^+ as an initiator (obtained from the reaction of $BF_3 + H_2O$).

71. In cationic polymerization, H^+ is the initiator, obtained from the reaction of $BF_3 + H_2O$. Why is HCl not used as the source of H^+?

CHAPTER 6

Isomers and Stereochemistry

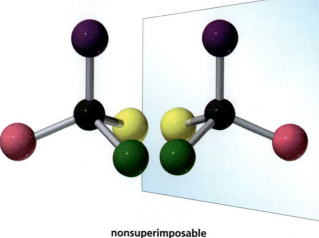

**nonsuperimposable
mirror images**

Compounds that have the same molecular formula but do not have identical structures are called **isomers**. Isomers fall into two main classes: *constitutional isomers* and *stereoisomers*. **Constitutional isomers** differ in the way their atoms are connected (Section 3.0). For example, ethyl alcohol and dimethyl ether are constitutional isomers with molecular formula C_2H_6O. The oxygen in ethyl alcohol is bonded to a carbon and to a hydrogen, whereas the oxygen in dimethyl ether is bonded to two carbons.

constitutional isomers

CH_3CH_2OH and CH_3OCH_3
ethyl alcohol   **dimethyl ether**

$CH_3CH_2CH_2CH_2Cl$ and $\overset{\displaystyle Cl}{\overset{|}{CH_3CH_2CHCH_3}}$
1-chlorobutane   **2-chlorobutane**

Unlike the atoms in constitutional isomers, the atoms in *stereoisomers* are connected in the same way. **Stereoisomers** differ in the way their atoms are arranged in space. There are two kinds of stereoisomers: **cis–trans isomers** and isomers that contain **asymmetric centers**.

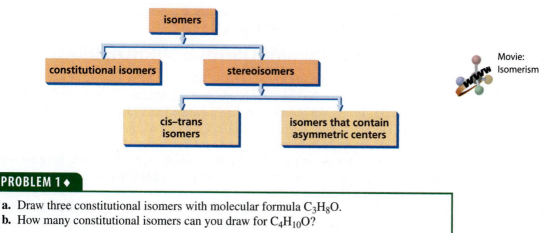

Movie:
Isomerism

PROBLEM 1 ♦

a. Draw three constitutional isomers with molecular formula C_3H_8O.
b. How many constitutional isomers can you draw for $C_4H_{10}O$?

6.1 CIS–TRANS ISOMERS RESULT FROM RESTRICTED ROTATION

Cis–trans isomers result from restricted rotation. Restricted rotation can be caused either by a *double bond* or by a *cyclic structure*. We have seen that because of the restricted rotation about its carbon–carbon double bond, an alkene such as 2-pentene can exist as cis and trans isomers (Section 4.4). The **cis isomer** has the hydrogens on the *same side* of the double bond, and the **trans isomer** has the hydrogens on *opposite sides* of the double bond.

3-D Molecules:
cis-2-Pentene;
trans-2-Pentene

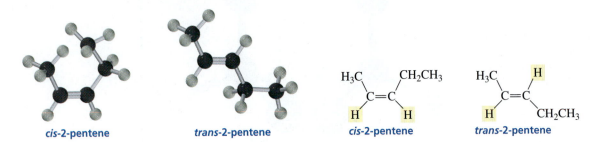

cis-2-pentene   *trans*-2-pentene   *cis*-2-pentene   *trans*-2-pentene

As a result of restricted rotation about the bonds in a ring, cyclic compounds also can have cis and trans isomers (Section 3.12). The cis isomer has the hydrogens on the same side of the ring, whereas the trans isomer has the hydrogens on opposite sides of the ring.

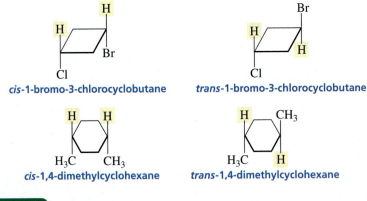

cis-1-bromo-3-chlorocyclobutane   *trans*-1-bromo-3-chlorocyclobutane

cis-1,4-dimethylcyclohexane   *trans*-1,4-dimethylcyclohexane

---

PROBLEM 2

Draw the cis and trans isomers for the following:
a. 1-ethyl-3-methylcyclobutane   **c.** 1-bromo-4-chlorocyclohexane
b. 3-hexene   **d.** 2-methyl-3-heptene

6.2 A CHIRAL OBJECT HAS A NONSUPERIMPOSABLE MIRROR IMAGE

Why can't you put your right shoe on your left foot? Why can't you put your right glove on your left hand? It is because hands, feet, gloves, and shoes have right-handed and left-handed forms. An object with a right-handed and a left-handed form is said to be **chiral** (ky-ral), a word derived from the Greek word *cheir*, which means "hand."

A chiral object has a *nonsuperimposable mirror image*. In other words, its mirror image is not the same as an image of the object itself. A hand is chiral because when you look at your right hand in a mirror, you see not a right hand but a left hand (Figure 6.1). In contrast, a chair is *not* chiral; the reflection of the chair in the mirror looks the same as the chair itself. Objects that are not chiral are said to be **achiral**. An achiral object has a *superimposable mirror image*. Some other achiral objects are a fork, a glass, and a balloon (assuming they are unadorned).

chiral objects

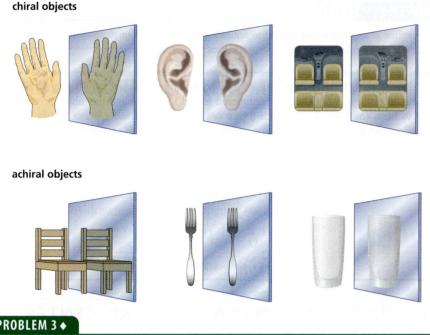

◀ **Figure 6.1**
Using a mirror to test for chirality.
A chiral object is not the same as its mirror image—they are nonsuperimposable.
An achiral object is the same as its mirror image—they are superimposable.

achiral objects

PROBLEM 3 ◆

a. Name five capital letters that are chiral.   **b.** Name five capital letters that are achiral.

6.3 AN ASYMMETRIC CENTER IS THE CAUSE OF CHIRALITY IN A MOLECULE

Objects are not the only things that can be chiral. Molecules can be chiral, too. The usual cause of chirality in a molecule is an *asymmetric center*.

An **asymmetric center** is an atom bonded to four different groups. Each of the compounds shown below has an asymmetric center indicated by a star. For example, the starred carbon in 4-octanol is an asymmetric center because it is bonded to four different groups (H, OH, $CH_2CH_2CH_3$, and $CH_2CH_2CH_2CH_3$). Notice that the atoms immediately bonded to the asymmetric center are not necessarily different from one another; the propyl and butyl groups are different even though the point at which they differ is several atoms away from the asymmetric center. The starred carbon in 2,4-dimethylhexane is an asymmetric center because it is bonded to four different groups (methyl, ethyl, isobutyl, and hydrogen).

A molecule with one asymmetric center is chiral.

$$\overset{*}{CH_3CH_2CH_2CHCH_2CH_2CH_2CH_3}$$
$$\underset{OH}{|}$$
4-octanol

$$\overset{*}{CH_3CHCH_2CH_3}$$
$$\underset{Br}{|}$$
2-bromobutane

$$\overset{CH_3}{\overset{|}{CH_3CHCH_2\overset{*}{C}HCH_2CH_3}}$$
$$\underset{CH_3}{|}$$
2,4-dimethylhexane

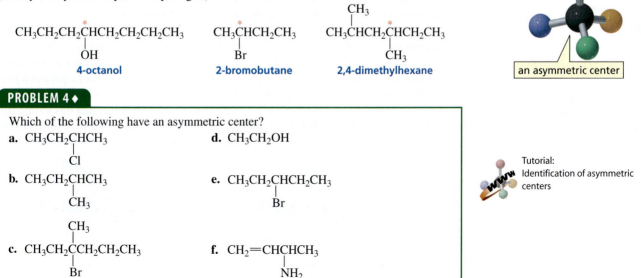

an asymmetric center

PROBLEM 4 ◆

Which of the following have an asymmetric center?

a. $CH_3CH_2CHCH_3$
$\quad\quad\quad\underset{Cl}{|}$

b. $CH_3CH_2CHCH_3$
$\quad\quad\quad\underset{CH_3}{|}$

$\quad\quad\underset{CH_3}{|}$
c. $CH_3CH_2CCH_2CH_2CH_3$
$\quad\quad\underset{Br}{|}$

d. CH_3CH_2OH

e. $CH_3CH_2CHCH_2CH_3$
$\quad\quad\quad\quad\underset{Br}{|}$

f. $CH_2{=}CHCHCH_3$
$\quad\quad\quad\quad\underset{NH_2}{|}$

Tutorial:
Identification of asymmetric centers

Eilhardt Mitscherlich (1794–1863), *a German chemist, studied medicine so he could travel to Asia in order to satisfy his interest in Oriental languages. He later became fascinated by chemistry. He was a professor of chemistry at the University of Berlin and wrote a successful chemistry textbook that was published in 1829.*

Later, chemists recognized how lucky Pasteur had been. Sodium ammonium tartrate forms asymmetric crystals only under the precise conditions that Pasteur happened to employ. Under other conditions, the symmetrical crystals that had fooled Mitscherlich are formed. But to quote Pasteur, "Chance favors the prepared mind."

Separating enantiomers by hand, as Pasteur did, is not a universally useful method because few compounds form asymmetric crystals. Fortunately enantiomers can now be separated relatively easily by a technique called **chromatography**. In this method, the mixture to be separated is dissolved in a solvent, and the solution is passed through a column packed with a chiral material that tends to absorb organic compounds. The two enantiomers can be expected to move through the column at different rates because they will have different affinities for the chiral material—just as a right hand prefers a right-handed glove to a left-handed glove—so one enantiomer will emerge from the column before the other.

6.12 RECEPTORS

A **receptor** is a protein that binds a particular molecule. Because a receptor is chiral, it will bind one enantiomer better than the other. In Figure 6.3, the receptor binds the *R* enantiomer, but it does not bind the *S* enantiomer.

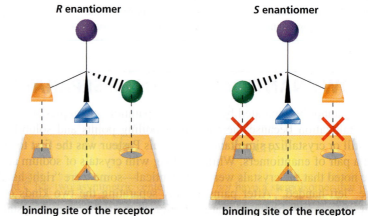

▶ **Figure 6.3**
A schematic diagram showing why only one enantiomer is bound by a receptor. One enantiomer fits into the binding site and one does not.

The fact that a receptor typically recognizes only one enantiomer causes enantiomers to have different physiological properties. For example, receptors located on the exteriors of nerve cells in the nose are able to perceive and differentiate the estimated 10,000 smells to which they are exposed. The reason that (*R*)-(−)-carvone (found in spearmint oil) and (*S*)-(+)-carvone (the main constituent of caraway seed oil) have such different odors is that each enantiomer fits into a different receptor.

(*R*)-(−)-carvone
spearmint oil
$[\alpha]_D^{20\,°C} = -62.5$

(*S*)-(+)-carvone
caraway seed oil
$[\alpha]_D^{20\,°C} = +62.5$

Frances O. Kelsey receives a medal from John F. Kennedy for preventing the sale of thalidomide (see page 161).

Many drugs exert their physiological activity by binding to cell-surface receptors. If the drug has an asymmetric center, the receptor can bind one of the enantiomers preferentially. Thus, enantiomers of a drug can have the same physiological activities, different degrees of the same activity, or very different activities, depending on the drug.

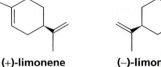

PROBLEM 26 ◆

Limonene exists as two different stereoisomers. The *R* enantiomer is found in oranges, and the *S* enantiomer is found in lemons. Which of the following is found in oranges?

(+)-limonene   (−)-limonene

THE ENANTIOMERS OF THALIDOMIDE

Thalidomide was developed in then West Germany and was first marketed in 1957 for insomnia and morning sickness. At that time it was available in more than 40 countries but had not been approved for use in the United States because Frances O. Kelsey, a physician for the Federal Food and Drug Administration (FDA), had insisted upon additional tests. (See the box on Drug Safety on page 559.)

The dextrorotatory isomer has stronger sedative properties, but the commercial drug was a racemic mixture. No one knew that the levorotatory isomer was highly teratogenic—causes horrible birth defects—until women who had been given the drug during the first three months of pregnancy gave birth to babies with a wide variety of defects, deformed limbs being the most common. About 10,000 children were damaged by the drug.

It was eventually determined that the dextrorotatory isomer also has mild teratogenic activity and that each of the enan-

tiomers can racemize (interconvert) in the body. Thus, it is not clear whether the birth defects would have been less severe if the women had been given the dextrorotatory isomer only. Because thalidomide was found to damage fast-growing cells in the developing fetus, it has recently has been approved—with restrictions—as an anticancer drug to discourage the production of cancer cells.

asymmetric center

thalidomide

CHIRAL DRUGS

Until relatively recently, most drugs have been marketed as racemic mixtures because of the high cost of separating enantiomers. In 1992, however, the Food and Drug Administration (FDA) issued a policy statement encouraging drug companies to use recent advances in separation techniques to develop single-enantiomer drugs. Now, one-third of all drugs sold are single enantiomers.

If a drug is sold as a racemate, the FDA requires that both enantiomers be tested, since the enantiomers can have similar or very different properties. Examples are numerous. The *S* isomer of Prozac, an antidepressant, is better at blocking serotonin; but it is used up faster than the *R* isomer. Testing has shown that the anesthetic (*S*)-(+)-ketamine is four times more potent than (*R*)-(−)-ketamine and the disturbing side effects are apparently only associated with the (*R*)-(−)-enantiomer.

Only the *S* isomer of the beta-blocker propranolol shows activity; the *R* isomer is inactive. The activity of ibuprofen, the popular analgesic marketed as Advil, Nuprin, and Motrin, resides primarily in the (*S*)-(+)-enantiomer. Heroin addicts can be maintained with (−)-α-acetylmethadol for a 72-hour period compared to 24 hours with racemic methadone. This means less frequent visits to the clinic; a single dose can get an addict through an entire weekend.

Prescribing a single enantiomer prevents the patient from having to metabolize the less potent enantiomer and decreases the chance of unwanted drug interactions. Drugs that could not be given as racemates because of the toxicity of one of the enantiomers can now be used. For example, (*S*)-penicillamine can be used to treat Wilson's disease even though (*R*)-penicillamine causes blindness.

6.13 THE STEREOCHEMISTRY OF REACTIONS

When we looked at the electrophilic addition reactions that alkenes undergo (Chapter 5), we examined the step-by-step process by which each reaction occurs (the mechanism of the reaction), and we determined what products are formed. However, we did not consider the stereochemistry of the reactions.

Stereochemistry is the field of chemistry that deals with the structures of molecules in three dimensions. When we study the stereochemistry of a reaction, we are

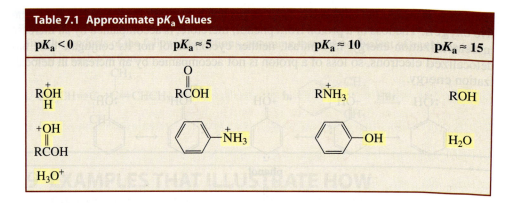

Table 7.1 Approximate pK_a Values

p$K_a < 0$	p$K_a \approx 5$	p$K_a \approx 10$	p$K_a \approx 15$

PROBLEM 17 ♦ SOLVED

Which is a stronger acid?
a. $CH_3CH_2CH_2OH$ or $CH_3CH{=}CHOH$

b.
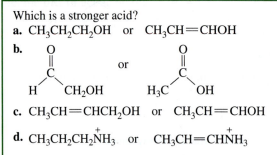

c. $CH_3CH{=}CHCH_2OH$ or $CH_3CH{=}CHOH$

d. $CH_3CH_2CH_2\overset{+}{N}H_3$ or $CH_3CH{=}CH\overset{+}{N}H_3$

Solution to 17a The conjugate base of propyl alcohol does not have delocalized electrons, but the conjugate base of allyl alcohol does. Because delocalized electrons stabilize a compound, the conjugate base of allyl alcohol is the weaker of the two bases. Therefore, allyl alcohol is a stronger acid than propyl alcohol, because the weaker the base, the stronger is its conjugate acid.

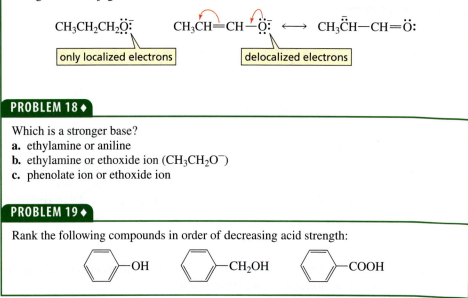

PROBLEM 18 ♦

Which is a stronger base?
a. ethylamine or aniline
b. ethylamine or ethoxide ion ($CH_3CH_2O^-$)
c. phenolate ion or ethoxide ion

PROBLEM 19 ♦

Rank the following compounds in order of decreasing acid strength:

⟨benzene⟩—OH   ⟨benzene⟩—CH₂OH   ⟨benzene⟩—COOH

7.10 ULTRAVIOLET AND VISIBLE SPECTROSCOPY

UV/Vis spectroscopy provides information about compounds that have conjugated double bonds. Ultraviolet (UV) light has wavelengths ranging from 180 to 400 nm (nanometers); visible (Vis) light has wavelengths ranging from 400 to 780 nm. The shorter the wavelength, the greater is the energy of the radiation. Ultraviolet light,

therefore, has greater energy than visible light. If a compound absorbs **ultraviolet light**, a UV spectrum is obtained; if it absorbs **visible light**, a visible spectrum is obtained.

The UV spectrum of acetone is shown in Figure 7.2. The λ_{max} (stated as "lambda max") is the wavelength corresponding to the highest point of the absorption band. The UV spectrum shows that acetone has a $\lambda_{max} = 195$ nm.

The shorter the wavelength, the greater is the energy of the radiation.

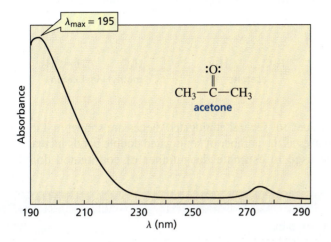

◀ **Figure 7.2**
The UV spectrum of acetone.

ULTRAVIOLET LIGHT AND SUNSCREENS

Exposure to ultraviolet light stimulates specialized cells in the skin to produce a black pigment known as melanin, which causes the skin to look tan. Melanin absorbs UV light, so it protects our bodies from the harmful effects of the sun. If more UV light reaches the skin than the melanin can absorb, the light will burn the skin and cause photochemical reactions that can result in skin cancer.

UV-A is the lowest-energy UV light (315 to 400 nm) and does the least biological damage. Fortunately, most of the more dangerous, higher-energy UV light, UV-B (290 to 315 nm) and UV-C (180 to 290 nm), is filtered out by the ozone layer in the stratosphere. That is why we need to worry about the apparent thinning of the ozone layer (Section 5.17).

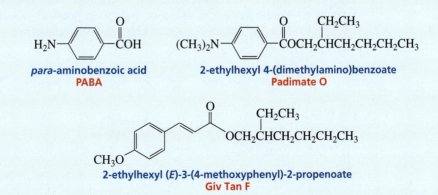

Applying a sunscreen can protect skin against UV light. The amount of protection provided by a particular sunscreen is indicated by its SPF (sun protection factor). The higher the SPF, the greater the protection. Some sunscreens contain an inorganic component, such as zinc oxide, that reflects the light as it reaches the skin. Others contain a compound that absorbs UV light. PABA was the first commercially available UV-absorbing sunscreen. PABA absorbs UV-B light, but is not very soluble in oily skin lotions. Less polar compounds, such as Padimate O, are now commonly used. Recent research has shown that sunscreens that absorb only UV-B light do not give adequate protection against skin cancer; both UV-A and UV-B protection are needed. Giv Tan F absorbs both UV-B and UV-A light, so it gives better protection.

7.11 THE λ_{max} INCREASES AS THE NUMBER OF CONJUGATED DOUBLE BONDS INCREASES

The more conjugated double bonds that a compound has, the longer is the wavelength at which the λ_{max} occurs. For example, methyl vinyl ketone has two conjugated double bonds, whereas acetone has only one; hence, the λ_{max} for methyl vinyl ketone is at 219 nm—a longer wavelength than the λ_{max} for acetone (195 nm).

3-D Molecule:
Methyl vinyl ketone

The λ_{max} increases as the number of conjugated double bonds increases.

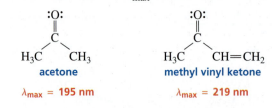

acetone
λ_{max} = **195 nm**

methyl vinyl ketone
λ_{max} = **219 nm**

The λ_{max} values for several conjugated dienes are shown in Table 7.2. Notice that the λ_{max} increases as the number of conjugated double bonds increases. Thus, the λ_{max} value can be used to estimate the number of conjugated double bonds in the compound.

Table 7.2 λ_{max} Values for Ethene and Conjugated Polyenes	
Compound	λ_{max} **(nm)**
$H_2C{=}CH_2$	165
	217
	256
	290
	334
	364

If a compound has enough conjugated double bonds, it will absorb visible light ($\lambda_{max} > 400$ nm), and the compound will be colored. Thus, β-carotene, a precursor of vitamin A—found in carrots, apricots, and sweet potatoes—is an orange substance. Lycopene—found in tomatoes, watermelon, and pink grapefruit—is red.

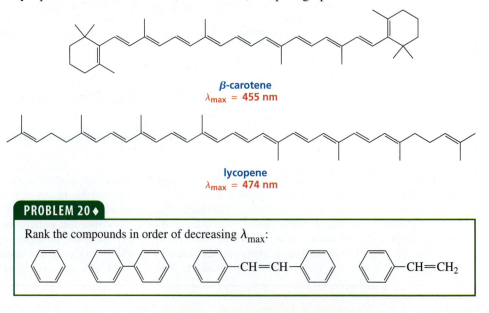

β-carotene
λ_{max} = **455 nm**

lycopene
λ_{max} = **474 nm**

PROBLEM 20 ◆

Rank the compounds in order of decreasing λ_{max}:

7.12 A COMPOUND THAT ABSORBS VISIBLE LIGHT IS COLORED

White light is a mixture of all wavelengths of visible light. If any of these wavelengths are removed from white light, the remaining light is colored. Therefore, a compound that absorbs visible light is colored, and its color depends on the color of the wavelengths of the absorbed light. The wavelengths that the compound does not absorb are reflected back to the viewer, producing the color the viewer sees.

The relationship between the wavelengths of the light that a substance absorbs and the substance's observed color is shown in Table 7.3. Notice that two absorption bands are necessary to produce green. Most colored compounds have fairly broad absorption bands, although vivid colors have narrow absorption bands. The human eye is able to distinguish more than a million different shades of color!

Table 7.3 Dependence of the Color Observed on the Wavelength of Light Absorbed	
Wavelengths absorbed (nm)	**Observed color**
380–460	yellow
380–500	orange
440–560	red
480–610	purple
540–650	blue
380–420 and 610–700	green

Lycopene, β-carotene, and anthocyanins are found in the leaves of trees, but their characteristic colors are usually obscured by the green color of chlorophyll (Section 8.3). In the fall, when chlorophyll degrades, the other colors become apparent.

Azobenzenes (benzene rings connected by an N=N bond) have an extended conjugated system that causes them to absorb light from the visible region of the spectrum. Some substituted azobenzenes, such as the two shown below, are used commercially as dyes. Varying the extent of conjugation and the substituents attached to the conjugated system creates a large number of different colors. Notice that the only difference between butter yellow and methyl orange is an $SO_3^- Na^+$ group. When margarine was first produced, it was colored with butter yellow to make it look more like butter. (White margarine would not have been very appetizing.) This dye was abandoned after it was found to be carcinogenic. β-Carotene is now used to color margarine (page 190). Methyl orange is a commonly used acid–base indicator (see Problem 47).

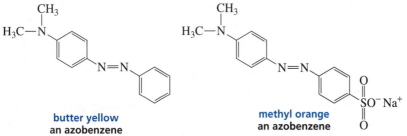

butter yellow **an azobenzene**	**methyl orange** **an azobenzene**

The lone-pair electrons on oxygen and nitrogen in the compounds shown below are available to interact with the π electron cloud of the ring; such an interaction increases the λ_{max}. Because the anilinium ion does not have a lone pair, its λ_{max} is similar to that of benzene.

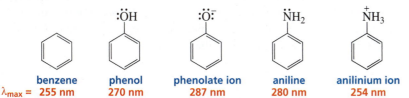

benzene	phenol	phenolate ion	aniline	anilinium ion
λ_{max} = **255 nm**	**270 nm**	**287 nm**	**280 nm**	**254 nm**

ANTHOCYANINS: A COLORFUL CLASS OF COMPOUNDS

A class of highly conjugated compounds called antho-cyanins is responsible for the red, purple, and blue colors of many flowers (poppies, peonies, cornflowers), fruits (cranberries, rhubarb, strawberries, blueberries), and vegetables (beets, radishes, red cabbage).

In a neutral or basic solution, the monocyclic fragment (on the right-hand side of the anthocyanin) is not conjugated with the rest of the molecule, so the anthocyanin does not absorb visible light and is therefore a colorless compound. In an acidic environment, however, the OH group becomes proto-

nated and water is eliminated. Loss of water results in the third ring's becoming conjugated with the rest of the molecule. As a result of the number of conjugated double bonds, the antho-cyanin absorbs visible light with wavelengths between 480 and 550 nm. The exact wavelength of light absorbed depends on the substituents (R and R′) on the anthocyanin. Thus, the flower, fruit, or vegetable appears red, purple, or blue, depending on what the R groups are. You can see this color change if you alter the pH of cranberry juice so that it is no longer acidic.

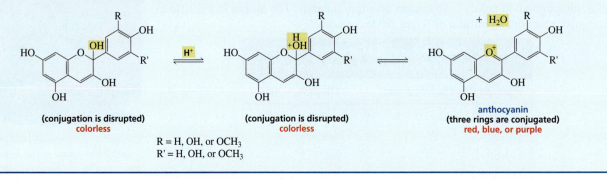

(conjugation is disrupted)
colorless

(conjugation is disrupted)
colorless

anthocyanin
(three rings are conjugated)
red, blue, or purple

R = H, OH, or OCH$_3$
R′ = H, OH, or OCH$_3$

PROBLEM 21 ◆

a. At pH = 7, one of the ions shown below is purple and the other is blue. Which is which?

b. What would be the difference in the colors of the compounds at pH = 3?

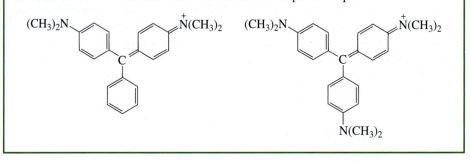

SUMMARY

Localized electrons belong to a single atom or are confined to a bond between two atoms. **Delocalized electrons** are shared by more than two atoms; they result when a *p* orbital overlaps the *p* orbitals of more than one adjacent atom. Electron delocalization occurs only if all the atoms sharing the delocalized electrons lie in or close to the same plane.

Each of benzene's six carbon atoms is sp^2 hybridized, with bond angles of 120°. A *p* orbital of each carbon overlaps the *p* orbitals of both adjacent carbons. The six π electrons are shared by all six carbons. Thus, benzene is a planar molecule with six delocalized π electrons.

Chemists use **resonance contributors**—structures with localized electrons—to approximate the actual structure of a compound that has delocalized electrons: the **resonance**

hybrid. To draw resonance contributors, move only π electrons or nonbonding electrons toward an sp^2 carbon. The total number of electrons and the numbers of paired and unpaired electrons do not change.

The greater the **predicted stability** of the resonance contributor, the more it contributes to the hybrid and the more similar its structure is to the real molecule. The extra stability a compound gains from having delocalized electrons is called **delocalization energy**. It tells us how much more stable a compound with delocalized electrons is than it would be if its electrons were localized. The greater the number of relatively stable resonance contributors and the more nearly equivalent they are, the greater is the delocalization energy of the compound.

Because electron delocalization increases the stability of a compound, allylic and benzylic cations are more stable, since they have delocalized electrons, than similarly substituted carbocations with only localized electrons. Delocalized electrons can affect the pK_a of a compound; a carboxylic acid and a phenol are more acidic than an alcohol such as ethanol, and a protonated aniline is more acidic than a protonated amine because in each case, the loss of a proton is accompanied by an increase in electron delocalization.

Conjugated double bonds are separated by one single bond. **Isolated double bonds** are separated by more than one single bond. Because dienes with conjugated double bonds have delocalized electrons, they are more stable than dienes with isolated double bonds.

An isolated diene, like an alkene, undergoes only 1,2-addition. If there is only enough electrophilic reagent to add to one of the double bonds, it will add preferentially to the one that forms the more stable carbocation. A conjugated diene reacts with a limited amount of electrophilic reagent to form a **1,2-addition product** and a **1,4-addition product**. The first step is addition of the electrophile to one of the sp^2 carbons at the end of the conjugated system.

Ultraviolet and **visible (UV/Vis) spectroscopy** provide information about compounds with conjugated double bonds. UV light is higher in energy than visible light; the shorter the wavelength, the greater is its energy. The more conjugated double bonds there are in a compound, the greater is its λ_{max} value.

SUMMARY OF REACTIONS

1. In the presence of excess electrophilic reagent, both double bonds of a *diene with isolated double bonds* will undergo electrophilic addition.

$$CH_2\!=\!CHCH_2CH_2\overset{\overset{\displaystyle CH_3}{|}}{C}\!=\!CH_2 \ + \ \underset{\text{excess}}{HBr} \ \longrightarrow \ CH_3\overset{}{C}HCH_2CH_2\overset{\overset{\displaystyle CH_3}{|}}{\underset{\underset{\displaystyle Br}{|}}{C}}CH_3$$

In the presence of only one equivalent of an electrophilic reagent, only the most reactive double bond of an isolated diene will undergo electrophilic addition (Section 7.8).

$$CH_2\!=\!CHCH_2CH_2\overset{\overset{\displaystyle CH_3}{|}}{C}\!=\!CH_2 \ + \ HBr \ \longrightarrow \ CH_2\!=\!CHCH_2CH_2\overset{\overset{\displaystyle CH_3}{|}}{\underset{\underset{\displaystyle Br}{|}}{C}}CH_3$$

2. *Conjugated dienes* undergo 1,2- and 1,4-addition with one equivalent of an electrophilic reagent (Section 7.8).

$$RCH\!=\!CHCH\!=\!CHR \ + \ HBr \ \longrightarrow \ \underset{\text{1,2-addition product}}{RCH_2\overset{}{\underset{\underset{\displaystyle Br}{|}}{C}}HCH\!=\!CHR} \ + \ \underset{\text{1,4-addition product}}{RCH_2CH\!=\!CH\overset{}{\underset{\underset{\displaystyle Br}{|}}{C}}HR}$$

PROBLEMS

22. Which of the following have delocalized electrons?

a. $CH_2\!=\!CH\overset{\overset{\displaystyle O}{\|}}{C}CH_3$

b.

c. $CH_3\overset{\overset{\displaystyle CH_3}{|}}{\underset{+}{C}}CH_2CH\!=\!CH_2$

d. $CH_3CH\!=\!CHOCH_2CH_3$

e. $CH_2\!=\!CHCH_2CH\!=\!CH_2$

f.

g. $CH_3CH_2NHCH_2CH\!=\!CHCH_3$

h. $CH_3CH_2NHCH\!=\!CHCH_3$

i. $CH_3CH_2\overset{}{\underset{+}{C}}HCH\!=\!CH_2$

j.

23. Draw resonance contributors for the following ions:

a. 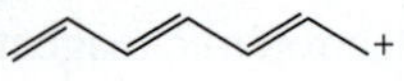 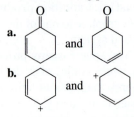   +

b.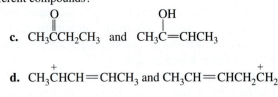

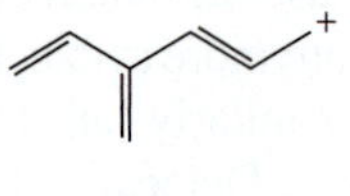

24. Are the following pairs of structures resonance contributors or different compounds?

a. [structure] and [structure]

b. [structure] and [structure]

c. $CH_3CCH_2CH_3$ and $CH_3C\!\!=\!\!CHCH_3$ (with O and OH)

d. $CH_3\overset{+}{C}HCH\!\!=\!\!CHCH_3$ and $CH_3CH\!\!=\!\!CHCH_2\overset{+}{C}H_2$

25. a. Draw resonance contributors for the following species. Indicate which are major contributors and which are minor contributors to the resonance hybrid.

1. $CH_3CH\!\!=\!\!CHOCH_3$

2. 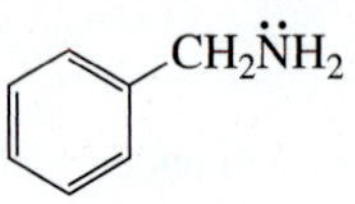 [benzene ring]—$CH_2\ddot{N}H_2$

3. $CH_3\overset{..}{\overset{..}{C}}HCH\!\!=\!\!NCH_3$

4. $CH_3CH\!\!=\!\!CH\overset{+}{C}H_2$

5.  [cyclopentadienyl cation structure] +

6. 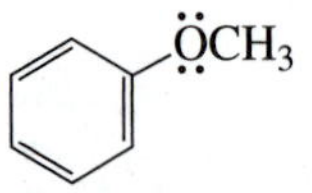 [benzene ring]—$\ddot{O}CH_3$

b. Do any of the species have resonance contributors that all contribute equally to the resonance hybrid?

26. Which resonance contributor makes the greater contribution to the resonance hybrid?

a. 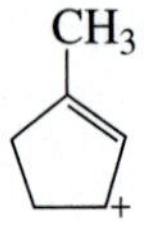 [cyclopentene structure with CH_3] or [cyclopentene structure with CH_3]

b.  or [structure]

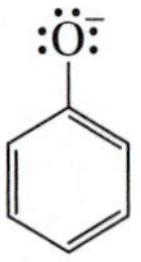

c. $CH_3\overset{+}{C}HCH\!\!=\!\!CH_2$ or $CH_3CH\!\!=\!\!CH\overset{+}{C}H_2$

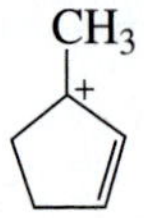

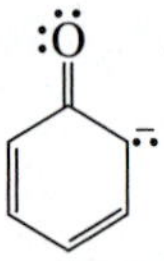

27. a. Which oxygen atom has the greater electron density?

$$CH_3COCH_3$$ (with O double bond)

b. Which compound has the greater electron density on its nitrogen atom? (Revisit Problem 25 in Chapter 2. Now you can understand why the two ring nitrogens have different basicities.)

[pyrrole-type structure] or [pyrrolidine-type structure]

c. Which compound has the greater electron density on its oxygen atom?

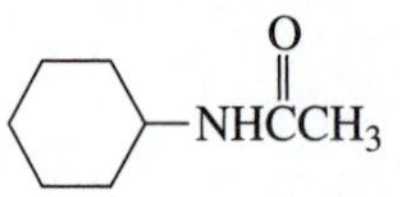

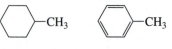

[cyclohexyl]—$NHCCH_3$ or [benzene]—$NHCCH_3$ (each with O)

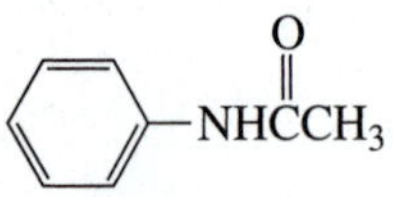

28. Which can lose a proton more readily, a methyl group bonded to cyclohexane or a methyl group bonded to benzene? Explain your answer.

[cyclohexane]—CH_3 [benzene]—CH_3

29. Rank the compounds in order of decreasing λ_{max}:

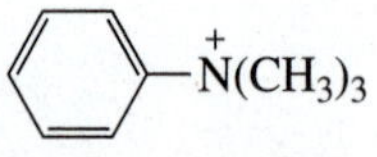

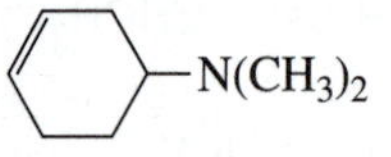

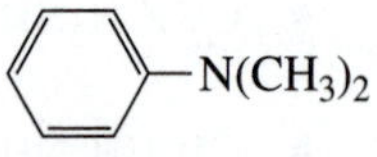

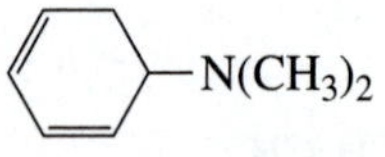

[benzene]—$\overset{+}{N}(CH_3)_3$ [cyclohexane]—$N(CH_3)_2$ [benzene]—$N(CH_3)_2$ [cyclohexane]—$N(CH_3)_2$

30. Rank the following compounds in order of decreasing acidity:

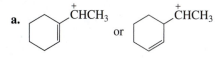

31. Which species in each pair is more stable?

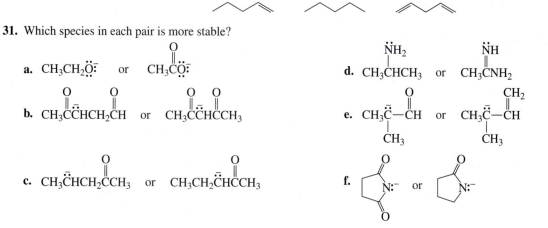

32. Which species in each of the pairs in Problem 31 is the stronger base?

33. Draw resonance contributors for the following species. Indicate which are major contributors to the resonance hybrid and which are minor contributors.

a. CH₃—N⁺(=Ö:)(—:Ö⁻)

b. HC(=O)N̈HCH₃

c. HC̈CH=CHC̈H₂

d. CH₃C̈H—N⁺(=O)(O⁻)

e. CH₂Ö̈H

f. CH₃C̈CHCCH₃

34. Rank the following compounds in order of decreasing acidity of the indicated hydrogen:

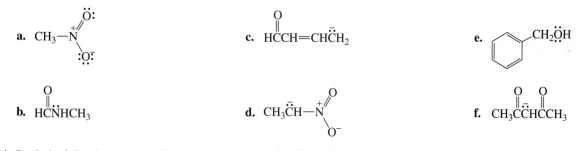

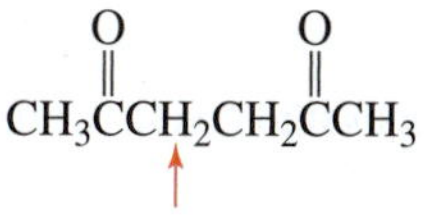

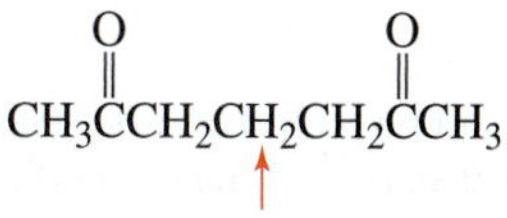

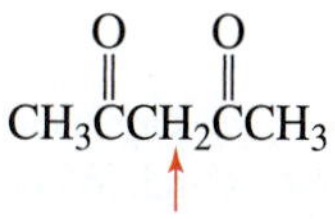

35. Draw resonance contributors for each of the following species, and rank the resonance contributors in order of decreasing contribution to the hybrid:

a. 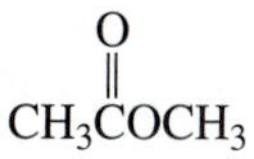 CH₃COCH₃   **b.** 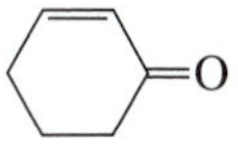 (cyclohexenone)   **c.** 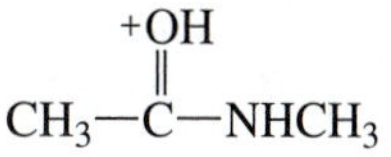 +OH / CH₃—C—NHCH₃

36. Draw the resonance hybrid for each of the species in Problem 35.

37. Name the following dienes and rank them in order of increasing stability. (Alkyl groups stabilize dienes in the same way that they stabilize alkenes; see Section 4.6.)

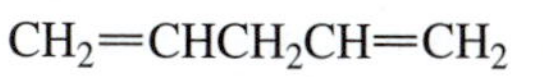

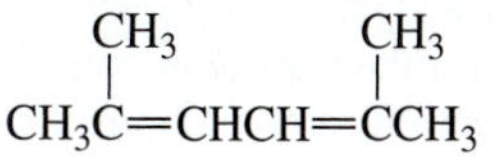

CH₃CH=CHCH=CHCH₃   CH₂=CHCH₂CH=CH₂   CH₃C(CH₃)=CHCH=C(CH₃)CH₃   CH₃CH=CHCH=CH₂

38. Which carbocation in each of the following pairs is more stable?

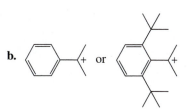

39. a. How many linear dienes have molecular formula C_6H_{10}? (Disregard cis–trans isomers.)
  b. How many of the linear dienes in part a are conjugated dienes?
  c. How many are isolated dienes?

40. What products would be obtained from the reaction of 1,3,5-hexatriene with one equivalent of HBr?

41. Give the major product of each of the following reactions, assuming the presence of one equivalent of each reagent:

a. + HBr ⟶

b. + HBr ⟶

42. How could you use UV spectroscopy to distinguish between the compounds in each of the following pairs?

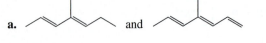

a. and

b. and

43. Give all the products of the following reaction:

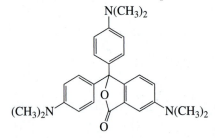

 + HO⁻ ⟶

44. Some credit card slips have a top sheet of "carbonless paper" that transfers an imprint of your signature to a sheet lying underneath. The paper contains tiny capsules filled with the colorless compound shown below:

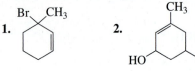

When you press on the paper, the capsules burst and the colorless compound comes into contact with the acid-treated bottom sheet, forming a highly colored compound. What is the structure of the colored compound?

45. a. How could each of the following compounds be prepared from a hydrocarbon in a single step?

1.   2.

  b. What other organic compound would be obtained from each synthesis?

46. Give the major products obtained from the reaction of one equivalent of HCl with the following:
  a. 2,3-dimethyl-1,3-pentadiene   **b.** 2,4-dimethyl-1,3-pentadiene

47. a. Methyl orange (whose structure is given in Section 7.12) is an acid–base indicator. In solutions of pH > 4, it is red; in solutions of pH < 4, it is yellow. Account for the change in color.
  b. Phenolphthalein, which is another indicator, exhibits a much more dramatic color change. In solutions of pH < 8.5, it is colorless; in solutions of pH > 8.5, it is deep red-purple. Account for the change in color.

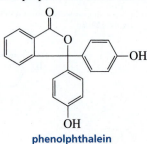

phenolphthalein

CHAPTER 8

Aromaticity
Reactions of Benzene and Substituted Benzenes

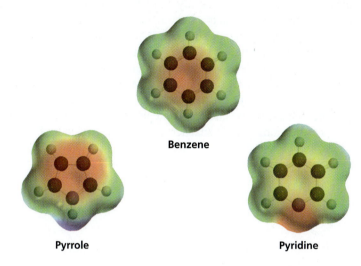

Benzene

Pyrrole

Pyridine

The compound we know as benzene was first isolated in 1825 by Michael Faraday. He extracted it from the liquid residue obtained after heating whale oil under pressure to produce the gas then being used to illuminate buildings in London.

Many substituted benzenes are found in nature. A few that have physiological activity are shown here:

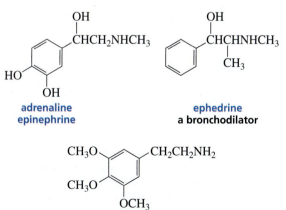

chloramphenicol
an antibiotic that is particularly
effective against typhoid fever

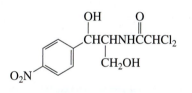

adrenaline
epinephrine

ephedrine
a bronchodilator

mescaline
active ingredient of
the peyote cactus

B I O G R A P H Y

Michael Faraday (1791–1867) *was born in England, a son of a blacksmith. At the age of 14, he was apprenticed to a bookbinder and educated himself by reading the books that he bound. He became an assistant to Sir Humphry Davy at the Royal Institution of Great Britain in 1812, where Faraday taught himself chemistry. In 1825, he became director of a laboratory there and, in 1833, a professor of chemistry. He is best known for his work on electricity and magnetism.*

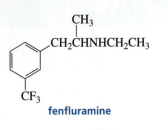
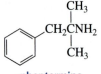

fenfluramine

phentermine

Many other physiologically active substituted benzenes are not found in nature, but exist because chemists have synthesized them. The now-banned diet drug fen-phen is a mixture of two synthetic substituted benzenes: fenfluramine and phentermine. Two other substituted benzenes, BHA and BHT, are preservatives found in a wide variety of packaged foods (Section 5.17).

When naturally occurring compounds are found to have desirable physiological activities, chemists will synthesize structurally similar compounds with the hope of developing them into useful products. For example, chemists have synthesized compounds with structures similar to adrenaline, producing amphetamine, a central nervous system stimulant, and the closely related methamphetamine, which are both used clinically as appetite suppressants. Methamphetamine, known as "speed," is also made and sold illegally because of its rapid and intense psychological effects. These compounds represent just a few of the many substituted benzenes that have been synthesized for commercial use by the chemical and pharmaceutical industries. The physical properties of several substituted benzenes are given in Appendix I.

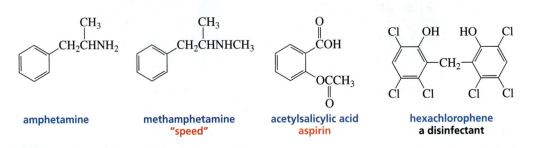

amphetamine   **methamphetamine "speed"**   **acetylsalicylic acid aspirin**   **hexachlorophene a disinfectant**

Compounds like benzene that have relatively few hydrogens in relation to the number of carbons, are typically found in oils produced by trees and other plants. Early chemists called such compounds **aromatic compounds** because of their pleasing fragrances. In this way, they were distinguished from **aliphatic compounds**, which have higher hydrogen-to-carbon ratios. Today, chemists use the word "aromatic" to signify certain kinds of chemical structures. We will now look at the features that cause a compound to be classified as aromatic.

MEASURING TOXICITY

Agent Orange, a defoliant widely used during the Vietnam War, is a mixture of two synthetic substituted benzenes: 2,4-D and 2,4,5-T. Dioxin (TCDD), a contaminant formed during the manufacture of Agent Orange, has been implicated as the causative agent of the various symptoms suffered by those exposed to this compound during the war.

2,4-dichlorophenoxyacetic acid 2,4-D   **2,4,5-trichlorophenoxyacetic acid 2,4,5-T**   **2,3,7,8-tetrachlorodibenzo[b,e][1,4]dioxin TCDD**

The toxicity of a compound is indicated by its LD_{50} value—the dosage found to kill 50% of the test animals exposed to it. Dioxin, with an LD_{50} value of 0.0006 mg/kg for guinea pigs, is an extremely toxic compound. Compare this with the LD_{50} values of some well-known but far less toxic poisons: 0.96 mg/kg for strychnine and 15 mg/kg for both arsenic trioxide and sodium cyanide. One of the most toxic agents known is the botulism toxin, with an LD_{50} value of about 1×10^{-8} mg/kg.

8.1 THE TWO CRITERIA FOR AROMATICITY

In Chapter 7, we saw that benzene is a planar, cyclic compound with a cyclic cloud of delocalized electrons above and below the plane of the ring. (See Figure 7.1 on page 172.) Because its π electrons are delocalized, all the C—C bonds in benzene have the same length.

Benzene is a particularly stable compound because its **delocalization energy**— the extra stability it gains from having delocalized electrons—is unusually large (Section 7.6). Compounds with unusually large delocalization energies are called **aromatic compounds**.

Aromatic compounds are particularly stable.

How can we tell whether a compound is aromatic by looking at its structure? In other words, what structural features do aromatic compounds have in common?

To be classified as aromatic, a compound must meet both of the following criteria:

1. *It must have an uninterrupted cyclic cloud of π electrons* (called a π cloud) *above and below the plane of the molecule.* Let's look a little more closely at what this means:

    For the π cloud to be cyclic, *the molecule must be cyclic.*
    For the π cloud to be uninterrupted, *every atom in the ring must have a p orbital.*
    For the π cloud to form, each *p* orbital must overlap with the *p* orbitals on either side of it. Therefore, *the molecule must be planar.*

    Tutorial: Aromaticity

2. *The π cloud must contain an odd number of pairs of π electrons.*

    Therefore, benzene is an aromatic compound because it is cyclic and planar, every carbon in the ring has a *p* orbital, and the π cloud contains *three* pairs of π electrons.

For a compound to be aromatic, it must be cyclic and planar, and have an uninterrupted cloud of π electrons. The π cloud must contain an odd number of pairs of π electrons.

8.2 APPLYING THE CRITERIA FOR AROMATICITY

Cyclobutadiene has two pairs of π electrons, and cyclooctatetraene has four pairs of π electrons. Unlike benzene, these compounds are *not* aromatic because they have an *even* number of pairs of π electrons. There is an additional reason why cyclooctatetraene is not aromatic—it is not planar, it is tub-shaped (Section 7.3). Because cyclobutadiene and cyclooctatetraene are not aromatic, they do not have the unusual stability of aromatic compounds.

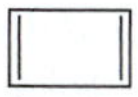

cyclobutadiene    **benzene**    **cyclooctatetraene**

3-D Molecules: Cyclobutadiene; Benzene; Cyclooctatetraene

Now let's look at some other compounds and determine whether they are aromatic. Cyclopentadiene is not aromatic because it does not have an uninterrupted ring of *p* orbital-bearing atoms. One of its ring atoms is sp^3 hybridized, and only sp^2 and sp carbons have *p* orbitals. Therefore, cyclopentadiene does not fulfill the first criterion for aromaticity.

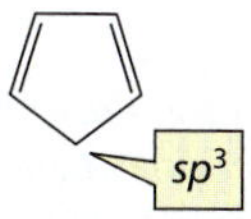

sp^3

cyclopentadiene

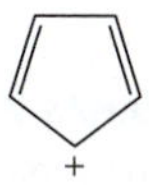

+

cyclopentadienyl cation

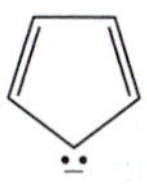

··

cyclopentadienyl anion

The cyclopentadienyl cation also *is not* aromatic because, although it is planar and has an uninterrupted ring of *p* orbital-bearing atoms, its π cloud has *two* (an even number) pairs of π electrons. The cyclopentadienyl anion *is* aromatic: it is planar and has an uninterrupted ring of *p* orbital-bearing atoms, and its π cloud contains *three* (an odd number) pairs of π electrons.

Notice that the negatively charged carbon in the cyclopentadienyl anion is sp^2 hybridized, because if it were sp^3 hybridized, the ion would not be aromatic. The resonance hybrid shows that all the carbons in the cyclopentadienyl anion are equivalent. Each carbon has exactly one-fifth of the negative charge associated with the anion.

When drawing resonance contributors, remember that only electrons move; atoms never move.

resonance contributors of the cyclopentadienyl anion

resonance hybrid

The criteria that determine whether a monocyclic hydrocarbon compound is aromatic can also be used to determine whether a polycyclic hydrocarbon compound is

BUCKYBALLS

Diamond and graphite are two familiar forms of pure carbon (Section 1.8). A third form was discovered unexpectedly in 1985, while scientists were conducting experiments designed to understand how long-chain molecules are formed in outer space. R. E. Smalley, R. F. Curl, Jr., and H. W. Kroto shared the 1996 Nobel Prize in chemistry for discovering this new form of carbon. They named the substance buckminsterfullerene (often shortened to fullerene) because its structure reminded them of the geodesic domes popularized by R. Buckminster Fuller, an American architect and philosopher. Its nickname is "buckyball."

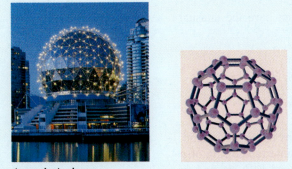

A geodesic dome

Consisting of a hollow cluster of 60 carbons, fullerene is the most symmetrical large molecule known. Like graphite, fullerene has only sp^2 carbons, but instead of being arranged in layers, the carbons are arranged in rings that fit together like the seams of a soccer ball. Each molecule has 32 interlocking rings (20 hexagons and 12 pentagons). At first glance, fullerene would appear to be aromatic because of its benzene-like rings. However, the curvature of the ball prevents the

molecule from fulfilling the first criterion for aromaticity—that it be planar.

Buckyballs have extraordinary chemical and physical properties. For example, they are exceedingly rugged, as shown by their ability to survive the extreme temperatures of outer space. Because they are essentially hollow cages, they can be manipulated to make materials never before known. For example, when a buckyball is "doped" by inserting potassium or cesium into its cavity, it becomes an excellent organic superconductor. These molecules are now being studied for use in many other applications, including the development of new polymers, new catalysts, and new drug delivery systems. The discovery of buckyballs is a strong reminder of the technological advances that can be achieved as a result of basic research.

Richard E. Smalley (1943–2005) *was born in Akron, Ohio. He received a B.S. from the University of Michigan and a Ph.D. from Princeton University. He was a professor of chemistry at Rice University.*

Robert F. Curl, Jr., *was born in Texas in 1933. He received a B.A. from Rice University and a Ph.D. from the University of California, Berkeley. He is a professor of chemistry at Rice University.*

Sir Harold W. Kroto *was born in 1939 in England and was a professor of chemistry at the University of Sussex. Now he is a professor of chemistry at Florida State University.*

aromatic. Naphthalene (five pairs of π electrons), phenanthrene (seven pairs of π electrons), and chrysene (nine pairs of π electrons) are aromatic.

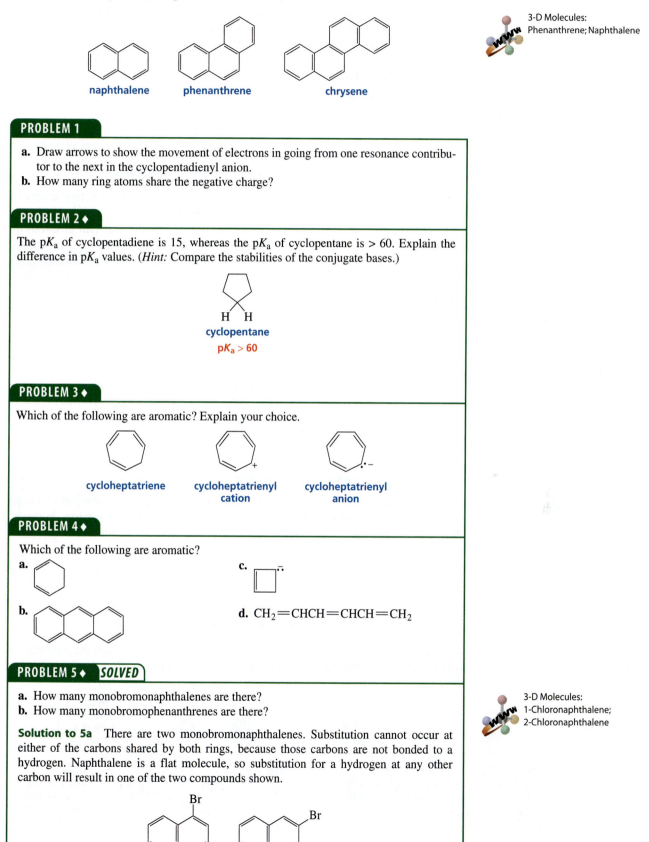

naphthalene   phenanthrene   chrysene

3-D Molecules:
Phenanthrene; Naphthalene

PROBLEM 1

a. Draw arrows to show the movement of electrons in going from one resonance contributor to the next in the cyclopentadienyl anion.
b. How many ring atoms share the negative charge?

PROBLEM 2 ♦

The pK_a of cyclopentadiene is 15, whereas the pK_a of cyclopentane is > 60. Explain the difference in pK_a values. (*Hint:* Compare the stabilities of the conjugate bases.)

H   H
cyclopentane
pK_a > 60

PROBLEM 3 ♦

Which of the following are aromatic? Explain your choice.

cycloheptatriene   cycloheptatrienyl   cycloheptatrienyl
                            cation   anion

PROBLEM 4 ♦

Which of the following are aromatic?
a.

c.

b.

d. CH_2=CHCH=CHCH=CH_2

PROBLEM 5 ♦ **SOLVED**

a. How many monobromonaphthalenes are there?
b. How many monobromophenanthrenes are there?

3-D Molecules:
1-Chloronaphthalene;
2-Chloronaphthalene

Solution to 5a   There are two monobromonaphthalenes. Substitution cannot occur at either of the carbons shared by both rings, because those carbons are not bonded to a hydrogen. Naphthalene is a flat molecule, so substitution for a hydrogen at any other carbon will result in one of the two compounds shown.

Br

Br

system. The metal atom in chlorophyll *a* is magnesium (Mg^{2+}). Vitamin B_{12} also has a ring system similar to a porphyrin ring system, but in this case, the metal ion is cobalt (Co^{3+}) (Section 17.10).

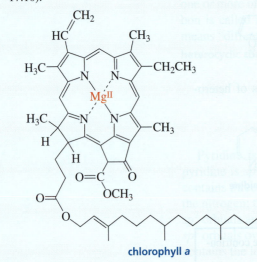

chlorophyll *a*

Chlorophyll a is the pigment that makes plants look green. This highly conjugated compound absorbs purple light, which causes plants to reflect green light (Section 7.12).

PORPHYRIN, BILIRUBIN, AND JAUNDICE

The average human breaks down about 6 g of hemoglobin each day. The protein portion (globin) and the iron are reutilized, but the porphyrin ring is broken down, being reduced first to biliverdin, a green compound, and then to bilirubin, a yellow compound. If more bilirubin is formed than can be excreted by the liver, it accumulates in the blood. When its concentration there reaches a certain level, it diffuses into the tissues, giving them a yellow appearance. This condition is known as jaundice.

8.4 THE NOMENCLATURE OF MONOSUBSTITUTED BENZENES

Some monosubstituted benzenes are named simply by attaching "benzene" to the name of the substituent.

bromobenzene	**chlorobenzene**	**nitrobenzene** **used as a solvent in shoe polish**	**ethylbenzene**

Some monosubstituted benzenes have names that incorporate the name of the substituent. Unfortunately, such names have to be memorized.

3-D Molecules:
Toluene; Bromobenzene

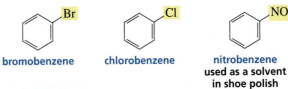

toluene	**phenol**	**aniline**	**benzenesulfonic acid**

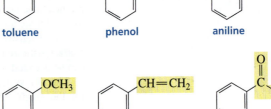

anisole	**styrene**	**benzaldehyde**	**benzoic acid**

When a benzene ring is a substituent, it is called a **phenyl group**. A benzene ring with a methylene group is called a **benzyl group**.

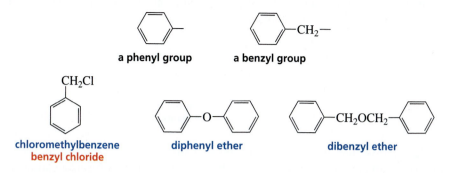

a phenyl group   a benzyl group

chloromethylbenzene   diphenyl ether   dibenzyl ether
benzyl chloride

An **aryl group** (Ar) is the general term for either a phenyl group or a substituted phenyl group, just as an alkyl group (R) is the general term for a group derived from an alkane. In other words, ArOH could be used to designate any of the following phenols:

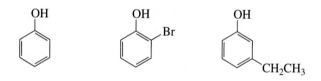

 THE TOXICITY OF BENZENE

Benzene, which has been widely used in chemical synthesis and has been frequently used as a solvent, is a toxic substance. The major adverse effects of chronic exposure are seen in the central nervous system and bone marrow; it causes leukemia and aplastic anemia. A higher than average incidence of leukemia, for example, has been found in industrial workers with long-term exposure to as little as 1 ppm benzene in the atmosphere. Toluene has replaced benzene as a solvent because, although it too is a central nervous system depressant, it does not cause leukemia or aplastic anemia. "Glue sniffing," a highly dangerous activity, produces narcotic central nervous system effects because glue contains toluene.

PROBLEM 8 ♦

Draw the structure of each of the following:
a. 2-phenylhexane   **b.** benzyl alcohol   **c.** 3-benzylpentane

8.5 HOW BENZENE REACTS

Aromatic compounds such as benzene undergo **electrophilic aromatic substitution reactions**: an electrophile substitutes for one of the hydrogens attached to the benzene ring.

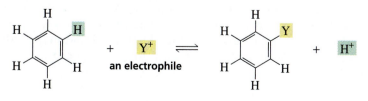

Now let's look at why this substitution reaction occurs. The cloud of π electrons above and below the plane of its ring causes benzene to be a nucleophile. It will, therefore, react with an electrophile (Y^+). When an electrophile attaches itself to a benzene ring, a carbocation intermediate is formed.

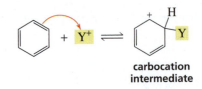

carbocation
intermediate

This description should remind you of the first step in an electrophilic addition reaction of an alkene: the alkene reacts with an electrophile and forms a carbocation intermediate (Section 5.1). In the second step of an electrophilic addition reaction of an alkene, the carbocation reacts with a nucleophile (Z^-) to form an addition product.

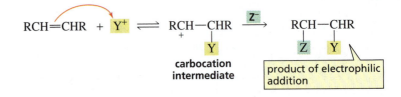

carbocation
intermediate

product of electrophilic
addition

If the carbocation intermediate formed from the reaction of benzene with an electrophile were to react similarly with a nucleophile (depicted as path *b* in Figure 8.1), the product of the reaction would not be aromatic. But if the carbocation instead were to lose a proton from the site of electrophilic attack (depicted as path *a* in Figure 8.1), the aromaticity of the benzene ring would be restored.

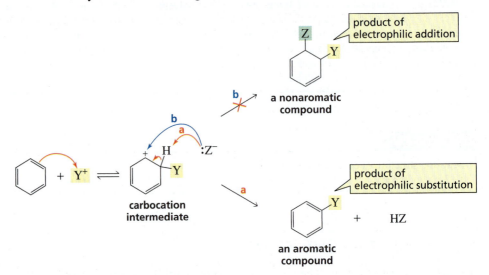

product of
electrophilic addition

a nonaromatic
compound

carbocation
intermediate

product of
electrophilic substitution

an aromatic
compound

▶ **Figure 8.1**
Reaction of benzene with an electrophile. Because the aromatic product is more stable, the reaction proceeds as (a) an electrophilic substitution reaction rather than (b) an electrophilic addition reaction.

Because the aromatic substitution product is much more stable than the nonaromatic addition product (Figure 8.2), benzene undergoes *electrophilic substitution reactions* that preserve aromaticity rather than *electrophilic addition reactions*—the reactions characteristic of alkenes—that would destroy aromaticity. The substitution reaction is more accurately called an **electrophilic aromatic substitution reaction**, since the electrophile substitutes for a hydrogen of an aromatic compound.

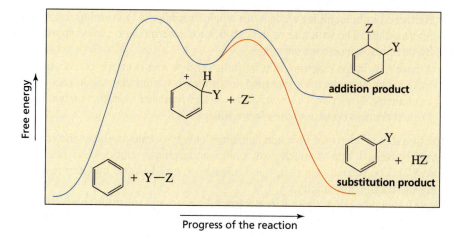

8.6 THE GENERAL MECHANISM FOR ELECTROPHILIC AROMATIC SUBSTITUTION REACTIONS

In an electrophilic aromatic substitution reaction, an electrophile becomes attached to a ring carbon, and an H^+ comes off the same ring carbon.

an electrophilic aromatic substitution reaction

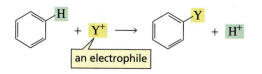

Tutorial:
Electrophilic aromatic substitution

The following are the five most common electrophilic aromatic substitution reactions:

1. **Halogenation:** A bromine (Br) or a chlorine (Cl) substitutes for a hydrogen.
2. **Nitration:** A nitro (NO_2) group substitutes for a hydrogen.
3. **Sulfonation:** A sulfonic acid (SO_3H) group substitutes for a hydrogen.
4. **Friedel–Crafts acylation:** An acyl ($RC=O$) group substitutes for a hydrogen.
5. **Friedel–Crafts alkylation:** An alkyl (R) group substitutes for a hydrogen.

All of these electrophilic aromatic substitution reactions take place by the same two-step mechanism.

general mechanism for electrophilic aromatic substitution

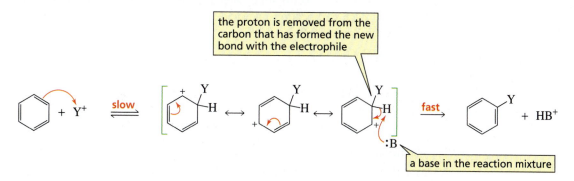

- Benzene (a nucleophile) reacts with an electrophile (Y^+), forming a carbocation intermediate. The structure of the carbocation intermediate can be approximated by three resonance contributors.
- A base (:B) in the reaction mixture pulls off a proton from the carbocation intermediate, and the electrons that held the proton move into the ring to reestablish its aromaticity. Notice that *the proton is always removed from the carbon that has formed the new bond with the electrophile.*

The first step is relatively slow and consumes energy because an aromatic compound is being converted into a much less stable nonaromatic carbocation intermediate (Figure 8.2). The second step is fast and releases energy because this step restores the stability-enhancing aromaticity.

We will look at each of these five electrophilic aromatic substitution reactions individually. As you study them, notice that they differ only in how the electrophile (Y^+) needed to start the reaction is generated. Once the electrophile is formed, all five reactions follow the same two-step mechanism for electrophilic aromatic substitution.

In an electrophilic aromatic substitution reaction, an electrophile (Y^+) is put on a ring carbon, and the H^+ comes off the same ring carbon.

8.7 HALOGENATION OF BENZENE

The bromination or chlorination of benzene requires a Lewis acid catalyst such as ferric bromide or ferric chloride. Recall that a *Lewis acid* is a compound that accepts a share in an electron pair (Section 2.9).

Donating a lone pair to the Lewis acid weakens the Br—Br (or Cl—Cl) bond, which makes Br_2 (or Cl_2) a better electrophile.

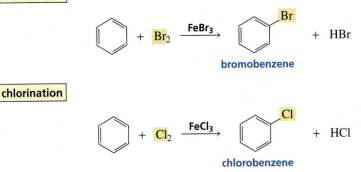

The mechanism for bromination is shown below. For the sake of clarity, only one of the three resonance contributors of the carbocation intermediate is shown in this and subsequent mechanisms for electrophilic aromatic substitution reactions. Bear in mind, however, that each carbocation intermediate actually has the three resonance contributors shown in Section 8.6.

mechanism for bromination

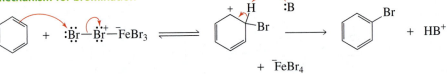

Movie:
Bromination of benzene

- The electrophile attaches to the benzene ring.
- A base (:B) from the reaction mixture (such as $^-FeBr_4$ or the solvent) removes a proton from the carbocation intermediate, thereby reforming the aromatic ring.

Chlorobenzene

Chlorination of benzene takes place by the same mechanism as bromination.

mechanism for chlorination

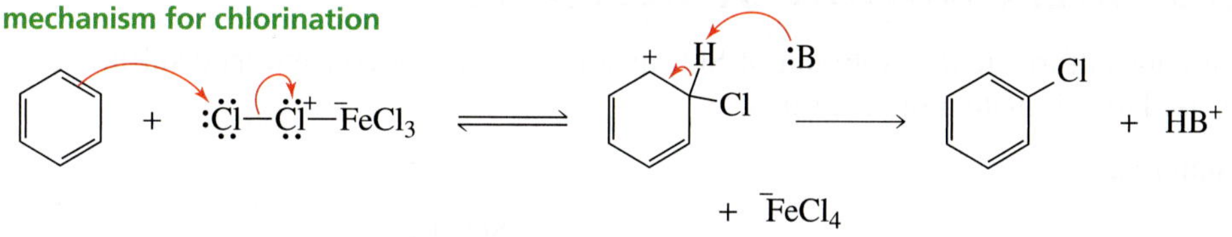

8.8 NITRATION OF BENZENE

The nitration of benzene with nitric acid requires sulfuric acid as a catalyst.

nitration

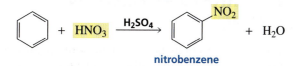

nitrobenzene

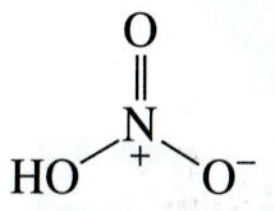

nitric acid

To generate the necessary electrophile, sulfuric acid protonates nitric acid. Protonated nitric acid loses water to form a nitronium ion, the electrophile required for nitration.

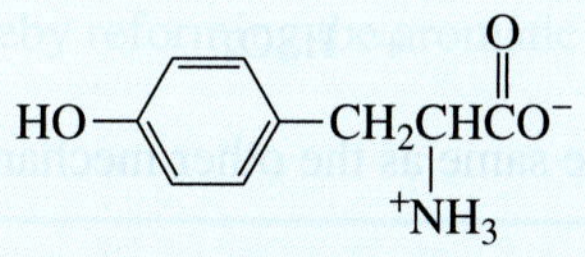

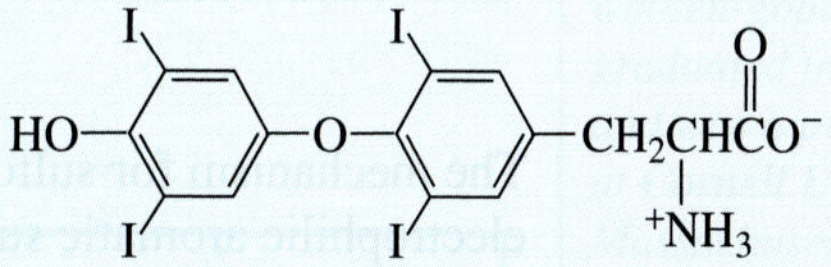

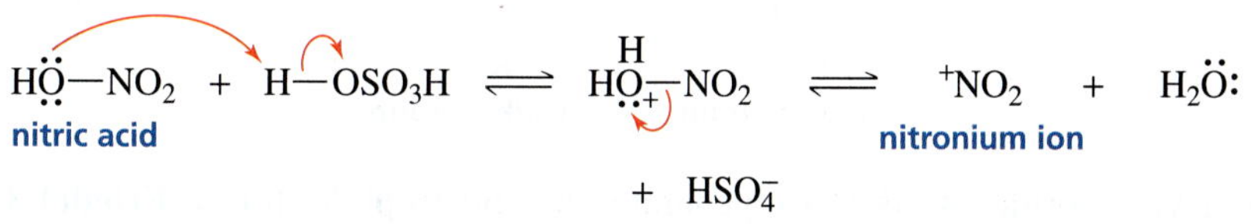

nitric acid   **nitronium ion**

$+ \ HSO_4^-$

nitronium ion

The mechanism for nitration is the same as the mechanisms described in Section 8.7.

mechanism for nitration

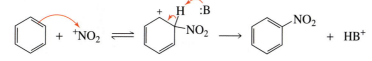

3-D Molecule:
Nitrobenzene

- The electrophile attaches to the benzene ring.
- A base (:B) from the reaction mixture (for example, H_2O, HSO_4^-, or solvent) removes a proton from the carbocation intermediate, thereby reforming the aromatic ring.

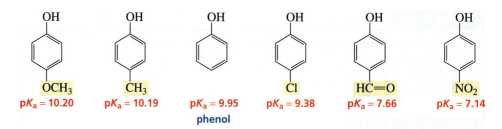

OH	OH	OH	OH	OH	OH
OCH$_3$	CH$_3$		Cl	HC=O	NO$_2$
pK_a = 10.20	pK_a = 10.19	pK_a = 9.95	pK_a = 9.38	pK_a = 7.66	pK_a = 7.14
		phenol			

Take a minute to compare the effect a substituent has on the reactivity of a benzene ring toward electrophilic aromatic substitution with the effect the substituent has on the pK_a of phenol. Notice that the more strongly deactivating the substituent, the lower the pK_a of the phenol; and the more strongly activating the substituent, the higher the pK_a of the phenol. In other words, *electron withdrawal decreases reactivity toward electrophilic aromatic substitution and increases acidity, whereas electron donation increases reactivity toward electrophilic aromatic substitution and decreases acidity.*

A similar substituent effect on pK_a is observed for substituted benzoic acids and substituted protonated anilines: electron-withdrawing substituents increase acidity; electron-donating substituents decrease acidity.

Tutorial:
Effect of substituents on pK_a

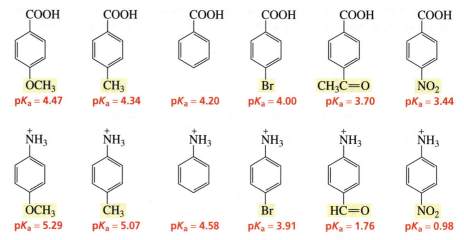

COOH	COOH	COOH	COOH	COOH	COOH
OCH$_3$	CH$_3$		Br	CH$_3$C=O	NO$_2$
pK_a = 4.47	pK_a = 4.34	pK_a = 4.20	pK_a = 4.00	pK_a = 3.70	pK_a = 3.44

$\overset{+}{N}H_3$	$\overset{+}{N}H_3$	$\overset{+}{N}H_3$	$\overset{+}{N}H_3$	$\overset{+}{N}H_3$	$\overset{+}{N}H_3$
OCH$_3$	CH$_3$		Br	HC=O	NO$_2$
pK_a = 5.29	pK_a = 5.07	pK_a = 4.58	pK_a = 3.91	pK_a = 1.76	pK_a = 0.98

The more deactivating (electron withdrawing) the substituent, the more it increases the acidity of a COOH, an OH, or an $^+$NH$_3$ group attached to a benzene ring.

The more activating (electron donating) the substituent, the more it decreases the acidity of a COOH, an OH, or an $^+$NH$_3$ group attached to a benzene ring.

PROBLEM 22 ◆

Which of the compounds in each of the following pairs is more acidic?

a. CH$_3\overset{O}{\overset{\|}{C}}$OH or ClCH$_2\overset{O}{\overset{\|}{C}}$OH

b. CH$_3$CH$_2\overset{O}{\overset{\|}{C}}$OH or H$_3\overset{+}{N}CH_2\overset{O}{\overset{\|}{C}}$OH

c. FCH$_2\overset{O}{\overset{\|}{C}}$OH or ClCH$_2\overset{O}{\overset{\|}{C}}$OH

d. H$\overset{O}{\overset{\|}{C}}$OH or CH$_3\overset{O}{\overset{\|}{C}}$OH

SUMMARY

To be classified as **aromatic**, a compound must have an uninterrupted cyclic cloud of π electrons that contains an *odd number of pairs* of π electrons.

A **heterocyclic compound** is a cyclic compound in which one or more of the ring atoms is a **heteroatom**—an atom other than carbon. Pyridine, pyrrole, furan, and thiophene are aromatic heterocyclic compounds.

Benzene's aromaticity causes it to undergo **electrophilic aromatic substitution reactions**. Electrophilic addition reactions that are characteristic of alkenes and dienes would lead to much less stable nonaromatic products. The most common electrophilic aromatic substitution reactions are halogenation, nitration, sulfonation, and Friedel–Crafts acylation and alkylation. Once the electrophile is generated,

all electrophilic aromatic substitution reactions take place by the same two-step mechanism: (1) the aromatic compound reacts with an electrophile, forming a carbocation intermediate; and (2) a base pulls off a proton from the carbon that formed the bond with the electrophile.

Some monosubstituted benzenes are named as substituted benzenes (for example, bromobenzene, nitrobenzene); others have names that incorporate the substituent (for example, toluene, phenol, aniline, anisole). The relative positions of two substituents on a benzene ring are indicated in the compound's name either by numbers or by the prefixes *ortho*, *meta*, and *para*.

Bromination or **chlorination** requires a Lewis acid catalyst. **Sulfonation** with sulfuric acid places an SO_3H group on the ring. **Nitration** with nitric acid requires sulfuric acid as a catalyst. An acyl chloride is used for a **Friedel–Crafts acylation**, a reaction that places an acyl group on a benzene ring; an alkyl halide is used for a **Friedel–Crafts** alkylation, a reaction that places an alkyl group on a benzene ring.

Benzene rings with substituents other than halo, nitro, sulfonic acid, alkyl, and acyl can be prepared by synthesiz-ing one of these substituted benzenes and then chemically changing the substituent.

The nature of the substituent affects both the reactivity of the benzene ring and the placement of an incoming substituent: the rate of electrophilic aromatic substitution is increased by electron-donating substituents and decreased by electron-withdrawing substituents. Substituents can donate or withdraw electrons **inductively** or by **resonance**.

The stability of the carbocation intermediate determines the position to which the substituent directs an incoming electrophile. All activating substituents and the weakly deactivating halogens are **ortho–para directors**; all substituents more deactivating than the halogens are **meta directors**.

In planning the synthesis of disubstituted benzenes, the order in which the substituents are placed on the ring and the point in a reaction sequence at which a substituent is chemically modified are important considerations.

The acidity of substituted benzoic acids, phenols, and anilinium ions is increased by electron-withdrawing substituents and decreased by electron-donating substituents.

SUMMARY OF REACTIONS

1. Electrophilic aromatic substitution reactions:

   a. Halogenation (Section 8.7)

   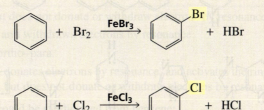

   b. Nitration and sulfonation (Sections 8.8 and 8.9)

   c. Friedel–Crafts acylation and alkylation (Sections 8.10 and 8.11)

Table 9.5	Relative Reactivities of Alkyl Halides		
In an S_N2 reaction:	$1° > 2° > 3°$	In an S_N1 reaction:	$3° > 2° > 1°$
In an E2 reaction:	$3° > 2° > 1°$	In an E1 reaction:	$3° > 2° > 1°$

A *secondary* alkyl halide forms both substitution and elimination products under S_N2/E2 conditions.

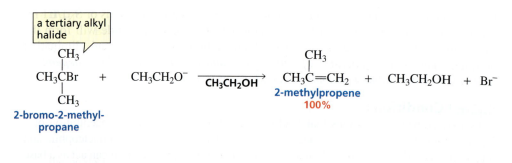

A *tertiary* alkyl halide is the least reactive of the alkyl halides in an S_N2 reaction and the most reactive in an E2 reaction (Table 9.5). Consequently, *only* the elimination product is formed when a tertiary alkyl halide reacts with a nucleophile/base under S_N2/E2 conditions.

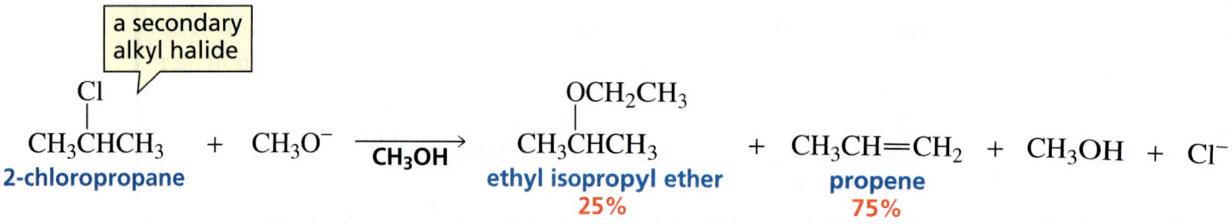

Tertiary alkyl halides undergo only elimination under S_N2/E2 conditions.

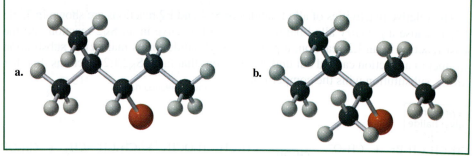

PROBLEM 22 ♦

Indicate whether the alkyl halides listed will give both substitution and elimination products, primarily substitution products, only elimination products, or no products when they react with methanol under S_N2/E2 conditions.

a. 1-bromobutane   **c.** 2-bromobutane
b. 1-bromo-2-methylpropane   **d.** 2-bromo-2-methylpropane

PROBLEM 23

Draw the stereoisomers that would be obtained in greatest yield from the reaction of the following alkyl chlorides with hydroxide ion:

a.   **b.**

S_N1/E1 Conditions

Now let's look at what happens when conditions favor $S_N1/E1$ reactions: a poor nucleophile/weak base. In $S_N1/E1$ reactions, the alkyl halide dissociates to form a carbocation, which can then either combine with the nucleophile to form the substitution product or lose a proton to form the elimination product.

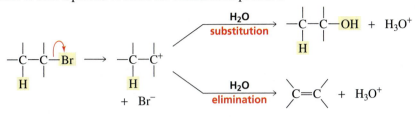

Alkyl halides have the same order of reactivity in S_N1 reactions as they do in E1 reactions (Table 9.5), because both reactions have the same rate-determining step—dissociation of the alkyl halide to form a carbocation. This means that all alkyl halides that react under $S_N1/E1$ conditions will give both substitution and elimination products. Remember that primary alkyl halides do not undergo $S_N1/E1$ reactions because primary carbocations are too unstable to be formed.

Primary alkyl halides do not form carbocations; therefore they cannot undergo S_N1 and E1 reactions.

Table 9.6 summarizes the products obtained when alkyl halides react with nucleophiles/bases under $S_N2/E2$ and $S_N1/E1$ conditions.

Table 9.6 Summary of the Products Expected in Substitution and Elimination Reactions		
Class of alkyl halide	**Products under $S_N2/E2$ Conditions**	**Products under $S_N1/E1$ Conditions**
Primary alkyl halide	Primarily substitution	Cannot undergo $S_N1/E1$ reactions
Secondary alkyl halide	Both substitution and elimination	Both substitution and elimination
Tertiary alkyl halide	Only elimination	Both substitution and elimination

The stereoisomers obtained from substitution and elimination reactions are summarized in Table 9.7.

Table 9.7 Stereochemistry of Substitution and Elimination Reactions	
Reaction	**Products**
S_N1	Both stereoisomers (R and S) are formed.
E1	Both E and Z stereoisomers are formed, but more of the stereoisomer with the bulkiest groups on opposite sides of the double bond is formed.
S_N2	Only the inverted product is formed.
E2	Both E and Z stereoisomers are formed, but more of the stereoisomer with the bulkiest groups on opposite sides of the double bond is formed.

PROBLEM 24 ◆

Indicate whether the alkyl halides listed will give both substitution and elimination products, primarily substitution products, only elimination products, or no products when they react with sodium methoxide under $S_N1/E1$ conditions.

1. 1-bromobutane
2. 1-bromo-2-methylpropane
3. 2-bromobutane
4. 2-bromo-2-methylpropane

PROBLEM 25

Which of the following reactions will go faster if the concentration of the nucleophile is increased?

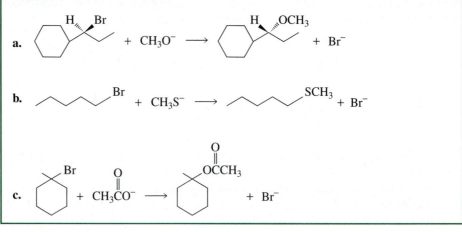

a.

b.

c.

9.12 SOLVENT EFFECTS

S_N1 and E1 reactions are faster in protic polar solvents. A polar solvent has a partial positive charge and a partial negative charge; a *protic* polar solvent is a polar solvent with a hydrogen bonded to an oxygen or a nitrogen. Water (H_2O) and alcohols (ROH) are examples of protic polar solvents. Protic polar solvents cluster around the carbocation intermediate, with the negative poles of the solvent surrounding the positive charge of the carbocation (Figure 9.7a).

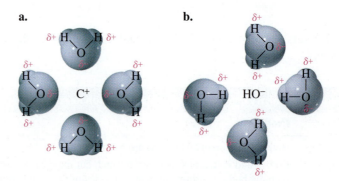

▶ **Figure 9.7**
a. The interactions of a protic polar solvent with a positively charged species.
b. The interactions of a protic polar solvent with a negatively charged species.

When a carbocation interacts with a protic polar solvent, the positive charge on the carbocation is spread out to the surrounding solvent molecules. Spreading out the charge stabilizes the charged species. Stabilizing the carbocation intermediate lowers the height of the energy barrier (the energy "hill") for the rate-determining step, causing the reaction to go faster.

In contrast, S_N2 and E2 reactions are slowed down by protic polar solvents. S_N2 and E2 reactions require a strong nucleophile/base. Most such reagents are negatively charged. The positive poles of a protic polar solvent surround the negatively charged nucleophile/base, thereby spreading out its charge and stabilizing it (Figure 9.7b). Stabilizing a reactant increases the height of the energy barrier for the reaction, causing the reaction to be slower. We would like, therefore, to carry out S_N2 and E2 reactions in a nonpolar solvent. However, negatively charged species generally will not dissolve in a nonpolar solvent. Therefore, an aprotic polar solvent is used. Because an aprotic polar solvent does not have a hydrogen bonded to an oxygen or a nitrogen, it interacts less effectively with an ion than does a polar protic solvent. In fact, aprotic polar solvents

such as DMSO and DMF interact very poorly with nucleophiles/bases because the partial positive charge on the solvent is located on the inside of the molecule, and is therefore less accessible to the negatively charged nucleophile/base.

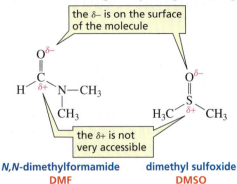

the $\delta-$ is on the surface of the molecule

the $\delta+$ is not very accessible

N,N-dimethylformamide   dimethyl sulfoxide
DMF   **DMSO**

 Tutorial:
Common terms

We have now seen that when a reaction can undergo both $S_N1/E1$ and $S_N2/E2$ reactions, the $S_N1/E1$ reactions will be favored by a poor (neutral) nucleophile/weak base in a protic polar solvent, whereas the $S_N2/E2$ reactions will be favored by a high concentration of a good nucleophile/strong base in an aprotic polar solvent.

S_N1 and E1 reactions of alkyl halides are favored by a poor nucleophile/weak base in a protic polar solvent.

S_N2 and E2 reactions of alkyl halides are favored by a high concentration of a good nucleophile/strong base in an aprotic polar solvent.

PROBLEM 26

Which reaction in each of the following pairs will take place more rapidly?

a. $CH_3Br + HO^- \xrightarrow{\text{DMSO}} CH_3OH + Br^-$

$CH_3Br + HO^- \xrightarrow{\text{EtOH}} CH_3OH + Br^-$

b. $CH_3Br + NH_3 \longrightarrow CH_3\overset{+}{N}H_3 + Br^-$

$CH_3Br + H_2O \longrightarrow CH_3OH + HBr$

c. $CH_3Br + NH_3 \xrightarrow{\text{Et}_2\text{O}} CH_3\overset{+}{N}H_3 + Br^-$

$CH_3Br + NH_3 \xrightarrow{\text{EtOH}} CH_3\overset{+}{N}H_3 + Br^-$

9.13 USING SUBSTITUTION REACTIONS TO SYNTHESIZE ORGANIC COMPOUNDS

In Section 9.4, you saw that nucleophilic substitution reactions of alkyl halides can lead to a wide variety of organic compounds. For example, ethers are synthesized when an *alkyl halide* reacts with an *alkoxide ion*. This reaction, discovered by Alexander Williamson in 1850, is still considered one of the best ways to synthesize an ether.

Williamson ether synthesis

R—Br   +   R—O⁻   ⟶   R—O—R   +   Br⁻
alkyl halide   alkoxide ion   ether

The alkoxide ion (RO^-) for the **Williamson ether synthesis** can be prepared using sodium metal.

$$2\,\text{ROH} + 2\,\text{Na} \longrightarrow 2\,\text{RO}^- + 2\,\text{Na}^+ + H_2$$

The Williamson ether synthesis is a nucleophilic substitution reaction. It requires a high concentration of a good nucleophile, which indicates that it is an S_N2 reaction. If you want to synthesize an ether such as butyl propyl ether, you have a choice of starting

materials: you can use either a propyl halide and butoxide ion or a butyl halide and propoxide ion.

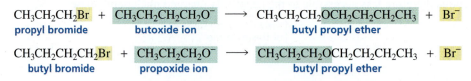

$CH_3CH_2CH_2Br$ + $CH_3CH_2CH_2CH_2O^-$ ⟶ $CH_3CH_2CH_2OCH_2CH_2CH_2CH_3$ + Br^-
propyl bromide   butoxide ion   butyl propyl ether

$CH_3CH_2CH_2CH_2Br$ + $CH_3CH_2CH_2O^-$ ⟶ $CH_3CH_2CH_2OCH_2CH_2CH_2CH_3$ + Br^-
butyl bromide   propoxide ion   butyl propyl ether

However, if you want to synthesize *tert*-butyl ethyl ether, the starting materials must be an ethyl halide and *tert*-butoxide ion. If you tried to use a *tert*-butyl halide and ethoxide ion as reactants, you would not obtain any ether, because the reaction of a tertiary alkyl halide under $S_N2/E2$ conditions forms only the elimination product. Consequently, a Williamson ether synthesis should be designed in such a way that *the less hindered alkyl group is provided by the alkyl halide and the more hindered alkyl group comes from the alkoxide ion.*

In ether synthesis, the less hindered group should be provided by the alkyl halide.

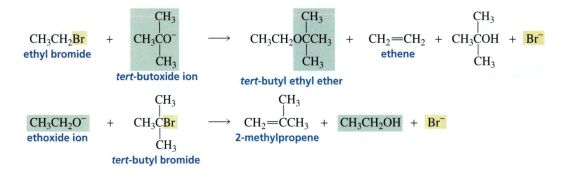

We saw in Section 5.14 that alkynes can be synthesized by the reaction of an acetylide anion with an alkyl halide.

$CH_3CH_2C≡C^-$ + $CH_3CH_2CH_2Br$ ⟶ $CH_3CH_2C≡CCH_2CH_2CH_3$ + Br^-

Now that you know that this is an S_N2 reaction (the alkyl halide reacts with a high concentration of a good nucleophile), you can understand why you were told that it is best to use primary alkyl halides and methyl halides in the reaction. These alkyl halides are the only ones that form primarily the desired substitution product.

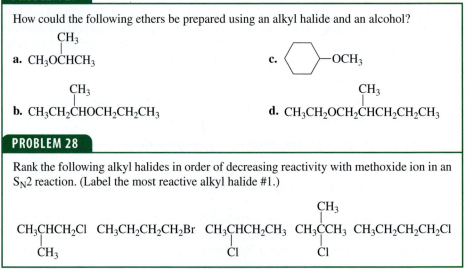

9.14 **BIOLOGICAL METHYLATING REAGENTS**

If an organic chemist wanted to put a methyl group on a nucleophile (Nu^-), methyl iodide would most likely be the methylating agent used. Of the methyl halides, methyl iodide has the most easily displaced leaving group because I^- is the weakest base of the halide ions. The reaction would be a simple S_N2 reaction.

In a living cell, however, methyl iodide is not available. Methyl halides are only slightly soluble in water, so they are not found in the predominantly aqueous environments of biological systems. Instead, biological systems use *S*-adenosylmethionine (SAM), a water-soluble compound, as a methylating agent. (A less common biological methylating agent is discussed in Section 17.11.) Although it looks much more complicated than methyl iodide, it performs the same function: it transfers a methyl group to a nucleophile. Notice that the methyl group of SAM is attached to a positively charged sulfur, which can readily accept the electrons left behind when the methyl group is transferred. In other words, the methyl group is attached to a very good leaving group, allowing biological methylation to take place at a reasonable rate.

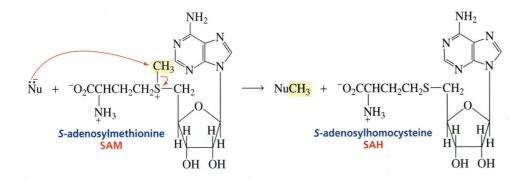

A specific example of a methylation reaction that takes place in biological systems is the conversion of noradrenaline (norepinephrine) to adrenaline (epinephrine), using SAM to provide the methyl group. Noradrenaline and adrenaline are hormones that control glycogen metabolism; they are released into the bloodstream in response to stress. Adrenaline is more potent than noradrenaline.

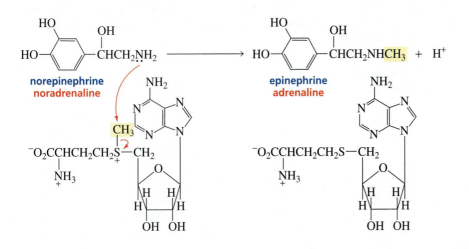

ERADICATING TERMITES

Alkyl halides can be very toxic to biological organisms. For example, methyl bromide is used to kill termites and other pests. Methyl bromide works by methylating the NH_2 and SH groups of enzymes, thereby destroying the enzyme's ability to catalyze necessary biologi-cal reactions. Unfortunately, methyl bromide has been found to deplete the ozone layer (Section 5.9), so its production has recently been banned in developed countries, and developing countries will have until 2015 to phase out its use.

S-ADENOSYLMETHIONINE: A NATURAL ANTIDEPRESSANT

Marketed under the name SAMe (pronounced Sammy), *S*-adenosylmethionine is sold in many health food and drug stores as a treatment for depression and arthritis. Although SAMe has been used clinically in Europe for more than two decades, it has not been rigorously evaluated in the United States and thus has not been approved by the FDA. It can be sold, however, because the FDA does not prohibit the sale of most naturally occurring substances as long as the marketer does not make therapeutic claims. SAMe has also been found to be effective in the treatment of liver diseases, such as diseases caused by alcohol and the hepatitis C virus. The attenuation of liver injuries is accompanied by increased levels of glutathione in the liver. Glutathione is an important biological antioxidant. SAM is required for the synthesis of cysteine, an amino acid that, in turn, is required for the synthesis of glutathione (Section 16.6).

SUMMARY

Alkyl halides undergo two kinds of **nucleophilic substitution reactions**: S_N2 and S_N1. In both reactions, a nucleophile substitutes for a halogen. An S_N2 reaction is bimolecular: two molecules are involved in the rate-limiting step; an S_N1 reaction is unimolecular: one molecule is involved in the rate-limiting step.

The rate of an **S_N2 reaction** depends on the concentration of both the alkyl halide and the nucleophile. An S_N2 reaction has a one-step mechanism: the nucleophile attacks the back side of the carbon that is attached to the halogen. The rate of an S_N2 reaction is affected by steric hindrance: the bulkier the groups at the back side of the carbon undergoing attack, the slower is the reaction. Tertiary alkyl halides, therefore, cannot undergo S_N2 reactions. An S_N2 reaction takes place with **inversion of configuration**.

The rate of an **S_N1 reaction** depends only on the concentration of the alkyl halide. The halogen departs in the first step, forming a carbocation that is attacked by a nucleophile in the second step. The rate of an S_N1 reaction depends on the ease of carbocation formation. Tertiary alkyl halides, therefore, are more reactive than secondary alkyl halides because tertiary carbocations are more stable than secondary carbocations. Primary carbocations are so unstable that primary alkyl halides cannot undergo S_N1 reactions. An S_N1 reaction forms both inverted and noninverted products.

The rates of both S_N2 and S_N1 reactions are influenced by the nature of the **leaving group**. Weak bases are the best leaving groups because weak bases form the weakest bonds. Thus, the weaker the basicity of the leaving group, the faster the reaction will occur. Therefore, the relative reactivities of alkyl halides that differ only in the halogen atom are $RI > RBr > RCl > RF$ in both S_N2 and S_N1 reactions.

Basicity is a measure of how well a species shares its lone pair with a proton. **Nucleophilicity** is a measure of how readily a species is able to attack an electron-deficient atom. In general, the stronger base is the better nucleophile.

In addition to undergoing nucleophilic substitution reactions, alkyl halides undergo elimination reactions, in which the halogen is removed from one carbon, a hydrogen is removed from an adjacent carbon, and a double bond is formed between the two carbons from which the atoms were eliminated. The product of an elimination reaction is therefore an alkene. There are two important elimination reactions, E1 and E2.

An **E2 reaction** is a one-step reaction; the proton and the halide ion are removed in the same step, so no intermediate is formed. In an **E1 reaction**, the alkyl halide dissociates, forming a carbocation intermediate. In a second step, a base removes a proton from a carbon that is adjacent to the positively charged carbon.

The major product of an elimination reaction is the more stable alkene—the one formed when a proton is removed from the β-carbon that is bonded to the fewest hydrogens. If both E and Z isomers are possible for the product, the one with the bulkiest groups on opposite sides of the double bond is more stable and, therefore, will be formed in greater yield.

Predicting which products are formed when an alkyl halide undergoes a reaction begins with determining whether the conditions favor $S_N2/E2$ or $S_N1/E1$ reactions. $S_N2/E2$ reactions are favored by a high concentration of a good nucleophile/strong base in an **aprotic polar solvent**, whereas $S_N1/E1$ reactions are favored by a poor nucleophile/weak base in a **protic polar solvent**.

When $S_N2/E2$ reactions are favored, primary alkyl halides form mainly substitution products, secondary alkyl halides form both substitution and elimination products, and tertiary alkyl halides form only elimination products. When $S_N1/E1$ conditions are favored, secondary and tertiary alkyl halides form both substitution and elimination products; primary alkyl halides do not undergo $S_N1/E1$ reactions.

SUMMARY OF REACTIONS

1. S_N2 reaction: a one-step mechanism

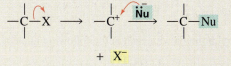

Relative reactivities of alkyl halides: $CH_3X > 1° > 2° > 3°$
Only the inverted product is formed.

2. S_N1 reaction: a two-step mechanism with a carbocation intermediate

Relative reactivities of alkyl halides: $3° > 2° > 1° > CH_3X$
Both the inverted and noninverted products are formed.

3. E2 reaction: a one-step mechanism

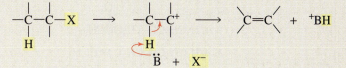

Relative reactivities of alkyl halides: $3° > 2° > 1°$
Both *E* and *Z* stereoisomers are formed; the isomer with the bulkiest groups on opposite sides of the double bond will be formed in greater yield.

4. E1 reaction: a two-step mechanism with a carbocation intermediate

Relative reactivities of alkyl halides: $3° > 2° > 1°$
Both *E* and *Z* stereoisomers are formed; the isomer with the bulkiest groups on opposite sides of the double bond will be formed in greater yield.

Competing S_N2 and E2 Reactions

Primary alkyl halides: mainly substitution
Secondary alkyl halides: substitution and elimination
Tertiary alkyl halides: only elimination

Competing S_N1 and E1 Reactions

Primary alkyl halides: cannot undergo S_N1 or E1 reactions
Secondary alkyl halides: substitution and elimination
Tertiary alkyl halides: substitution and elimination

PROBLEMS

29. Which reaction in each of the following pairs will take place more rapidly?

a. $CH_3Br + CH_3O^- \longrightarrow CH_3OCH_3 + Br^-$

$CH_3Br + CH_3OH \longrightarrow CH_3OCH_3 + HBr$

b. $CH_3I + NH_3 \longrightarrow CH_3\overset{+}{N}H_3 + I^-$

$CH_3Cl + NH_3 \longrightarrow CH_3\overset{+}{N}H_3 + Cl^-$

c. $CH_3Br + CH_3NH_2 \longrightarrow CH_3\overset{+}{N}H_2CH_3 + Br^-$

$CH_3Br + CH_3OH \longrightarrow CH_3OCH_3 + HBr$

30. Give the product of the reaction of methyl bromide with each of the following nucleophiles:

a. HO^-

b. $^-NH_2$

c. H_2S

d. HS^-

e. $CH_3CH_2O^-$

f. CH_3NH_2

31. Which is a better nucleophile?

a. H_2O or HO^-

b. NH_3 or $^-NH_2$

c. $CH_3\overset{\displaystyle O}{\overset{\|}{C}}O^-$ or $CH_3CH_2O^-$

d. 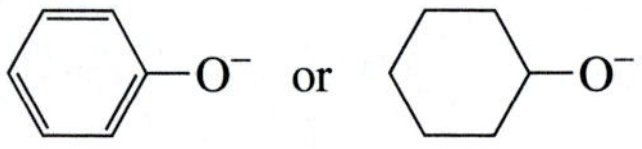 ⬡—O^- or ⬡—O^-

32. For each of the pairs in Problem 31, indicate which is a better leaving group.

33. What nucleophiles could be used to react with butyl bromide to prepare the following compounds?

a. $CH_3CH_2CH_2CH_2OH$

b. $CH_3CH_2CH_2CH_2OCH_3$

c. $CH_3CH_2CH_2CH_2SCH_2CH_3$

d. $CH_3CH_2CH_2CH_2C\equiv N$

e. $CH_3CH_2CH_2CH_2O\overset{\displaystyle O}{\overset{\|}{C}}CH_3$

f. $CH_3CH_2CH_2CH_2C\equiv CCH_3$

34. Which alkyl halide in each pair would you expect to be more reactive in an S_N2 reaction with a given nucleophile?

a. $CH_3CH_2\underset{\underset{\displaystyle I}{|}}{C}HCH_3$ or $CH_3CH_2\underset{\underset{\displaystyle Br}{|}}{C}HCH_3$

b. $CH_3CH_2\underset{\underset{\displaystyle CH_3}{|}}{\overset{}{C}}HBr$ or $CH_3CH_2\underset{\underset{\displaystyle CH_2CH_3}{|}}{\overset{}{C}}HBr$

c. $CH_3CH_2CH_2\underset{\underset{\displaystyle Br}{}}{\overset{\overset{\displaystyle CH_3}{|}}{C}H}Br$ or $CH_3CH_2\underset{}{\overset{\overset{\displaystyle CH_3}{|}}{C}H}CH_2Br$

d. 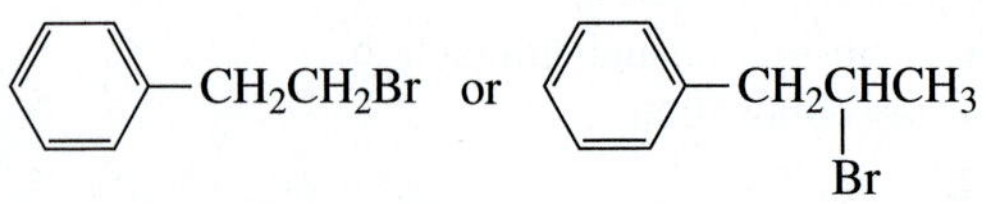 ⬡—CH_2CH_2Br or ⬡—$CH_2\underset{\underset{\displaystyle Br}{|}}{C}HCH_3$

35. For each of the pairs in Problem 34, which compound would be more reactive in an S_N1 reaction?

36. For each of the following reactions, give the substitution products; if the products can exist as stereoisomers, show what stereoisomers are obtained:

a. (*R*)-2-bromopentane + high concentration of CH_3O^-

b. (*R*)-2-bromopentane + CH_3OH

c. *trans*-1-bromo-4-methylcyclohexane + high concentration of CH_3O^-

d. *trans*-1-bromo-4-methylcyclohexane + CH_3OH

e. 3-bromo-3-methylpentane + CH_3OH

37. Give the major product obtained when each of the following alkyl halides undergoes an E2 reaction:

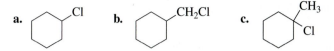

a. [cyclohexane with Cl] **b.** [cyclohexane with CH₂Cl] **c.** [cyclohexane with CH₃ and Cl]

38. Give the stereoisomer that would be obtained in greater yield when each of the following alkyl halides undergoes an E2 reaction:

a. $CH_3CHCH_2CH_3$ **b.** $CH_3CHCH_2CH_3$ **c.** $CH_3CHCH_2CH_2CH_3$
    |
    Br   Cl   Cl

39. Which reactant in each of the following pairs will undergo an elimination reaction more rapidly?

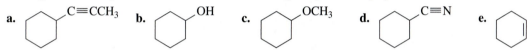

a. $(CH_3)_3CCl$ $\xrightarrow[H_2O]{HO^-}$ **b.** $(CH_3)_3CBr$ $\xrightarrow[H_2O]{HO^-}$

               or   or

     $(CH_3)_3CI$ $\xrightarrow[H_2O]{HO^-}$ $(CH_3)_2CHBr$ $\xrightarrow[H_2O]{HO^-}$

40. a. Identify the three products that are formed when 2-bromo-2-methylpropane is dissolved in a mixture of 80% ethanol and 20% water.
   b. Explain why the same products are obtained when 2-chloro-2-methylpropane is dissolved in a mixture of 80% ethanol and 20% water.

41. Starting with bromocyclohexane, how could the following compounds be prepared?

a. [cyclohexane with C≡CCH₃] **b.** [cyclohexane with OH] **c.** [cyclohexane with OCH₃] **d.** [cyclohexane with C≡N] **e.** [cyclohexene]

42. For each of the following reactions, give the major elimination product; if the product can exist as stereoisomers, indicate which stereoisomer is obtained in greater yield:

a. (R)-2-bromohexane + high concentration of HO^-
b. (R)-2-bromohexane + H_2O
c. 3-bromo-3-methylpentane + high concentration of HO^-
d. 3-bromo-3-methylpentane + H_2O

43. The rate of reaction of methyl iodide with quinuclidine was measured in nitrobenzene, and then the rate of reaction of methyl iodide with triethylamine was measured in the same solvent. The concentration of the reagents was the same in both experiments.
   a. Which reaction was faster?
   b. Which reaction had the larger rate constant?

                      CH_2CH_3
                        |
               $CH_3CH_2NCH_2CH_3$
    quinuclidine   **triethylamine**

44. Which substitution reaction in each of the following pairs will occur more rapidly?

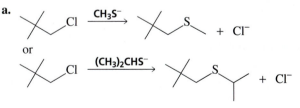

a. [(CH₃)₃C–CH₂Cl] $\xrightarrow{CH_3S^-}$ [(CH₃)₃C–CH₂S–] + Cl⁻

   or

   [(CH₃)₃C–CH₂Cl] $\xrightarrow{(CH_3)_2CHS^-}$ [(CH₃)₃C–CH₂S–CH(CH₃)₂] + Cl⁻

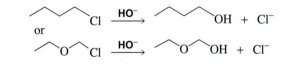

b. [CH₃CH₂CH₂CH₂Cl] $\xrightarrow{HO^-}$ [CH₃CH₂CH₂CH₂OH] + Cl⁻

   or

   [CH₃CH₂–O–CH₂CH₂Cl] $\xrightarrow{HO^-}$ [CH₃CH₂–O–CH₂CH₂OH] + Cl⁻

45. Which of the following is more reactive in an E2 reaction?

a. 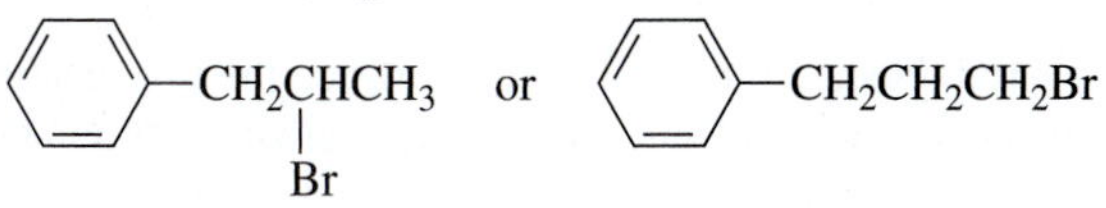 [benzene]–CH_2CHCH_3 or [benzene]–$CH_2CH_2CH_2Br$
                |
                Br

b. $CH_3CH_2CHCH_3$ or $CH_2{=}CHCH_2CHCH_3$
        | |
        Br   Br

46. Would you expect methoxide ion to be a better nucleophile if it were dissolved in CH_3OH or in DMSO? Why?

47. a. Explain why 1-bromo-2,2-dimethylpropane has difficulty undergoing either S_N2 or S_N1 reactions.
   b. Can it undergo E2 and E1 reactions?

48. Which stereoisomer would be obtained in greater yield from an E2 reaction of each of the following alkyl halides?

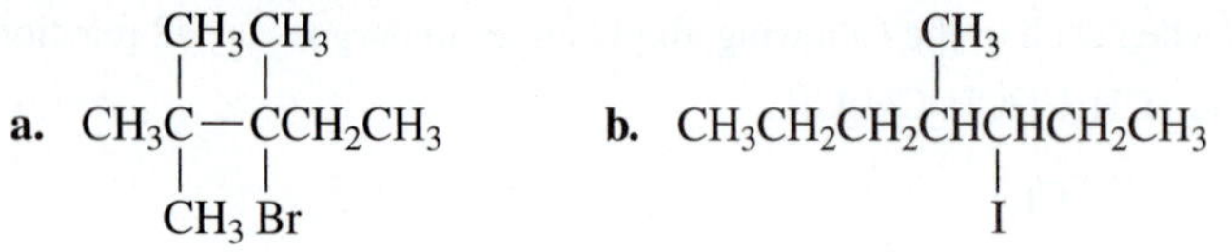

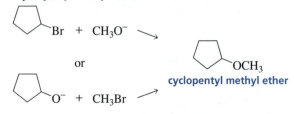

49. An ether can be prepared by an S_N2 reaction of an alkyl halide with an alkoxide ion (RO^-). Which set of alkyl halide and alkoxide ion would give you a better yield of cyclopentyl methyl ether?

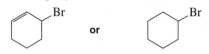

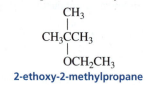

cyclopentyl methyl ether

50. Dr. Don T. Doit wanted to synthesize the anesthetic 2-ethoxy-2-methylpropane. He used ethoxide ion and 2-chloro-2-methylpropane for his synthesis and ended up with no ether. What was the product of his synthesis? What reagents should he have used?

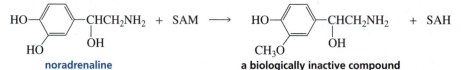

2-ethoxy-2-methylpropane

51. Which alkyl halide undergoes an E1 reaction more rapidly?

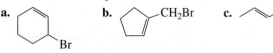

52. In Section 9.14, we saw that *S*-adenosylmethionine (SAM) methylates the nitrogen atom of noradrenaline to form adrenaline, a more potent hormone. If SAM methylates an OH group on the benzene ring instead, it completely destroys noradrenaline's activity. Give the mechanism for the methylation of the OH group by SAM.

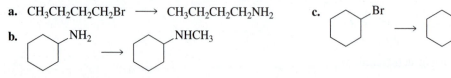

53. Give the substitution products obtained from the reaction of each of the following alkyl halides with ethanol:

a. [cyclohexene with Br]   **b.** [cyclopentene-CH$_2$Br]   **c.** [allyl bromide]

54. Show how the following compounds could be synthesized using the given starting materials:

a. $CH_3CH_2CH_2CH_2Br \longrightarrow CH_3CH_2CH_2CH_2NH_2$ **c.** [cyclohexane-Br $\longrightarrow$ cyclohexene]

b. [cyclohexane-NH$_2$ $\longrightarrow$ cyclohexane-NHCH$_3$]

55. A cyclic compound can be formed by an intramolecular reaction. An intramolecular reaction is one in which the two reacting groups are in the same molecule. Give the structure of the ether that would be formed from each of the following intramolecular reactions.

a. $BrCH_2CH_2CH_2CH_2O^- \longrightarrow$ ether   **b.** $ClCH_2CH_2CH_2CH_2CH_2O^- \longrightarrow$ ether

CHAPTER 10

Reactions of Alcohols, Amines, Ethers, and Epoxides

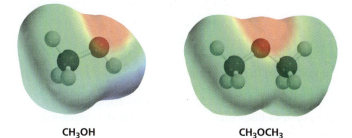

CH₃OH   CH₃OCH₃

e have seen that alkyl halides undergo substitution and elimination reactions because of their electron-withdrawing halogen atoms (Chapter 9). Compounds with other electron-withdrawing groups also undergo substitution and elimination reactions. The relative reactivity of these compounds depends on the electron-withdrawing group. For example, an alcohol (ROH) has an electron-withdrawing OH group. An OH group, however, is much more basic than a halogen, so we will see that it is much harder to displace.

10.1 THE NOMENCLATURE OF ALCOHOLS

Before we look at the reactions of alcohols, we need to learn how to name them. An **alcohol** is a compound in which a hydrogen of an alkane has been replaced by an OH group. We have seen that alcohols are classified as **primary**, **secondary**, or **tertiary**, depending on whether the OH group is bonded to a primary, secondary, or tertiary carbon—the same way alkyl halides are classified (Section 3.5).

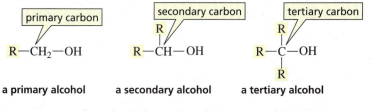

a primary alcohol   a secondary alcohol   a tertiary alcohol

The common name of an alcohol consists of the name of the alkyl group to which the OH group is attached, followed by the word "alcohol."

CH_3CH_2OH $CH_3CH_2CH_2OH$ CH_3CHOH
ethyl alcohol   propyl alcohol |
                                                    CH_3
                                          isopropyl alcohol

The **functional group** is the center of reactivity in an organic compound. In an alcohol, the OH is the functional group. The IUPAC system uses the suffix "ol" to denote the OH group. Thus, the systematic name of an alcohol is obtained by replacing

methyl alcohol

ethyl alcohol

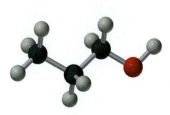

propyl alcohol

the "e" at the end of the name of the parent hydrocarbon with the suffix "ol." This should remind you of the use of the suffix "ene" to denote the functional group of an alkene (Section 4.2).

$$CH_3OH \qquad CH_3CH_2OH$$
methanol **ethanol**

When necessary, the position of the functional group is indicated by a number.

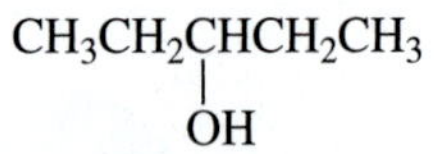

$$CH_3CH_2CHCH_2CH_3$$
$$|$$
$$OH$$
3-pentanol

Let's review the rules used to name a compound that has a functional group suffix:

1. The parent hydrocarbon is the longest chain containing the functional group. The parent chain is numbered in the direction that gives the *functional group suffix the lowest possible number.*

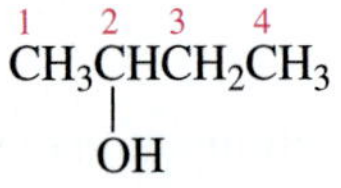

2-butanol 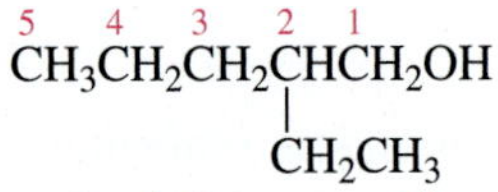 **2-ethyl-1-pentanol**

> The longest continuous chain has six carbons, but the longest continuous chain containing the OH functional group has five carbons so the compound is named as a pentanol.

2. If there is a functional group suffix and a substituent, the functional group suffix gets the lowest possible number.

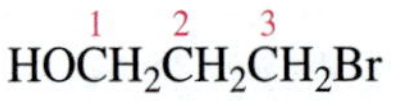

3-bromo-1-propanol 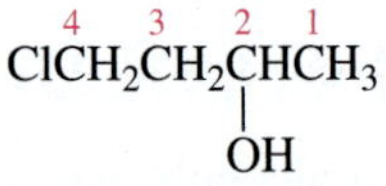 **4-chloro-2-butanol** 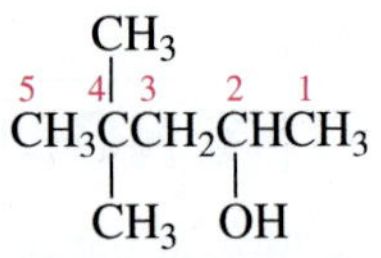 **4,4-dimethyl-2-pentanol**

3. If counting in either direction gives the same number for the functional group suffix, the chain is numbered in the direction that gives a substituent the lowest possible number. Notice that a number is not needed to designate the position of a functional group suffix in a cyclic compound, because it is assumed to be at the 1-position.

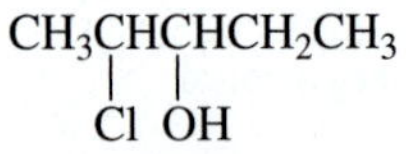

2-chloro-3-pentanol **2-methyl-4-heptanol** **3-methylcyclohexanol**
not **not** **not**
4-chloro-3-pentanol **6-methyl-4-heptanol** **5-methylcyclohexanol**

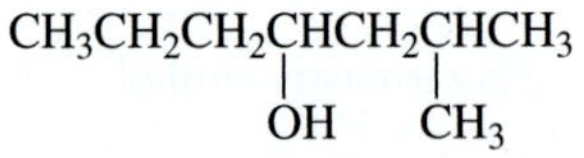

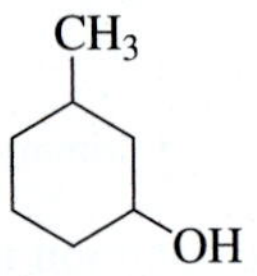

4. If there is more than one substituent, the substituents are stated in alphabetical order.

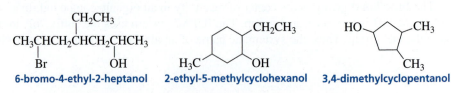

6-bromo-4-ethyl-2-heptanol **2-ethyl-5-methylcyclohexanol** **3,4-dimethylcyclopentanol**

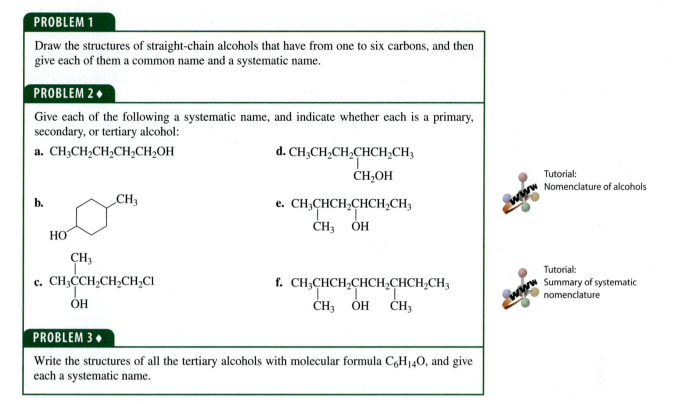

PROBLEM 1

Draw the structures of straight-chain alcohols that have from one to six carbons, and then give each of them a common name and a systematic name.

PROBLEM 2 ♦

Give each of the following a systematic name, and indicate whether each is a primary, secondary, or tertiary alcohol:

a. $CH_3CH_2CH_2CH_2CH_2OH$

b.

c. $CH_3CCH_2CH_2CH_2Cl$ (with CH_3 above and OH below)

d. $CH_3CH_2CH_2CHCH_2CH_3$ with CH_2OH below

e. $CH_3CHCH_2CHCH_2CH_3$ with CH_3 and OH below

f. $CH_3CHCH_2CHCH_2CHCH_2CH_3$ with CH_3, OH, CH_3 below

Tutorial:
Nomenclature of alcohols

Tutorial:
Summary of systematic nomenclature

PROBLEM 3 ♦

Write the structures of all the tertiary alcohols with molecular formula $C_6H_{14}O$, and give each a systematic name.

10.2 SUBSTITUTION REACTIONS OF ALCOHOLS

An **alcohol** has a strongly basic group (HO^-) that cannot be displaced by a nucleophile. Therefore, an alcohol cannot undergo a nucleophilic substitution reaction.

a strongly basic leaving group

$$CH_3-\overset{..}{\underset{..}{O}}H + Br^- \xrightarrow{\times\!\!\!\!\!\rightarrow} CH_3-Br + HO^-$$
strong base

However, if the alcohol's OH group is converted into a group that is a weaker base (and therefore a better leaving group), a nucleophilic substitution reaction can occur. One way to convert an OH group into a weaker base is to protonate it by adding acid to the solution. Protonation changes the leaving group from HO^- to H_2O, which is a weak enough base to be displaced by a nucleophile. The substitution reaction is slow and requires heat (except in the case of tertiary alcohols) if it is to take place in a reasonable period of time.

The weaker the base, the more easily it can be displaced.

a weakly basic leaving group

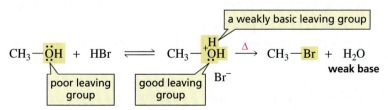

$$CH_3-\overset{..}{\underset{..}{O}}H + HBr \rightleftharpoons CH_3-\overset{H}{\underset{..}{\overset{+}{O}}H} \xrightarrow{\Delta} CH_3-Br + H_2O$$
weak base

poor leaving group   good leaving group Br^-

Because the OH group of the alcohol has to be protonated before it can be displaced by a nucleophile, only weakly basic nucleophiles (I^-, Br^-, Cl^-) can be used in the substitution reaction. Moderately and strongly basic nucleophiles (NH_3, RNH_2, CH_3O^-) cannot be used because they would also be protonated in the acidic solution and, once protonated, would no longer be nucleophiles ($^+NH_4$, RNH_3^+) or would be poor nucleophiles (CH_3OH).

PROBLEM 4◆

Why are NH_3 and CH_3NH_2 no longer nucleophiles when they are protonated?

Primary, secondary, and tertiary alcohols all undergo nucleophilic substitution reactions with HI, HBr, and HCl to form alkyl halides.

$$CH_3CH_2CH_2OH + HI \xrightarrow{\Delta} CH_3CH_2CH_2I + H_2O$$

1-propanol
a primary alcohol
1-iodopropane

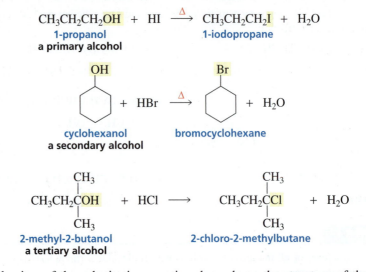

cyclohexanol
a secondary alcohol
bromocyclohexane

$$CH_3CH_2\overset{\overset{\textstyle CH_3}{|}}{\underset{\underset{\textstyle CH_3}{|}}{C}}OH + HCl \longrightarrow CH_3CH_2\overset{\overset{\textstyle CH_3}{|}}{\underset{\underset{\textstyle CH_3}{|}}{C}}Cl + H_2O$$

2-methyl-2-butanol
a tertiary alcohol
2-chloro-2-methylbutane

The mechanism of the substitution reaction depends on the structure of the alcohol. Secondary and tertiary alcohols undergo S_N1 reactions—an S_N1 reaction of a protonated alcohol.

mechanism for the S_N1 reaction of an alcohol

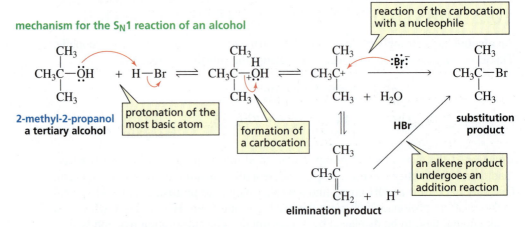

2-methyl-2-propanol
a tertiary alcohol

protonation of the most basic atom

formation of a carbocation

reaction of the carbocation with a nucleophile

substitution product

HBr

an alkene product undergoes an addition reaction

elimination product

- An acid always reacts with an organic molecule in the same way: it protonates the most basic atom in the reactant.

- Weakly basic water is the leaving group that is expelled, forming a carbocation.

- The carbocation has two possible fates: it can combine with a nucleophile and form a substitution product, or it can lose a proton and form an elimination product.

Secondary and tertiary alcohols undergo S_N1 reactions with hydrogen halides.

Although the reaction can form both a substitution product and an elimination product, only the substitution product is actually obtained because any alkene formed in an elimination reaction will undergo a subsequent addition reaction with HBr to form more of the substitution product (Section 5.1).

Tertiary alcohols undergo substitution reactions with hydrogen halides faster than secondary alcohols do, because tertiary carbocations are easier to form than secondary carbocations (Section 9.4). Thus, the reaction of a tertiary alcohol with a hydrogen halide proceeds readily at room temperature, whereas the reaction of a secondary alcohol with a hydrogen halide has to be heated to have the reaction occur at the same rate.

Carbocation stability: 3°>2°>1°

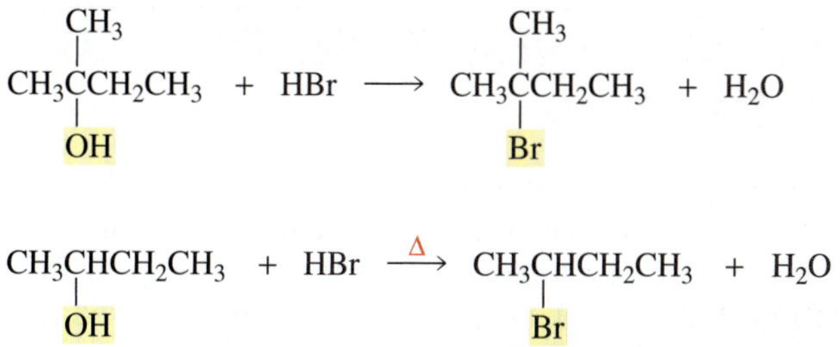

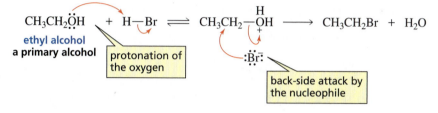

Primary alcohols cannot undergo S_N1 reactions because primary carbocations are too unstable to be formed (Section 9.4). Therefore, when a primary alcohol reacts with a hydrogen halide, it must do so in an S_N2 reaction.

Primary alcohols undergo S_N2 reactions with hydrogen halides.

mechanism for the S_N2 reaction of an alcohol

$$CH_3CH_2\overset{..}{\underset{..}{O}}H \; + \; H\!-\!Br \; \rightleftharpoons \; CH_3CH_2\!-\!\overset{H}{\underset{+}{O}}H \; \longrightarrow \; CH_3CH_2Br \; + \; H_2O$$

ethyl alcohol
a primary alcohol | protonation of the oxygen |

:Br:⁻

| back-side attack by the nucleophile |

- The acid protonates the most basic atom in the reactant.
- The nucleophile hits the back side of the carbon and displaces the leaving group.

PROBLEM 5   *SOLVED*

Using the pK_a values of the conjugate acids of the leaving groups (the pK_a of HBr is −9; the pK_a of H_2O is 15.7; the pK_a of H_3O^+ is −1.7), explain the difference in reactivity in substitution reactions between:

a. CH_3Br and CH_3OH

b. $CH_3\overset{+}{O}H_2$ and CH_3OH

Solution to 5a   The conjugate acid of the leaving group of CH_3Br is HBr; its pK_a is −9; the conjugate acid of the leaving group of CH_3OH is H_2O; its pK_a is 15.7. Because HBr is a much stronger acid than H_2O, Br^- is a much weaker base than HO^-. (Recall the stronger the acid, the weaker is its conjugate base.) Therefore, Br^- is a much better leaving group than HO^-, causing CH_3Br to be much more reactive in a substitution reaction than CH_3OH.

PROBLEM 6 ◆

Give the major product of each of the following reactions:

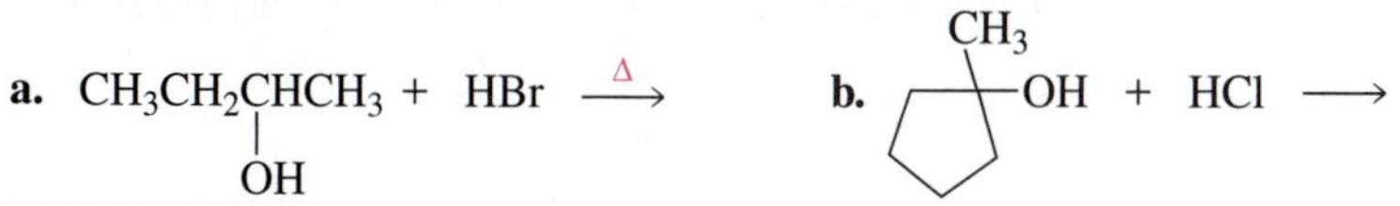

a. $CH_3CH_2\underset{\underset{OH}{|}}{C}HCH_3 + HBr \overset{\Delta}{\longrightarrow}$ **b.** [cyclopentane ring with CH_3 and OH] $+ HCl \longrightarrow$

PROBLEM 7   *SOLVED*

Show how 1-butanol can be converted into the following compounds:

a. $CH_3CH_2CH_2CH_2OCH_3$ **c.** $CH_3CH_2CH_2CH_2NHCH_2CH_3$

b. $CH_3CH_2CH_2CH_2O\overset{\overset{O}{\|}}{C}CH_2CH_3$ **d.** $CH_3CH_2CH_2CH_2C\equiv N$

Solution to 7a   Because the OH group of 1-butanol is too basic to be substituted, the alcohol must first be converted into an alkyl halide. The alkyl halide has a leaving group that can be substituted by CH_3O^-, the nucleophile required to obtain the desired product.

$$CH_3CH_2CH_2CH_2OH \; \overset{HBr}{\underset{\Delta}{\longrightarrow}} \; CH_3CH_2CH_2CH_2Br \; \overset{CH_3O^-}{\longrightarrow} \; CH_3CH_2CH_2CH_2OCH_3$$

10.3 ELIMINATION REACTIONS OF ALCOHOLS: DEHYDRATION

An alcohol can undergo an elimination reaction by losing an OH from one carbon and an H from an adjacent carbon. The product of the reaction is an alkene. Overall, this amounts to the elimination of a molecule of water. Loss of water from a molecule is called **dehydration**. Dehydration of an alcohol requires an acid catalyst and heat. Sulfuric acid (H_2SO_4) is a commonly used acid catalyst.

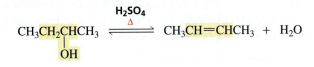

The dehydration of a secondary or a tertiary alcohol is an E1 reaction.

mechanism for the E1 dehydration of an alcohol

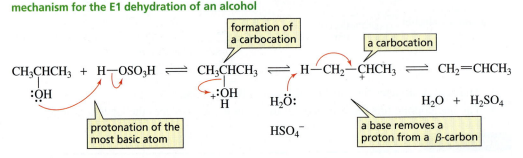

- The acid protonates the most basic atom in the reactant. As we saw earlier, protonation converts the very poor leaving group (HO^-) into a good leaving group (H_2O).
- Water departs, leaving behind a carbocation.
- A base in the reaction mixture (water is the base present in the highest concentration) removes a proton from a β-carbon (a carbon adjacent to the positively charged carbon), forming an alkene and regenerating the acid catalyst.

An acid protonates the most basic atom in a molecule.

Secondary and tertiary alcohols undergo dehydration by an E1 pathway.

Notice that the dehydration reaction is an E1 reaction of a protonated alcohol.

When more than one elimination product can be formed, the major product is the more stable alkene—the one obtained by removing a proton from the β-carbon bonded to the fewest hydrogens (Figure 10.1).

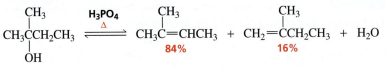

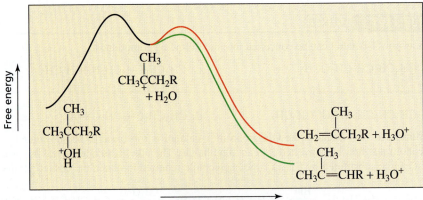

▶ **Figure 10.1**
The reaction coordinate diagram for the dehydration of a protonated alcohol. The major product is the more substituted alkene because the transition state leading to its formation is more stable, allowing that alkene to be formed more rapidly.

Because the rate-determining step in the dehydration of a secondary or a tertiary alcohol is formation of a carbocation intermediate, the rate of dehydration parallels the ease with which the carbocation is formed. Tertiary alcohols are the easiest to dehydrate because tertiary carbocations are more stable and are, therefore, easier to form than secondary and primary carbocations (Section 9.4).

relative ease of dehydration

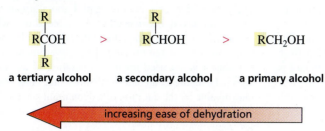

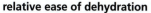

While the dehydration of a tertiary or a secondary alcohol is an E1 reaction, the dehydration of a primary alcohol is an E2 reaction because primary carbocations are too unstable to be formed.

Primary alcohols undergo dehydration by an E2 pathway.

mechanism for the E2 dehydration and the competing substitution (S_N2) reaction

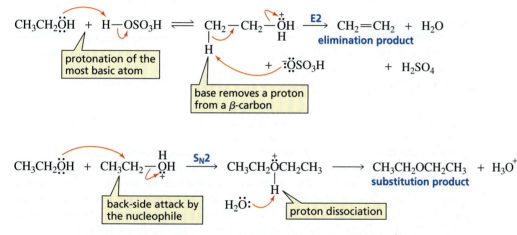

- The acid protonates the most basic atom in the reactant.
- A base removes a proton in the elimination reaction.
- An ether is also obtained; it is the product of a competing S_N2 reaction, since primary alkyl halides are the ones most likely to form substitution products in S_N2/E2 reactions (Section 9.11).

PROBLEM 8 ◆

List the following alcohols in order of decreasing rate of dehydration in the presence of acid:

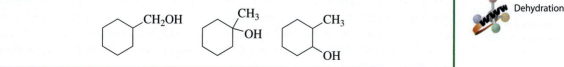

Movie: Dehydration

GRAIN ALCOHOL AND WOOD ALCOHOL

When ethanol is ingested, it acts on the central nervous system. Moderate amounts affect judgment and lower inhibitions. Higher amounts interfere with motor coordination and cause slurred speech and amnesia. Still higher amounts cause nausea and loss of consciousness. Ingesting very large amounts of ethanol interferes with spontaneous respiration and can be fatal.

The ethanol in alcoholic beverages is produced by the fermentation of glucose, generally obtained from grapes or from grains such as corn, rye, and wheat (which is why ethanol is also known as grain alcohol). Grains are cooked in the presence of malt (sprouted barley) to convert much of their starch into glucose. Yeast is added to convert the glucose into ethanol and carbon dioxide (Section 18.5).

$$C_6H_{12}O_6 \xrightarrow{\text{yeast enzymes}} 2\ CH_3CH_2OH\ +\ 2\ CO_2$$
glucose   ethanol

The kind of beverage produced (white or red wine, beer, scotch, bourbon, champagne) depends on the plant species

providing the glucose, whether the CO_2 formed in the fermentation is allowed to escape, whether other substances are added, and how the beverage is purified (by sedimentation, for wines; by distillation, for scotch and bourbon).

The tax imposed on liquor would make ethanol a prohibitively expensive laboratory reagent. Laboratory alcohol, therefore, is not taxed, because ethanol is needed in a wide variety of commercial processes. Although this alcohol is not taxed, it is carefully regulated by the federal government to make certain that it is not used for the preparation of alcoholic beverages. Denatured alcohol—ethanol that has been made undrinkable by the addition of a denaturant such as benzene or methanol—is not taxed, but the added impurities make it unfit for many laboratory uses.

Methanol, also known as wood alcohol because at one time it was obtained by heating wood in the absence of oxygen, is highly toxic. Ingesting even very small amounts can cause blindness, and ingesting as little as an ounce can be fatal.

PROBLEM 9

Heating an alcohol with H_2SO_4 is a good way to prepare a symmetrical ether such as diethyl ether.
a. Explain why it is not a good way to prepare an unsymmetrical ether such as ethyl propyl ether.
b. How would you synthesize ethyl propyl ether?

The products obtained from the acid-catalyzed elimination (dehydration) of an alcohol are identical to those obtained from the elimination reaction of an alkyl halide. That is, both the *E* and *Z* stereoisomers are obtained as products. The reaction produces more of the stereoisomer in which the bulkier group on each of the sp^2 carbons are on opposite sides of the double bond; that stereoisomer, being more stable, is formed more rapidly, since the transition state leading to its formation is more stable.

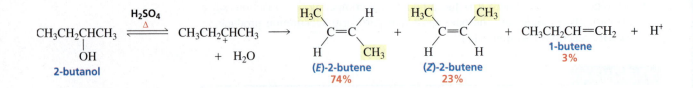

Alcohols and ethers undergo S$_N$1 reactions unless they would have to form a methyl or primary carbocation, in which case they undergo S$_N$2 reactions.

We can summarize what we have learned about the mechanisms by which alcohols undergo substitution and elimination reactions: they react by S$_N$1 and E1 pathways, unless they cannot do so. In other words, 3° and 2° alcohols undergo S$_N$1 and E1 reactions; 1° alcohols, because they cannot form primary carbocations, have to undergo S$_N$2 and E2 reactions.

BIOLOGICAL DEHYDRATIONS

Dehydration reactions occur in many important biological processes. Instead of being catalyzed by strong acids, which would not be available to a cell, they are catalyzed by enzymes. Fumarase, for example, is the enzyme that catalyzes the dehydration of malate in the citric acid cycle. The citric acid cycle is a series of reactions that oxidize compounds derived from carbohydrates, fatty acids, and amino acids (Section 18.7).

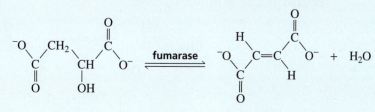

Enolase, another enzyme, catalyzes the dehydration of α-phosphoglycerate in glycolysis (Section 18.4). Glycolysis is a series of reactions that prepare glucose for entry into the citric acid cycle.

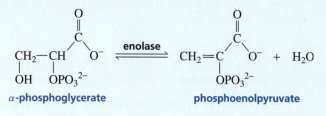

PROBLEM 10 ◆

Give the major product formed when each of the following alcohols is heated in the presence of H_2SO_4:

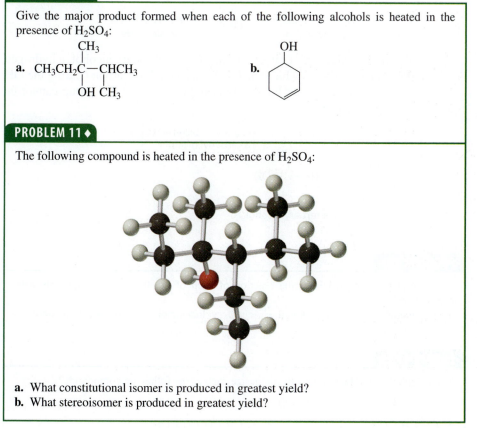

a.   CH₃CH₂C—CHCH₃ (with CH₃ above C, OH and CH₃ below)

b. (cyclohexenol structure)

PROBLEM 11 ◆

The following compound is heated in the presence of H_2SO_4:

a. What constitutional isomer is produced in greatest yield?
b. What stereoisomer is produced in greatest yield?

10.4 OXIDATION OF ALCOHOLS

We have seen that a **reduction reaction** *increases* the number of C—H bonds in a compound (Section 5.12). Oxidation is the reverse of reduction. Therefore, an **oxidation reaction** *decreases* the number of C—H bonds (or increases the number of C—O bonds).

Secondary alcohols are oxidized to *ketones*. Chromic acid (H_2CrO_4) is the reagent commonly used to oxidize alcohols.

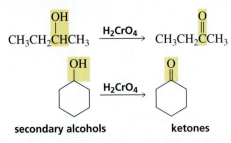

secondary alcohols   ketones

Secondary alcohols are oxidized to ketones.

Primary alcohols are initially oxidized to *aldehydes* by this reagent. The reaction, however, does not stop at the aldehyde. Instead, the aldehyde is further oxidized to a *carboxylic acid*.

a primary alcohol   an aldehyde   a carboxylic acid

The oxidation of a primary alcohol will stop at the aldehyde if pyridinium chlorochromate (PCC) is used as the oxidizing agent in a solvent such as dichloromethane (CH_2Cl_2).

Primary alcohols are oxidized to aldehydes and carboxylic acids.

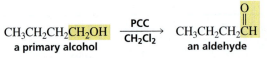

a primary alcohol   an aldehyde

Notice that in the oxidation of either a primary or a secondary alcohol, a hydrogen is removed from the carbon to which the OH is attached. The carbon bearing the OH group in a tertiary alcohol is not bonded to a hydrogen, so its OH group cannot be oxidized to a carbonyl group.

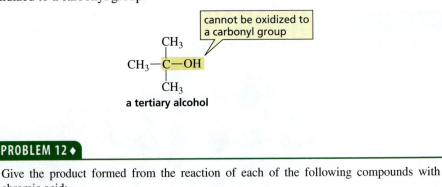

cannot be oxidized to a carbonyl group

a tertiary alcohol

PROBLEM 12 ◆

Give the product formed from the reaction of each of the following compounds with chromic acid:

a. 3-pentanol   **b.** 1-pentanol   **c.** cyclohexanol   **d.** benzyl alcohol

---

PROBLEM 13 ◆

What alcohol would be required to synthesize each of the following compounds?

a. CH₃CH₂CCH₃   **b.**   **c.** CH₃CH₂CH₂COH

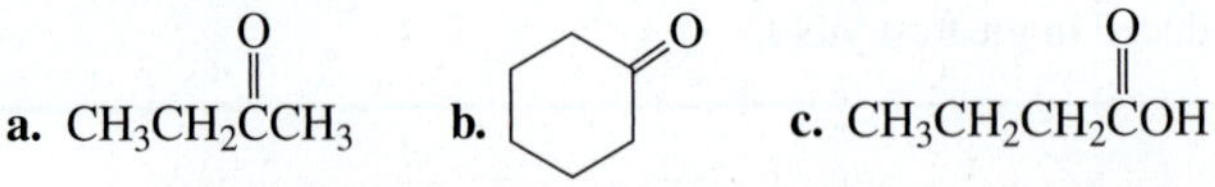

BLOOD ALCOHOL CONTENT

As blood passes through the arteries in our lungs, an equilibrium is established between alcohol in the blood and alcohol in the breath. Therefore, if the concentration of one is known, the concentration of the other can be estimated.

The test that law enforcement agencies use to approximate a person's blood alcohol level is based on the oxidation of breath ethanol. An oxidizing agent impregnated onto an inert material is enclosed within a sealed glass tube. When the test is to be administered, the ends of the tube are broken off and replaced with a mouthpiece at one end and a balloon-type bag at the other. The person being tested blows into the mouthpiece until the bag is filled with air.

Any ethanol in the breath is oxidized as it passes through the column. When ethanol is oxidized, the red-orange dichromate ion is reduced to green chromic ion. The greater the con-

centration of alcohol in the breath, the farther the green color spreads through the tube.

$$CH_3CH_2OH \ + \ \underset{\textbf{red orange}}{Cr_2O_7{}^{2-}} \ \xrightarrow{\ H^+\ } \ \underset{}{CH_3\overset{\displaystyle O}{\overset{\|}{C}}OH} \ + \ \underset{\textbf{green}}{Cr^{3+}}$$

If the person fails this test—determined by the extent to which the green color spreads through the tube—a more accurate Breathalyzer test is administered. The Breathalyzer test also depends on the oxidation of breath ethanol, but it provides more accurate results because it is quantitative. In the test, a known volume of breath is bubbled through an acidic solution of sodium dichromate, and the concentration of the green chromic ion is measured precisely with a UV/Vis spectrophotometer (Section 7.12).

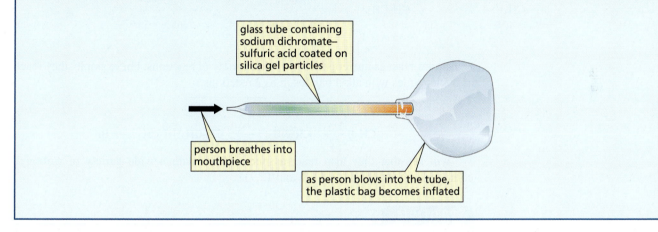

glass tube containing sodium dichromate–sulfuric acid coated on silica gel particles

person breathes into mouthpiece

as person blows into the tube, the plastic bag becomes inflated

10.5 AMINES DO NOT UNDERGO SUBSTITUTION OR ELIMINATION REACTIONS

We have just seen that alcohols are much less reactive than alkyl halides in substitution and elimination reactions. Amines are even *less reactive* than alcohols. The relative reactivities of an alkyl fluoride (the least reactive of the alkyl halides because it has the poorest leaving group), an alcohol, and an amine can be appreciated by comparing the pK_a values of the conjugate acids of their leaving groups, recalling that the weaker the acid, the stronger is its conjugate base and the poorer it is as a leaving group. The leaving group of an amine ($^-NH_2$) is such a strong base that amines cannot undergo substitution or elimination reactions.

relative reactivities

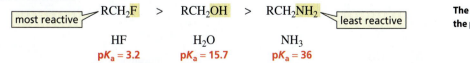

| most reactive | RCH$_2$F | > | RCH$_2$OH | > | RCH$_2$NH$_2$ | least reactive |

| | HF | H$_2$O | NH$_3$ |
| | $pK_a = 3.2$ | $pK_a = 15.7$ | $pK_a = 36$ |

The stronger the base, the poorer it is as a leaving group.

Protonation of the amino group makes it a better leaving group, but not nearly as good a leaving group as a protonated alcohol, which is ~13 pK_a units more acidic than

a protonated amine. Therefore, unlike the leaving group of a protonated alcohol, the leaving group of a protonated amine cannot be replaced by a halide ion or dissociate to form a carbocation.

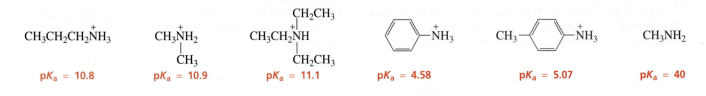

$$CH_3CH_2\overset{+}{O}H_2 \quad > \quad CH_3CH_2\overset{+}{N}H_3$$
$$pK_a = -2.4 \qquad\qquad pK_a = 11.2$$

Although amines cannot undergo substitution or elimination reactions, they are extremely important organic compounds. (The lone pair on the nitrogen atom allows it to act as both a base and a nucleophile.)

Amines are the most common organic bases. We have seen that protonated amines have pK_a values of about 11 (Section 2.3) and that protonated anilines have pK_a values of about 5 (Section 7.9). Neutral amines have very high pK_a values. For example, the pK_a of methylamine is 40.

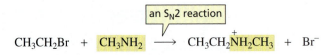

$$CH_3CH_2CH_2\overset{+}{N}H_3 \qquad CH_3\overset{+}{N}H_2 \qquad CH_3CH_2\overset{+}{N}H \qquad \qquad \qquad CH_3NH_2$$

$pK_a = 10.8$ $pK_a = 10.9$ $pK_a = 11.1$ $pK_a = 4.58$ $pK_a = 5.07$ $pK_a = 40$

Amines react as nucleophiles in a wide variety of reactions. For example, they react as nucleophiles with alkyl halides in S_N2 reactions.

an S_N2 reaction

$$CH_3CH_2Br \; + \; CH_3NH_2 \; \longrightarrow \; CH_3CH_2\overset{+}{N}H_2CH_3 \; + \; Br^-$$

We will see that they also react as nucleophiles with a wide variety of carbonyl compounds (Sections 11.7, 11.8, and 12.7).

PROBLEM 14

Why can protonated amino groups not be displaced by strongly basic nucleophiles such as HO^-?

ALKALOIDS

Alkaloids are amines found in the leaves, bark, roots, or seeds of many plants. Examples include caffeine (found in tea leaves, coffee beans, and cola nuts), nicotine (found in tobacco leaves), and cocaine (obtained from the coca bush in the rainforest areas of Colombia, Peru, and Bolivia). Ephedrine, a bronchodilator, is an alkaloid obtained from *Ephedra sinica*, a plant found in China. Morphine is an alkaloid obtained from opium, the juice derived from a species of poppy (Section 21.3).

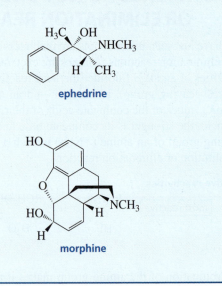

ephedrine

morphine

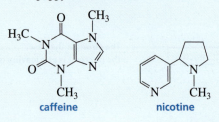

caffeine   nicotine

10.6 NOMENCLATURE OF ETHERS

An **ether** is a compound in which an oxygen is bonded to two alkyl substituents. The common name of an ether consists of the names of the two alkyl substituents (in alphabetical order), followed by the word "ether." The smallest ethers are almost always named by their common names.

$CH_3OCH_2CH_3$
ethyl methyl ether

$CH_3CH_2OCH_2CH_3$
diethyl ether

$$CH_3CHCH_2OCCH_3$$
with CH_3 above and CH_3 CH_3 below
tert-butyl isobutyl ether

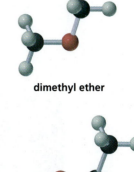
dimethyl ether

The IUPAC system names an ether as an alkane with an RO substituent. The substituents are named by replacing the "yl" ending in the name of the alkyl substituent with "oxy."

CH_3O-
methoxy

CH_3CH_2O-
ethoxy

CH_3CHO- with CH_3
isopropoxy

CH_3CH_2CHO- with CH_3
sec-butoxy

CH_3CO- with CH_3 above and CH_3 below
tert-butoxy

diethyl ether

$CH_3CHCH_2CH_3$ with OCH_3
2-methoxybutane

$CH_3CH_2CHCH_2CH_2OCH_2CH_3$ with CH_3
1-ethoxy-3-methylpentane

PROBLEM 15 ◆

a. Give the systematic name for the following ethers:

1. $CH_3OCH_2CH_3$

2. $CH_3CH_2OCH_2CH_3$

3. $CH_3CH_2CH_2CH_2CHCH_2CH_2CH_3$ with OCH_3

4. $CH_3CH_2CH_2OCH_2CH_2CH_2CH_3$

b. Do all of these ethers have common names?

c. What are their common names?

Tutorial:
Nomenclature of ethers

ANESTHETICS

Because diethyl ether (commonly known simply as ether) is a short-lived muscle relaxant, it was at one time widely used as an inhalation anesthetic. However, it takes effect slowly and has a slow and unpleasant recovery period, so other compounds, such as enflurane, isoflurane, and halothane, have replaced it as an anesthetic.

Even so, diethyl ether is still used where trained anesthesiologists are scarce, because it is the safest anesthetic for an untrained person to administer. Anesthetics interact with the nonpolar molecules of cell membranes, causing the membranes to swell, which interferes with their permeability.

$CH_3CH_2OCH_2CH_3$
"ether"

$CF_3CHClOCHF_2$
isoflurane

$CHClFCF_2OCHF_2$
enflurane

$CF_3CHClBr$
halothane

Sodium pentothal (also called thiopental sodium) is commonly used as an intravenous anesthetic. The onset of anesthesia and the loss of consciousness occur within seconds of its administration. Care must be taken when administering sodium pentothal because the dose for effective anesthesia is 75% of the lethal dose. Because of its toxicity, it cannot be used as

the sole anesthetic. It is generally used to induce anesthesia before an inhalation anesthetic is administered. Propofol is an anesthetic that has all the properties of the "perfect anesthetic": it can be used as the sole anesthetic by intravenous drip, it has a rapid and pleasant induction period and a wide margin of safety, and recovery from the drug also is rapid and pleasant.

(continued...)

sodium pentothal   **propofol**

Amputation of a leg without anesthetic in 1528.

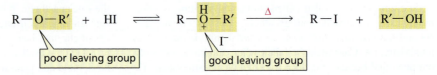

A painting showing the first use of anesthesia in 1846 at Massachusetts General Hospital by surgeon John Collins Warren.

10.7 NUCLEOPHILIC SUBSTITUTION REACTIONS OF ETHERS

The OR group of an **ether** and the OH group of an alcohol have nearly the same basicity because the conjugate acids of these two groups have similar pK_a values. (The pK_a of CH_3OH is 15.5, and the pK_a of H_2O is 15.7.) Both groups are strong bases, so both are very poor leaving groups. Consequently, ethers, like alcohols, need to be activated before they can undergo nucleophilic substitution.

$$R-\ddot{O}-H \qquad R-\ddot{O}-R$$
an alcohol   **an ether**

Ethers, like alcohols, can be activated by protonation. Ethers, therefore, can undergo nucleophilic substitution reactions with HBr or HI. As with alcohols, the reaction of ethers with hydrogen halides is slow. The reaction mixture must be heated to cause the reaction to occur at a reasonable rate.

What happens *after* the ether is protonated depends on the structure of the ether. If departure of ROH creates a relatively stable carbocation (such as a tertiary carbocation), an S_N1 reaction occurs.

ether cleavage: an S_N1 reaction

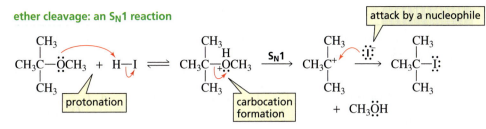

- Protonation converts the very basic RO^- leaving group into the less basic ROH leaving group.
- The leaving group departs.
- The halide ion combines with the carbocation.

However, if departure of the leaving group would create an unstable carbocation (such as a methyl or primary carbocation), the leaving group cannot depart. It has to be displaced by the halide ion. In other words, an S_N2 reaction occurs.

ether cleavage: an S_N2 reaction

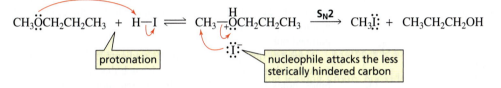

- Protonation converts the very basic RO^- leaving group into the less basic ROH leaving group.
- The halide ion preferentially attacks the less sterically hindered of the two alkyl groups.

In summary, ethers are cleaved by an S_N1 pathway, unless the instability of the carbocation causes the reaction to follow an S_N2 pathway.

Because the only reagents that react with ethers are hydrogen halides, ethers are frequently used as solvents. Some common ether solvents are shown in Table 10.1.

Table 10.1 Some Ethers That Are Used as Solvents

$CH_3CH_2OCH_2CH_3$				$CH_3OCH_2CH_2OCH_3$	$CH_3OC(CH_3)_3$
diethyl ether "ether"	tetrahydrofuran **THF**	tetrahydropyran **THP**	1,4-dioxane	1,2-dimethoxyethane **DME**	*tert*-butyl methyl ether **MTBE**

3-D Molecules:
Diethyl ether; Tetrahydrofuran

PROBLEM 16 **SOLVED**

Explain why methyl propyl ether forms both methyl iodide and propyl iodide when it is heated with excess HI.

Solution   We have just seen that the S_N2 reaction of methyl propyl ether with an equivalent of HI forms methyl iodide and propyl alcohol because the methyl group is less sterically hindered to attack by the iodide ion. When there is excess HI, the alcohol product of this first reaction can react with HI in another S_N2 reaction (Section 10.1). Thus, the major products are methyl iodide and propyl iodide.

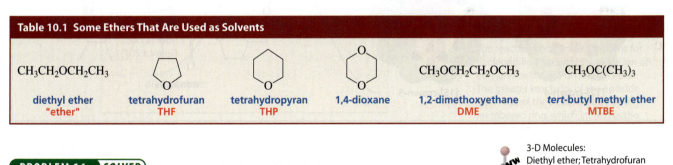

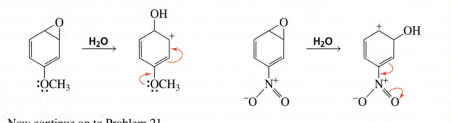

Now continue on to Problem 21.

Now continue on to Problem 21.

PROBLEM 21 ◆

Which compound is more likely to be carcinogenic? (*Hint:* Read the box on benzo[*a*]pyrene to see why the 4,5-epoxide is harmful.)

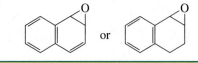

BENZO[*a*]PYRENE AND CANCER

Benzo[*a*]pyrene is one of the most carcinogenic of the aromatic hydrocarbons. It is formed whenever an organic compound is not completely burned. For example, benzo[*a*]pyrene is found in cigarette smoke, automobile exhaust, and charcoal-broiled meat. Several arene oxides

can be formed from benzo[*a*]pyrene. The two most harmful are the 4,5-oxide and the 7,8-oxide. It has been suggested that people who develop lung cancer as a result of smoking may have a higher than normal concentration of cytochrome P_{450} in their lung tissue.

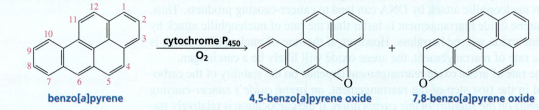

The 4,5-oxide is harmful because it forms a carbocation that cannot be stabilized by electron delocalization without destroying the aromaticity of an adjacent benzene ring. Thus, the carbocation is relatively unstable, so the epoxide tends not to open until it is attacked by a nucleophile (the carcinogenic pathway).

The 7,8-oxide is harmful because it reacts with water to form a diol, which then forms a diol epoxide. The diol epoxide does not readily undergo rearrangement (the harmless pathway), because it opens to a carbocation that is destabilized by the electron-withdrawing OH groups. Therefore, the diol epoxide can exist long enough to be attacked by nucleophiles (the carcinogenic pathway).

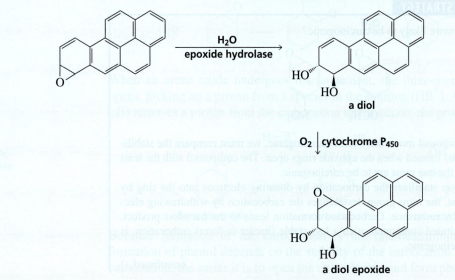

CHIMNEY SWEEPS AND CANCER

In 1775, British physician Percival Potts became the first to recognize that environmental factors can cause cancer when he observed that chimney sweeps had a higher incidence of scrotum cancer than the male population as a whole. He theorized that something in the chimney soot was causing cancer. We now know that it was benzo[*a*]pyrene.

Titch Cox, the chimney sweep responsible for cleaning the 800 chimneys at Buckingham Palace.

SUMMARY

The leaving groups of **alcohols** and **ethers** are stronger bases than halide ions are, so alcohols and ethers are less reactive than alkyl halides and have to be protonated before they can undergo a substitution or an elimination reaction. **Epoxides** do not have to be activated, because ring strain increases their reactivity. Amines cannot undergo substitution or elimination reactions because their leaving groups ($^-NH_2$, ^-NHR, $^-NR_2$) are very strong bases.

Primary, secondary, and tertiary alcohols undergo nucleophilic substitution reactions with HI, HBr, and HCl to form alkyl halides. These are S_N1 reactions in the case of tertiary and secondary alcohols and S_N2 reactions in the case of primary alcohols.

An alcohol undergoes an elimination reaction if heated with an acid. **Dehydration** (elimination of a water molecule) is an E1 reaction in the case of tertiary and secondary alcohols and an E2 reaction in the case of primary alcohols. Tertiary alcohols are the easiest to dehydrate, and primary alcohols are the hardest. The major product is the more substituted alkene. If the alkene has stereoisomers, both the *E* and *Z* stereoisomers are formed, but the one with the bulkiest groups on opposite sides of the double bond predominates.

Chromic acid oxidizes secondary alcohols to ketones and primary alcohols to carboxylic acids. PCC oxidizes primary alcohols to aldehydes.

Ethers can undergo nucleophilic substitution reactions with HBr or HI. If departure of the leaving group creates a relatively stable carbocation, an S_N1 reaction occurs; otherwise, an S_N2 reaction occurs.

Epoxides undergo nucleophilic substitution reactions. Under basic conditions, the least sterically hindered ring-carbon is attacked; under acidic conditions, the most substituted ring-carbon is attacked. **Arene oxides** undergo rearrangement to form phenols and nucleophilic attack to form addition products. An arene oxide's cancer-causing potential depends on the stability of the carbocation formed during rearrangement.

SUMMARY OF REACTIONS

1. Nucleophilic substitution reactions of *alcohols* (Section 10.2).

$$\text{ROH} + \text{HBr} \xrightarrow{\Delta} \text{RBr}$$

$$\text{ROH} + \text{HI} \xrightarrow{\Delta} \text{RI}$$

$$\text{ROH} + \text{HCl} \xrightarrow{\Delta} \text{RCl}$$

relative rate: tertiary > secondary > primary

2. Elimination reactions of *alcohols*: dehydration (Section 10.3).

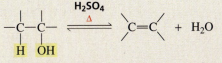

relative rate: tertiary > secondary > primary

3. Oxidation of *alcohols* (Section 10.4).

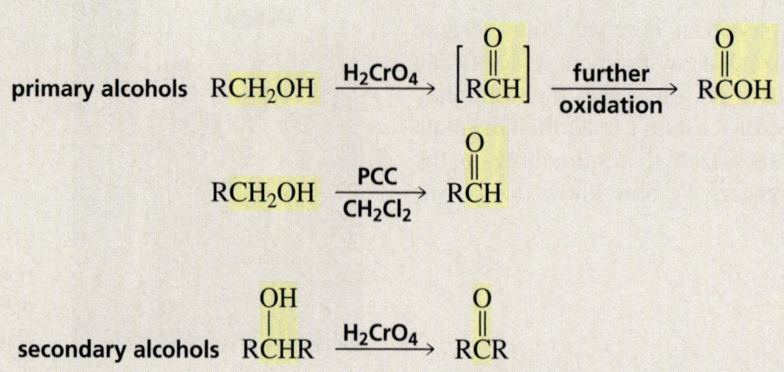

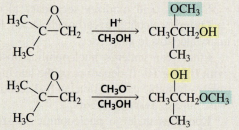

4. Nucleophilic substitution reactions of *ethers* (Section 10.7).

$$ROR' + HX \xrightarrow{\Delta} ROH + R'X$$

HX = HBr or HI

5. Nucleophilic substitution reactions of *epoxides* (Section 10.8)

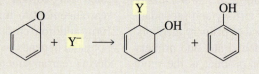

under acidic conditions, the nucleophile attacks the more substituted ring-carbon

under basic conditions, the nucleophile attacks the less sterically hindered ring-carbon

6. Reactions of *arene oxides*: ring opening and rearrangement (Section 10.9).

PROBLEMS

22. Give the product of each of the following reactions:

a. 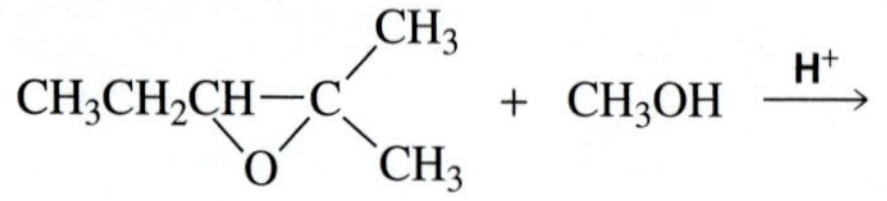 CH₃CH₂CH—C(CH₃)(CH₃)O + CH₃OH $\xrightarrow{H^+}$

b. CH₃CHCH₂OCH₃ + HI $\xrightarrow{\Delta}$
    CH₃

c. cyclohexyl-CH₂CH₂OH $\xrightarrow{H_2CrO_4}$

d. CH₃CH₂CH—C(CH₃)(CH₃)O + CH₃OH $\xrightarrow{CH_3O^-}$

e. CH₃CH—CCH₃(CH₃)(OH)CH₃ $\xrightarrow[\Delta]{H_2SO_4}$

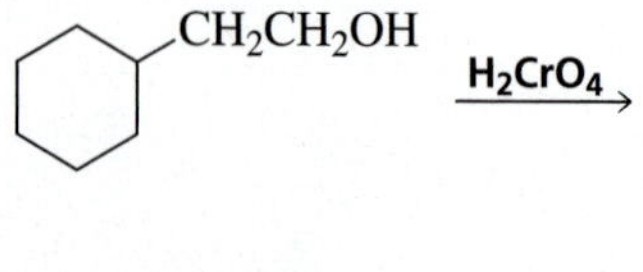

f. 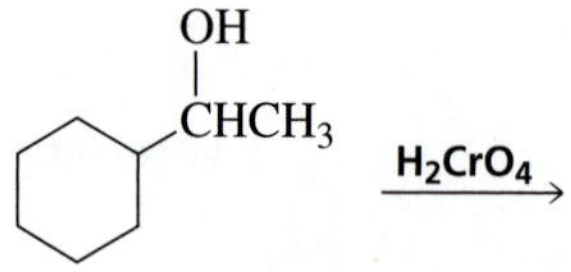 cyclohexyl-CH(OH)CH₃ $\xrightarrow{H_2CrO_4}$

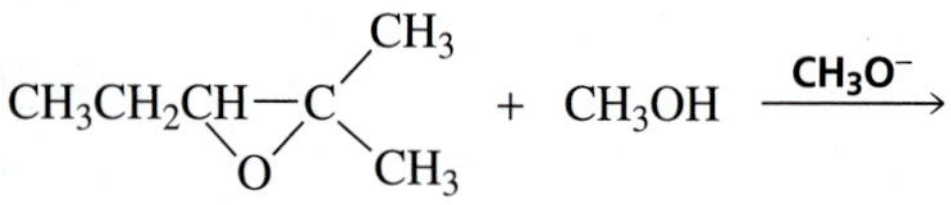

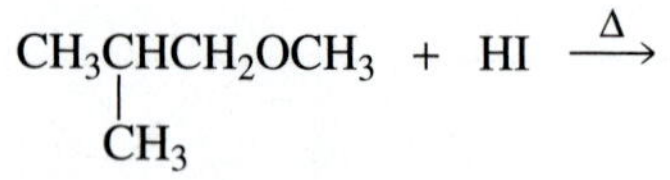

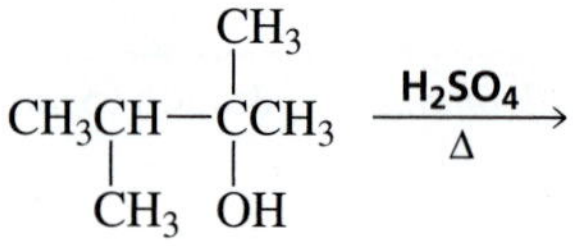

23. Give common and systematic names for each of the following:
   a. CH$_3$CHOCH$_2$CH$_2$CH$_3$ **b.** CH$_3$CH$_2$CH$_2$CH$_2$OCH$_2$CH$_3$ **c.** CH$_3$CH$_2$CHOCH$_3$ **d.** CH$_3$CHOCHCH$_3$
         | | | |
         CH$_3$ CH$_3$ CH$_3$ CH$_3$

24. Which alcohol in each pair will undergo dehydration more rapidly when heated with H$_2$SO$_4$?

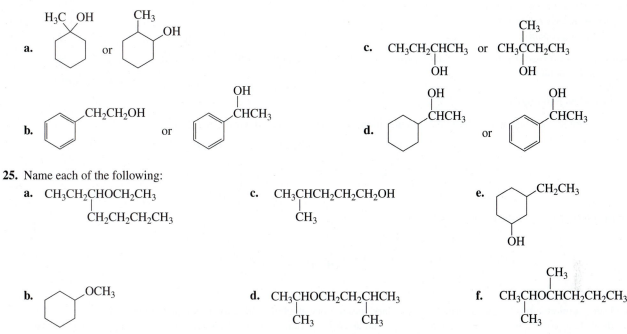

25. Name each of the following:
   a. CH$_3$CH$_2$CHOCH$_2$CH$_3$ **c.** CH$_3$CHCH$_2$CH$_2$CH$_2$OH   **e.** [cyclohexane with CH$_2$CH$_3$ and OH]
              | |
              CH$_2$CH$_2$CH$_2$CH$_3$ CH$_3$

   b. [cyclohexane with OCH$_3$]   **d.** CH$_3$CHOCH$_2$CH$_2$CHCH$_3$ **f.** CH$_3$CHOCHCH$_2$CH$_2$CH$_3$
                                                                 | | | |
                                                                 CH$_3$ CH$_3$ CH$_3$ CH$_3$

26. Using the given starting material, any necessary inorganic reagents, and any carbon-containing compounds with no more than two carbon atoms, indicate how the following syntheses could be carried out:
   a. [cyclohexanol] ⟶ [cyclohexene]

   b. CH$_3$CH$_2$C≡CH ⟶ CH$_3$CH$_2$C≡CCH$_2$CH$_3$

   c. CH$_3$CH$_2$C≡CH ⟶ CH$_3$CH$_2$C≡CCH$_2$CH$_2$OH

27. Draw structures for the following:
   a. diisopropyl ether   **c.** *sec*-butyl isobutyl ether
   b. allyl vinyl ether   **d.** benzyl phenyl ether

28. If any of the ethers in Problem 27 can exist as stereoisomers, draw the stereoisomers.

29. Give the product of each of the following reactions:

   a. [spiro epoxide on cyclohexane] $\xrightarrow[\text{CH}_3\text{OH}]{\text{CH}_3\text{O}^-}$

   b. CH$_3$COCH$_2$CH$_3$ + HBr $\xrightarrow{\Delta}$
              |
              CH$_3$
         (with CH$_3$ on top)

   c. CH$_3$CH$_2$CHCHCH$_3$ $\xrightarrow[\Delta]{\text{H}_2\text{CrO}_4}$
              | |
              OH CH$_3$
         (CH$_3$ on top)

   d. CH$_3$CHCH$_2$OCH$_3$ + HI $\xrightarrow{\Delta}$
              |
              CH$_3$

   e. [spiro epoxide on cyclohexane] $\xrightarrow[\text{CH}_3\text{OH}]{\text{H}^+}$

   f. CH$_3$CH$_2$CHOCCH$_3$ + HI $\xrightarrow{\Delta}$
              | |
              CH$_3$ CH$_3$
         (CH$_3$ on top)

g. 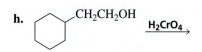 $\xrightarrow{\text{H}_2\text{CrO}_4}$

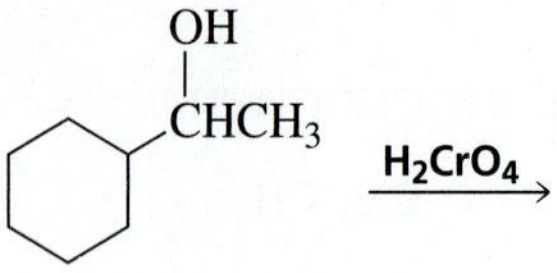

h. $\xrightarrow{\text{H}_2\text{CrO}_4}$

30. Draw structures for the following:
  a. *trans*-4-methylcyclohexanol
  b. 3-ethoxy-1-propanol

31. Give the product formed from the reaction of each of the following compounds with chromic acid:
  a. 3-methyl-2-pentanol
  b. butanol
  c. 2-methylcyclohexanol

32. Which alcohol in each pair will undergo dehydration more rapidly when heated with H_2SO_4?

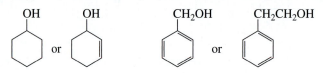

33. Propose a mechanism for the following reaction:

$$\text{CH}_3\text{CHCH}-\text{CH}_2 \ + \ \text{CH}_3\text{O}^- \ \xrightarrow{\text{CH}_3\text{OH}} \ \text{CH}_3\text{CH}-\text{CHCH}_2\text{OCH}_3 \ + \ \text{Cl}^-$$

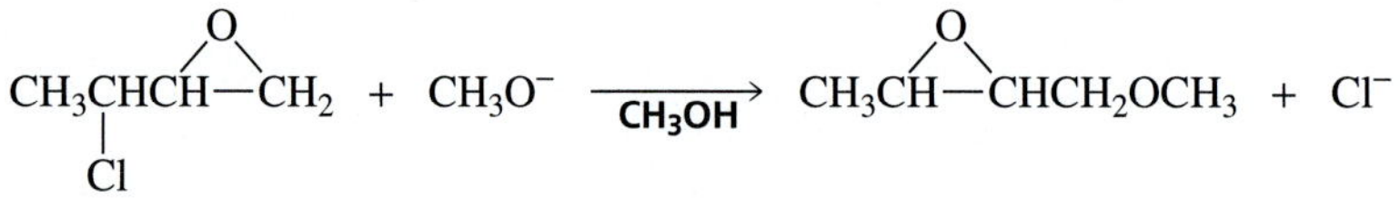

34. The observed relative reactivities of primary, secondary, and tertiary alcohols with a hydrogen halide are 3° > 2° > 1°. If a secondary alcohol underwent an S_N2 reaction rather than an S_N1 reaction with a hydrogen halide, what would be the relative reactivities of the three classes of alcohols?

35. Give the major product expected from the reaction of 1,2-epoxybutane with each of the following reagents:
  a. 0.1 M HCl
  b. CH_3OH/H^+
  c. CH_3OH/CH_3O^-
  d. 0.1 M NaOH

36. Name each of the following:

a.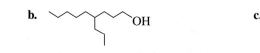

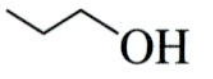

b.

c.

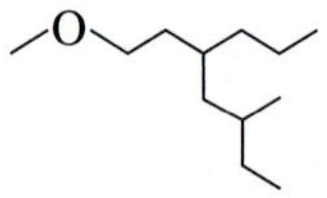

37. When ethyl ether is heated with excess HI for several hours, the only organic product obtained is ethyl iodide. Explain why ethyl alcohol is not obtained as a product.

38. Ethylene oxide reacts readily with HO^- because of the strain in the three-membered ring. Explain why cyclopropane, with approximately the same amount of strain, does not react with HO^-.

39. Propose a mechanism for each of the following reactions:

  a. 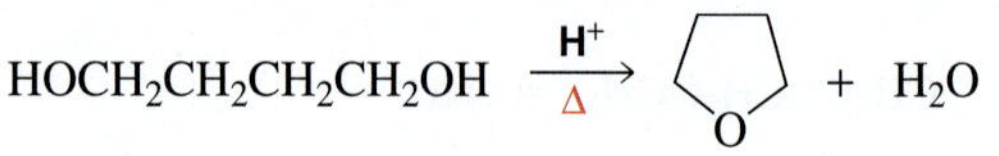 $\text{HOCH}_2\text{CH}_2\text{CH}_2\text{CH}_2\text{OH} \ \xrightarrow[\Delta]{\text{H}^+} \ $ + H_2O

  b. 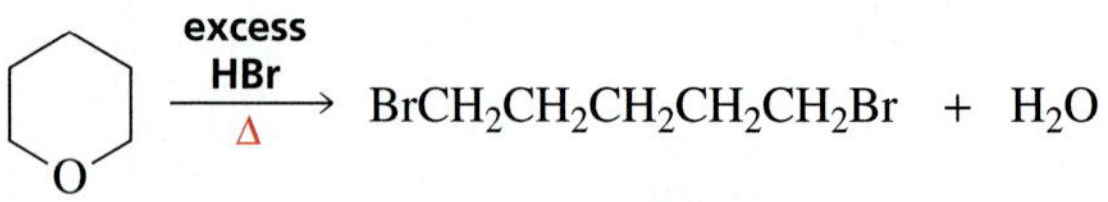 $\xrightarrow[\Delta]{\substack{\textbf{excess} \\ \textbf{HBr}}} \ \text{BrCH}_2\text{CH}_2\text{CH}_2\text{CH}_2\text{CH}_2\text{Br} \ + \ H_2O$

40. Triethylene glycol is one of the products obtained from the reaction of ethylene oxide and hydroxide ion. Propose a mechanism for its formation.

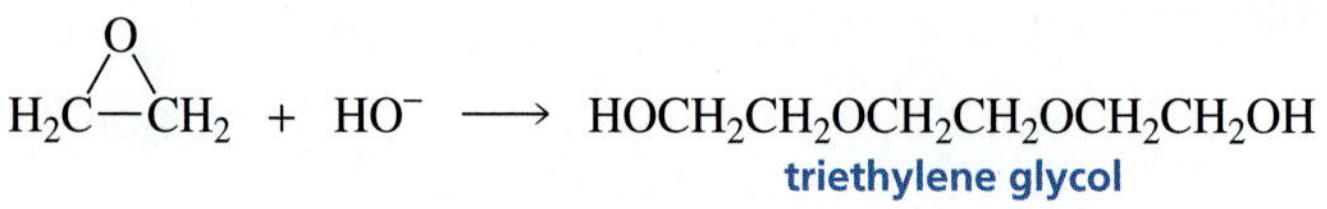

$$\text{H}_2\text{C}-\text{CH}_2 \ + \ \text{HO}^- \ \longrightarrow \ \text{HOCH}_2\text{CH}_2\text{OCH}_2\text{CH}_2\text{OCH}_2\text{CH}_2\text{OH}$$
$$\text{triethylene glycol}$$

41. Explain why the major product obtained from the acid-catalyzed dehydration of 1-butanol is 2-butene.

42. What alkenes would you expect to be obtained from the acid-catalyzed dehydration of 1-hexanol?

43. Propose a mechanism for the following reaction:

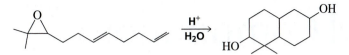

44. Which of the following ethers would be obtained in greatest yield directly from alcohols?

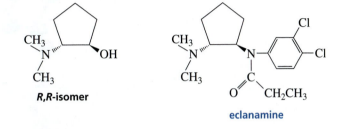

$CH_3OCH_2CH_2CH_3$ $CH_3CH_2OCH_2CH_2CH_3$ $CH_3CH_2OCH_2CH_3$

45. Explain why (S)-2-butanol forms a racemic mixture when it is heated in sulfuric acid.

46. Two stereoisomers are obtained from the reaction of cyclopentene oxide and dimethylamine. The R,R-isomer is used in the manufacture of eclanamine, an antidepressant. What other stereoisomer is obtained?

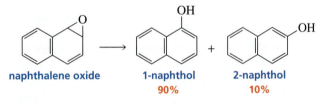

47. Explain why more 1-naphthol than 2-naphthol is obtained from the rearrangement of naphthalene oxide.

48. Three arene oxides can be obtained from phenanthrene.

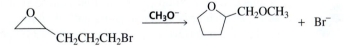

phenanthrene

  a. Give the structures of the three phenanthrene oxides.
  b. What phenols can be obtained from each phenanthrene oxide?
  c. If a phenanthrene oxide can lead to the formation of more than one phenol, which phenol is obtained in greater yield?
  d. Which of the three phenanthrene oxides is most likely to be carcinogenic?

49. a. Propose a mechanism for the following reaction:

b. A small amount of a product containing a six-membered ring is also formed. Give the structure of that product.
  c. Why is so little six-membered ring product formed?

50. Show how each of the following compounds could be prepared from bromocyclohexane.

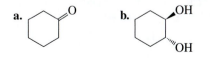

CHAPTER 11

Carbonyl Compounds I
Nucleophilic Acyl Substitution

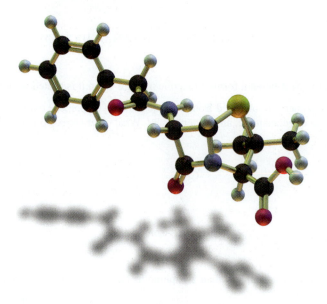

Penicillin G

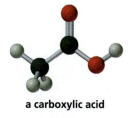

a carboxylic acid

an acyl chloride

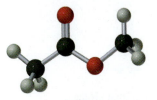

an ester

an amide

The **carbonyl group**—a carbon double bonded to an oxygen—is probably the most important functional group. Compounds containing carbonyl groups—called **carbonyl compounds**—are abundant in nature. Many play important roles in biological processes. Hormones, vitamins, amino acids, proteins, drugs, and flavorings are just a few of the carbonyl compounds that affect us daily.

An **acyl group** consists of a carbonyl group attached to an alkyl group (R) or an aryl group (Ar).

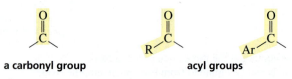

a carbonyl group   acyl groups

The group (or atom) attached to the acyl group strongly affects the reactivity of the carbonyl compound. In fact, carbonyl compounds can be divided into two classes determined by that group: Class I carbonyl compounds are those in which the acyl group is attached to a group (or atom) that *can* be replaced by another group. Carboxylic acids, acyl chlorides, esters, and amides belong to this class. All of these compounds contain a group (OH, Cl, OR, NH_2, NHR, NR_2) that can be replaced by a nucleophile.

carbonyl compounds with groups that can be replaced by a nucleophile

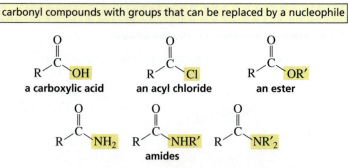

a carboxylic acid   an acyl chloride   an ester

amides

Acyl chlorides, esters, and amides are all called **carboxylic acid derivatives** because they differ from a carboxylic acid only in the nature of the group that has replaced the OH group of the carboxylic acid.

Class II carbonyl compounds are those in which the acyl group is attached to a group that *cannot* be readily replaced by another group. Aldehydes and ketones belong to this class. The H bonded to the acyl group of an aldehyde and the R bonded to the acyl group of a ketone cannot be readily replaced by a nucleophile.

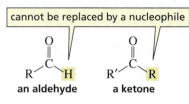

This chapter discusses the reactions of Class I carbonyl compounds. We will see that these compounds undergo substitution reactions, because they have an acyl group attached to a group that can be replaced (substituted) by a nucleophile. The reactions of Class II carbonyl compounds—aldehydes and ketones—will be considered in Chapter 12, where we will see that these compounds *do not* undergo substitution reactions because their acyl group is attached to a group that *cannot* be replaced by a nucleophile.

11.1 THE NOMENCLATURE OF CARBOXYLIC ACIDS AND CARBOXYLIC ACID DERIVATIVES

We will look at the names of carboxylic acids first because their names form the basis of the names of other carbonyl compounds.

Naming Carboxylic Acids

The functional group of a carboxylic acid is called a **carboxyl group**.

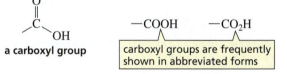

In systematic (IUPAC) nomenclature, a **carboxylic acid** is named by replacing the terminal "e" of the alkane name with "oic acid." For example, the one-carbon alkane is methan*e*, so the one-carbon carboxylic acid is methan*oic acid*.

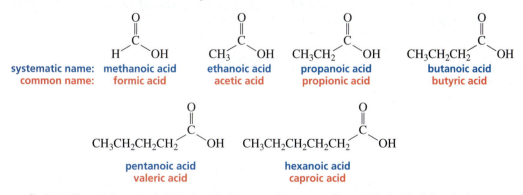

Carboxylic acids containing six or fewer carbons are frequently called by their common names. These names were chosen by early chemists to describe some feature of the compound, usually its origin. For example, formic acid is found in ants, bees, and other stinging insects; its name comes from *formica*, which is Latin for "ant." Acetic acid—contained in vinegar—got its name from *acetum*, the Latin word for

"vinegar." Propionic acid is the smallest acid that shows some of the characteristics of the larger fatty acids (Section 19.1); its name comes from the Greek words *pro* ("the first") and *pion* ("fat"). Butyric acid is found in rancid butter; the Latin word for "butter" is *butyrum*. Caproic acid is found in goat's milk; if you have the occasion to smell both a goat and caproic acid, you will find that they have similar odors. *Caper* is the Latin word for "goat."

In systematic (IUPAC) nomenclature, the position of a substituent is designated by a number. The carbonyl carbon is always the C-1 carbon. In common nomenclature, the position of a substituent is designated by a lowercase Greek letter, and the carbonyl carbon is not given a designation. The carbon adjacent to the carbonyl carbon is the **α-carbon**, the carbon adjacent to the α-carbon is the β-carbon, and so on.

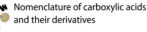

α = alpha
β = beta
γ = gamma
δ = delta
ε = epsilon

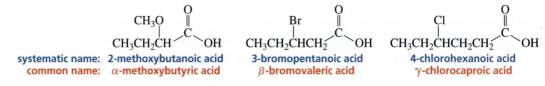

Tutorial:
Nomenclature of carboxylic acids and their derivatives

Take a careful look at the following examples to make sure that you understand the difference between systematic and common nomenclature:

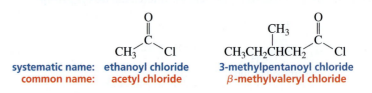

systematic name: **2-methoxybutanoic acid**   **3-bromopentanoic acid**   **4-chlorohexanoic acid**
common name: **α-methoxybutyric acid**   **β-bromovaleric acid**   **γ-chlorocaproic acid**

Naming Acyl Chlorides

Acyl chlorides have a Cl in place of the OH group of a carboxylic acid. They are named by taking the acid name and replacing "ic acid" with "yl chloride."

systematic name: **ethanoyl chloride**   **3-methylpentanoyl chloride**
common name: **acetyl chloride**   **β-methylvaleryl chloride**

Naming Esters

An **ester** has an OR group in place of the OH group of a carboxylic acid. In naming an ester, the name of the group (R) attached to the **carboxyl oxygen** is stated first, followed by the name of the acid with "ic acid" replaced by "ate."

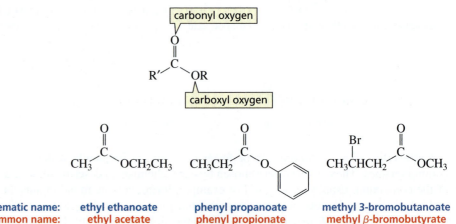

carbonyl oxygen

carboxyl oxygen

systematic name: **ethyl ethanoate**   **phenyl propanoate**   **methyl 3-bromobutanoate**
common name: **ethyl acetate**   **phenyl propionate**   **methyl β-bromobutyrate**

Salts of carboxylic acids are named in the same way. The cation is named first, followed by the name of the acid with "ic acid" replaced by "ate."

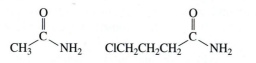

systematic name: sodium methanoate	**potassium ethanoate**
common name: sodium formate	**potassium acetate**

Naming Amides

An **amide** has an NH_2, NHR, or NR_2 group in place of the OH group of a carboxylic acid. Amides are named by taking the acid name and replacing "oic acid" or "ic acid" with "amide."

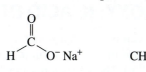

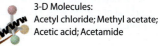
3-D Molecules:
Acetyl chloride; Methyl acetate;
Acetic acid; Acetamide

systematic name: ethanamide	**4-chlorobutanamide**
common name: acetamide	**γ-chlorobutyramide**

If a substituent is bonded to the nitrogen, the name of the substituent is stated first (if there is more than one substituent bonded to the nitrogen, they are stated alphabetically), followed by the name of the amide. The name of each substituent is preceded by a capital *N* to indicate that the substituent is bonded to a nitrogen.

3-D Molecule:
N-Methylbenzamide

N-cyclohexylpropanamide   *N*-ethyl-*N*-methylpentanamide   *N,N*-diethylbutanamide

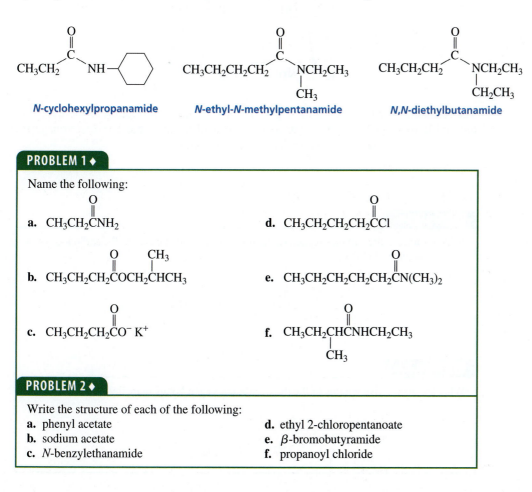

PROBLEM 1 ◆

Name the following:

a. $CH_3CH_2\overset{O}{\overset{\|}{C}}NH_2$

b. $CH_3CH_2CH_2\overset{O}{\overset{\|}{C}}OCH_2\overset{CH_3}{\overset{|}{C}}HCH_3$

c. $CH_3CH_2CH_2\overset{O}{\overset{\|}{C}}O^-\ K^+$

d. $CH_3CH_2CH_2CH_2\overset{O}{\overset{\|}{C}}Cl$

e. $CH_3CH_2CH_2CH_2CH_2\overset{O}{\overset{\|}{C}}N(CH_3)_2$

f. $CH_3CH_2\overset{O}{\overset{|}{C}}H\overset{\|}{C}NHCH_2CH_3$
     $\overset{|}{C}H_3$

PROBLEM 2 ◆

Write the structure of each of the following:
a. phenyl acetate
b. sodium acetate
c. *N*-benzylethanamide

d. ethyl 2-chloropentanoate
e. β-bromobutyramide
f. propanoyl chloride

11.2 THE STRUCTURES OF CARBOXYLIC ACIDS AND CARBOXYLIC ACID DERIVATIVES

The **carbonyl carbon** in carboxylic acids and carboxylic acid derivatives is sp^2 hybridized. It uses its three sp^2 orbitals to form σ bonds to the carbonyl oxygen, the α-carbon, and a substituent (Y). The three atoms attached to the carbonyl carbon are in the same plane, and the bond angles are each approximately 120°.

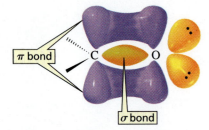

▲ **Figure 11.1**
Bonding in a carbonyl group. The π bond is formed by side-to-side overlap of a p orbital of carbon with a p orbital of oxygen.

The **carbonyl oxygen** is also sp^2 hybridized. One of its sp^2 orbitals forms a σ bond with the carbonyl carbon, and each of the other two sp^2 orbitals contains a lone pair. The remaining p orbital of the carbonyl oxygen overlaps the remaining p orbital of the carbonyl carbon to form a π bond (Figure 11.1).

Esters, carboxylic acids, and amides each have two major resonance contributors.

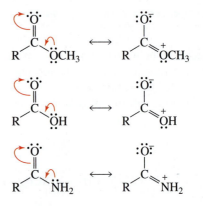

The resonance contributor on the right makes a greater contribution to the hybrid in the amide than in the ester or the carboxylic acid because the less electronegative nitrogen atom can better accommodate a positive charge.

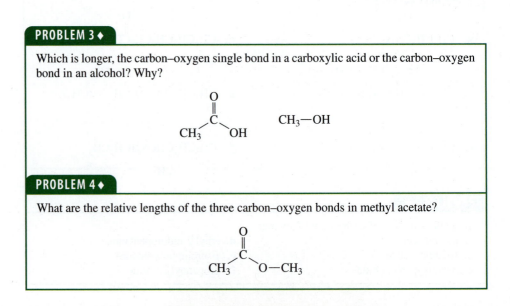

PROBLEM 3 ♦

Which is longer, the carbon–oxygen single bond in a carboxylic acid or the carbon–oxygen bond in an alcohol? Why?

PROBLEM 4 ♦

What are the relative lengths of the three carbon–oxygen bonds in methyl acetate?

11.3 **THE PHYSICAL PROPERTIES OF CARBONYL COMPOUNDS**

The acid properties of carboxylic acids were discussed in Sections 2.3 and 7.9. Recall that carboxylic acids have pK_a values of approximately 3–5 (Appendix II). The boiling points and other physical properties of carbonyl compounds are listed in Appendix I. Carbonyl compounds have the following relative boiling points:

relative boiling points

amide > carboxylic acid $\gg$ ester ~ acyl chloride ~ aldehyde ~ ketone

The boiling points of an ester, acyl chloride, ketone, and aldehyde are similar and are lower than the boiling point of an alcohol with a comparable molecular weight because only the alcohol molecules can form hydrogen bonds with each other. The boiling points of these four carbonyl compounds are higher than the boiling point of the same-size ether because of the polar carbonyl group.

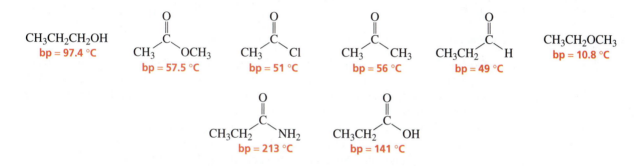

Carboxylic acids have relatively high boiling points because each molecule can form two hydrogen bonds.

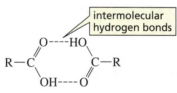

Amides have the highest boiling points, because they have strong dipole–dipole interactions, since the resonance contributor with separated charges contributes significantly to the overall structure of the compound (Section 11.2). In addition, if the nitrogen of an amide is bonded to a hydrogen, hydrogen bonds will form between the molecules.

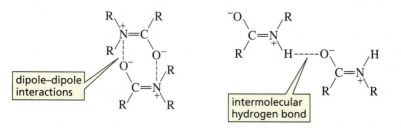

Carboxylic acid derivatives are soluble in solvents such as ethers, chlorinated alkanes, and aromatic hydrocarbons. Like alcohols and ethers, carbonyl compounds with fewer than about four carbons per oxygen atom are soluble in water (Section 3.7).

11.4 CARBOXYLIC ACIDS AND CARBOXYLIC ACID DERIVATIVES FOUND IN NATURE

Acyl halides are much more reactive than carboxylic acids and esters, which, in turn, are more reactive than amides. We will see the reason for these differences in reactivity in Section 11.6.

Because of their high reactivity, acyl halides are not found in nature. Carboxylic acids, on the other hand, being less reactive *are* found widely in nature. For example, glucose is metabolized to pyruvic acid (Section 18.4). (*S*)-(+)-Lactic acid is the compound responsible for the burning sensation felt in muscles during anaerobic exercise, and it is also found in sour milk. Spinach and other leafy green vegetables are rich in oxalic acid. Succinic acid and citric acid are important intermediates in the citric acid cycle, a series of reactions in biological systems that oxidize carbohydrates, fatty acids, and amino acids to CO_2 (Section 18.7). Citrus fruits are rich in citric acid; the concentration is greatest in lemons, less in grapefruit, and still less in oranges.

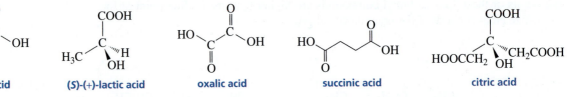

pyruvic acid   (*S*)-(+)-lactic acid   oxalic acid   succinic acid   citric acid

(*S*)-(-)-Malic acid is responsible for the sharp taste of unripe apples and pears. As these fruits ripen, the amount of malic acid decreases and the amount of sugar increases. The inverse relationship between the levels of malic acid and sugar is important for the propagation of the plant: malic acid prevents animals from eating the fruit until it becomes ripe, at which time its seeds are mature enough to germinate when they are scattered about. Prostaglandins are locally acting hormones that have several different physiological functions (Section 11.9), such as regulating inflammation, blood pressure, pain, and fever.

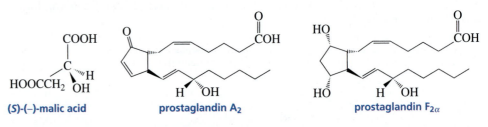

(*S*)-(–)-malic acid   prostaglandin A$_2$ prostaglandin F$_{2\alpha}$

Esters are also commonly found in nature. The aromas of many flowers and fruits are due to esters. (See Problems 19 and 30.)

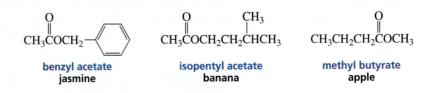

benzyl acetate
jasmine

isopentyl acetate
banana

methyl butyrate
apple

Carboxylic acids with an amino group on the α-carbon are commonly called **amino acids**. Amino acids are linked together by amide bonds to form peptides and proteins (Section 16.0).

an amino acid

general structure for a peptide or a protein

Caffeine, another naturally occurring amide, is found in cocoa and in coffee beans. Penicillin G has two amide bonds; the four-membered ring amide is the reactive part of the molecule (Section 11.13).

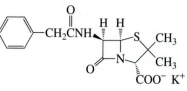

caffeine	**piperine** **the major component of black pepper**	**penicillin G**

THE DISCOVERY OF PENICILLIN

Sir Alexander Fleming (1881–1955), born in Scotland, was a professor of bacteriology at University College, London. The story is told that one day Fleming was about to throw away a culture of staphylococcal bacteria that had been contaminated by a rare strain of the mold *Penicillium notatum*. He noticed that the bacteria had disappeared wherever there was a particle of mold. This suggested to him that the mold must have produced an antibacterial substance. Ten years later, Howard Florey and Ernest Chain isolated the active substance—penicillin G (Section 11.13)—but the delay allowed sulfa drugs to be the first antibiotics (Section 21.4).

After penicillin G was found to cure bacterial infections in mice, it was used successfully in 1941 on nine cases of human bacterial infections. By 1943, penicillin G was being produced for the military and was first used for war casualties in Sicily and Tunisia. The drug became available to the civilian population in 1944. The pressure of the war made determination of penicillin G's structure a priority because once its structure was determined, large quantities of the drug could conceivably be synthesized.

Fleming, Florey, and Chain shared the 1945 Nobel Prize in physiology or medicine. Chain also discovered penicillinase, the enzyme that destroys penicillin (Section 11.13). Although Fleming is generally given credit for the discovery of penicillin, there is clear evidence that the germicidal activity of the mold was recognized in the nineteenth century by Lord Joseph Lister (1827–1912), the English physician renowned for the introduction of aseptic surgery.

Sir Alexander Fleming (1881–1955) *was born in Scotland, the seventh of eight children of a farmer. In 1902, he received a legacy from an uncle that, together with a scholarship, allowed him to study medicine at the University of London. He subsequently became a professor there in 1928. He was knighted in 1944.*

Sir Howard W. Florey (1898–1968) *was born in Australia and received a medical degree from the University of Adelaide. He went to England as a Rhodes Scholar and studied at both Oxford and Cambridge Universities. He became a professor of pathology at the University of Sheffield in 1931 and then at Oxford in 1935. Knighted in 1944, he was given a peerage in 1965 that made him Baron Florey of Adelaide.*

Ernest B. Chain (1906–1979) *was born in Germany and received a Ph.D. from Friedrich-Wilhelm University in Berlin. In 1933, he left Germany for England because Hitler had come to power. He studied at Cambridge, and in 1935 Florey invited him to Oxford. In 1948, he became the director of an institute in Rome, but he returned to England in 1961 to become a professor at the University of London.*

DALMATIANS: DON'T TRY TO FOOL MOTHER NATURE

When amino acids are metabolized, the excess nitrogen is concentrated into uric acid, a compound with five amide bonds. A series of enzyme-catalyzed reactions degrades uric acid all the way to ammonium ion. The extent to which uric acid is degraded in animals depends on the species. Birds, reptiles, and insects excrete excess nitrogen as uric acid. Mammals excrete excess nitrogen as allantoin. Excess nitrogen in aquatic animals is excreted as allantoic acid, urea, or ammonium salts.

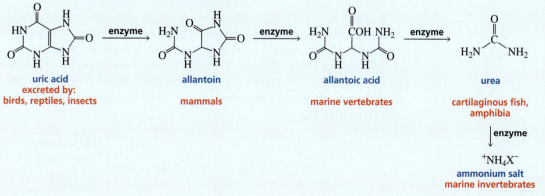

uric acid **excreted by:** **birds, reptiles, insects**	**allantoin** **mammals**	**allantoic acid** **marine vertebrates**	**urea** **cartilaginous fish,** **amphibia**

$^+NH_4X^-$
ammonium salt
marine invertebrates

(continued...)

Dalmatians, unlike other mammals, excrete high levels of uric acid. This is because breeders of Dalmatians have selected dogs that have no white hairs in their black spots, and the gene that causes the white hairs is linked to the gene that causes uric acid to be converted to allantoin. Dalmatians, therefore, are susceptible to gout, a painful buildup of uric acid in joints.

11.5 HOW CARBOXYLIC ACIDS AND CARBOXYLIC ACIDS COMPOUNDS REACT

The reactivity of carbonyl compounds is due to the polarity of the carbonyl group that results from oxygen being more electronegative than carbon. The carbonyl carbon is, therefore, electron deficient (an electrophile), so we can safely predict that it will be attacked by nucleophiles.

When a nucleophile attacks the carbonyl carbon of a carboxylic acid derivative, the weakest bond in the molecule—the carbon–oxygen π bond—breaks, and an intermediate is formed. The intermediate is called a **tetrahedral intermediate** because the sp^2 carbon in the reactant has become an sp^3 carbon in the intermediate. Generally, *a compound that has an sp^3 carbon bonded to an oxygen atom will be unstable if the sp^3 carbon is bonded to another electronegative atom.* The tetrahedral intermediate, therefore, is unstable because Y and Z are both electronegative atoms. A lone pair on the oxygen reforms the π bond, and either Y^- or Z^- is expelled with its bonding electrons.

> **A compound that has an sp^3 carbon bonded to an oxygen atom generally will be unstable if the sp^3 carbon is bonded to a second electronegative atom.**

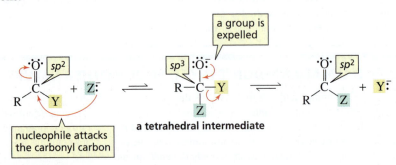

a tetrahedral intermediate

Whether Y^- or Z^- is expelled from the tetrahedral intermediate depends on their relative basicities. The weaker base is expelled preferentially, making this another example of the principle we first saw in Section 9.3: *the weaker the base, the better it is as a leaving group.* Because a weak base does not share its electrons as well as a strong base does, a weaker base forms a weaker bond—one that is easier to break. If Z^- is a much weaker base than Y^-, Z^- will be expelled and the reaction can be written as follows:

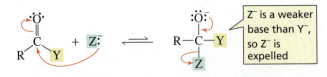

In this case, no new product is formed. The nucleophile attacks the carbonyl carbon, but the tetrahedral intermediate expels the attacking nucleophile and reforms the reactants.

On the other hand, if Y^- is a much weaker base than Z^-, Y^- will be expelled and a new product will be formed.

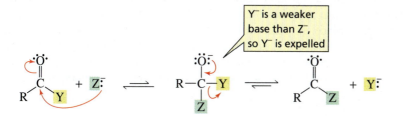

This reaction is called a **nucleophilic acyl substitution reaction** because a nucleophile (Z^-) has replaced the substituent (Y^-) that was attached to the acyl group in the reactant.

Movie:
Nucleophilic acyl substitution

If the basicities of Y^- and Z^- are similar, some molecules of the tetrahedral intermediate will expel Y^- and others will expel Z^-. When the reaction is over, both reactant and product will be present.

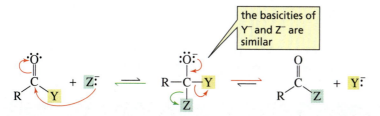

We can, therefore, make the following general statement about the reactions of carboxylic acid derivatives: *a carboxylic acid derivative will undergo a nucleophilic acyl substitution reaction, provided that the newly added group in the tetrahedral intermediate is not a weaker base than the group that is attached to the acyl group in the reactant.*

Tutorial:
Free energy diagrams for nucleophilic acyl substitution reactions

PROBLEM-SOLVING STRATEGY

The pK_a of HCl is -7; the pK_a of CH_3OH is 15.5. What is the product of the reaction of acetyl chloride with CH_3O^-?

In order to determine what the product of the reaction will be, we need to compare the basicities of the two groups that will be in the tetrahedral intermediate in order to see which one will be eliminated. Because HCl is a stronger acid than CH_3OH, Cl^- is a weaker base than CH_3O^-. Cl^-, therefore, will be eliminated from the tetrahedral intermediate, so the product of the reaction will be methyl acetate.

$$\underset{\textbf{acetyl chloride}}{\overset{O}{\underset{CH_3}{\overset{\|}{C}}Cl}} + CH_3O^- \longrightarrow CH_3-\overset{O^-}{\underset{OCH_3}{\overset{|}{\underset{|}{C}}}}-Cl \longrightarrow \underset{\textbf{methyl acetate}}{\overset{O}{\underset{CH_3}{\overset{\|}{C}}OCH_3}} + Cl^-$$

Now continue on to Problem 5.

PROBLEM 5

a. The pK_a of HCl is -7; the pK_a of H_2O is 15.7. What is the product of the reaction of acetyl chloride with HO^-?

b. The pK_a of NH_3 is 36; the pK_a of H_2O is 15.7. What is the product of the reaction of acetamide with HO^-?

11.6 RELATIVE REACTIVITIES OF CARBOXYLIC ACIDS AND CARBOXYLIC ACID DERIVATIVES

We have just seen that there are two steps in a nucleophilic acyl substitution reaction: formation of a tetrahedral intermediate and collapse of the tetrahedral intermediate. The weaker the base attached to the acyl group, the easier it is for *both steps* of the reaction to take place. In other words, the reactivity of a carboxylic acid derivative depends on the basicity of the substituent attached to the acyl group: the less basic the substituent, the more reactive is the carboxylic acid derivative. (The pK_a values of the conjugate acids of the leaving groups of Class I carbonyl compounds are shown in Table 11.1.)

The weaker the base, the better it is as a leaving group.

relative basicities of the leaving groups

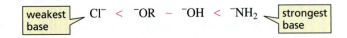

relative reactivities of carboxylic acid derivatives

relative reactivity: acyl chloride > ester ~ carboxylic acid > amide

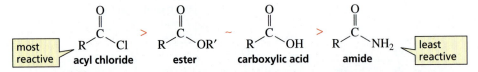

Table 11.1 The pK_a Values of the Conjugate Acids of the Leaving Groups of Carbonyl Compounds			
Carbonyl compound	Leaving group	Conjugate acid of the leaving group	pK_a
Class I			
$R-\overset{\overset{O}{\|}}{C}-Cl$	Cl^-	HCl	−7
$R-\overset{\overset{O}{\|}}{C}-OR'$	$^-OR'$	R'OH	~15–16
$R-\overset{\overset{O}{\|}}{C}-OH$	^-OH	H_2O	15.7
$R-\overset{\overset{O}{\|}}{C}-NH_2$	$^-NH_2$	NH_3	36
Class II			
$R-\overset{\overset{O}{\|}}{C}-H$	H^-	H_2	~40
$R-\overset{\overset{O}{\|}}{C}-R$	R^-	RH	~60

How does having a weak base attached to the acyl group make the *first* step of the nucleophilic acyl substitution reaction easier? A weaker base is a more electronegative base (Section 2.6). Therefore, it is better at withdrawing electrons inductively from the carbonyl carbon, and electron withdrawal increases the carbonyl carbon's susceptibility to nucleophilic attack.

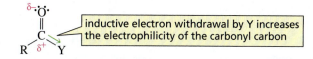

inductive electron withdrawal by Y increases the electrophilicity of the carbonyl carbon

A weak base attached to the acyl group also makes the *second* step of the nucleophilic acyl substitution reaction easier because weak bases—which form weak bonds—are easier to expel when the tetrahedral intermediate collapses.

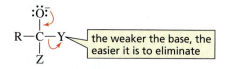

the weaker the base, the easier it is to eliminate

In Section 11.5, we saw that in a nucleophilic acyl substitution reaction, the nucleophile that forms the tetrahedral intermediate must not be a weaker base than the base that is already there. This means that *a carboxylic acid derivative can be converted into a less reactive carboxylic acid derivative, but not into one that is more reactive.* For example, an acyl chloride can be converted into an ester because an alkoxide ion, such as methoxide ion, is a stronger base than a chloride ion.

> For a carboxylic acid derivative to undergo a nucleophilic acyl substitution reaction, the incoming nucleophile must not be a much weaker base than the group that is to be replaced.

$$
\underset{R}{\overset{O}{\underset{\|}{C}}}\!\!-\!Cl \;+\; CH_3O^- \;\longrightarrow\; \underset{R}{\overset{O}{\underset{\|}{C}}}\!\!-\!OCH_3 \;+\; Cl^-
$$

An ester, however, cannot be converted into an acyl chloride because a chloride ion is a weaker base than an alkoxide ion.

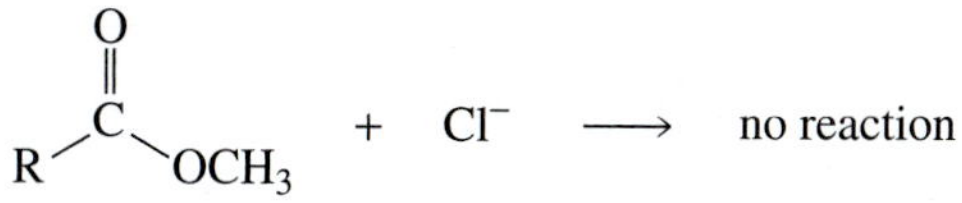

$$
\underset{R}{\overset{O}{\underset{\|}{C}}}\!\!-\!OCH_3 \;+\; Cl^- \;\longrightarrow\; \text{no reaction}
$$

PROBLEM 6 ◆

What will be the product of a nucleophilic acyl substitution reaction—a new carboxylic acid derivative, a mixture of two carboxylic acid derivatives, or no reaction—if the new group in the tetrahedral intermediate is the following?
a. a stronger base than the group that is already there
b. a weaker base than the group that is already there
c. similar in basicity to the group that is already there

PROBLEM 7 ◆

Using the pK_a values in Table 11.1, predict the products of the following reactions:

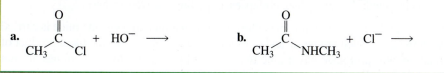

11.7 THE REACTIONS OF ACYL CHLORIDES

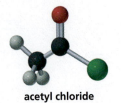

acetyl chloride

3-D Molecule:
Benzoyl chloride

Acyl chlorides react with alcohols to form esters, with water to form carboxylic acids, and with amines to form amides because in each case the incoming nucleophile is a stronger base than the departing chloride ion.

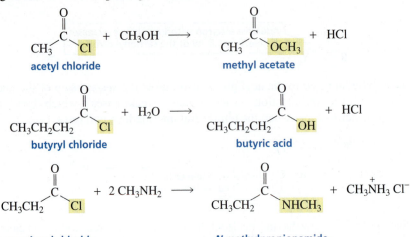

All carboxylic acid derivatives undergo nucleophilic acyl substitution reactions by one of the two following mechanisms. The only difference in the two mechanism is whether the nucleophile is neutral or charged.

mechanism for the conversion of an acyl chloride into an ester (with a neutral nucleophile)

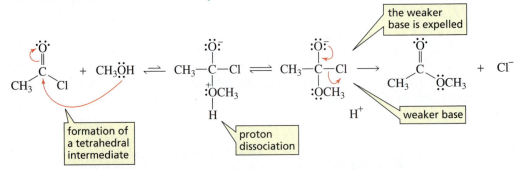

- The nucleophile attacks the carbonyl carbon, forming a tetrahedral intermediate.
- The proton dissociates before the tetrahedral intermediate collapses.
- The tetrahedral intermediate collapses, expelling the weaker base.

If the nucleophile is negatively charged, there is no need for the proton dissociation step.

mechanism for the conversion of an acyl chloride into an ester (with a negatively charged nucleophile)

The tetrahedral intermediate eliminates the weakest base.

- The nucleophile attacks the carbonyl carbon, forming a tetrahedral intermediate.
- The tetrahedral intermediate collapses, expelling the weaker base.

Notice that the reaction of an acyl chloride with an amine to form an amide is carried out with twice as much amine as acyl chloride, because the HCl formed as a product of the reaction will protonate any amine that has yet to react. Protonated amines are not nucleophiles, so they cannot react with the acyl chloride. Using twice as much amine guarantees that there is enough unprotonated amine to react with all the acyl chloride.

PROBLEM 8   SOLVED

Two amides are obtained from the reaction of acetyl chloride with a mixture of ethylamine and propylamine. Identify the amides.

Solution   Either of the amines can react with acetyl chloride, so both *N*-ethylacetamide and *N*-propylacetamide are formed.

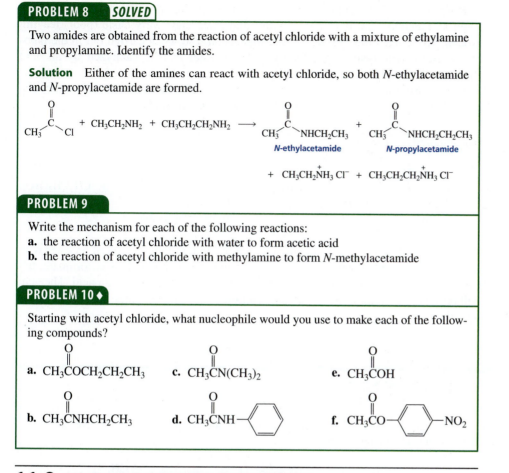

PROBLEM 9

Write the mechanism for each of the following reactions:
a. the reaction of acetyl chloride with water to form acetic acid
b. the reaction of acetyl chloride with methylamine to form *N*-methylacetamide

PROBLEM 10 ◆

Starting with acetyl chloride, what nucleophile would you use to make each of the following compounds?

a. $CH_3COCH_2CH_2CH_3$ **c.** $CH_3CN(CH_3)_2$ **e.** CH_3COH

b. $CH_3CNHCH_2CH_3$ **d.** CH_3CNH—⬡   **f.** CH_3CO—⬡—NO_2

11.8 THE REACTIONS OF ESTERS

Esters do not react with Cl^- because it is a much weaker base than the RO^- leaving group of the ester (Table 11.1).

An ester reacts with water to form a carboxylic acid and an alcohol. This is an example of a *hydrolysis* reaction. A **hydrolysis reaction** is a reaction with water that converts one compound into two compounds (*lysis* is Greek for "breaking down").

methyl acetate

a hydrolysis reaction

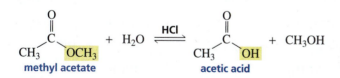

An ester reacts with an alcohol to form a new ester and a new alcohol. This is an example of an **alcoholysis** reaction—a reaction with an alcohol that converts one compound into two compounds. This particular alcoholysis reaction is also called a **transesterification reaction** because one ester is converted to another ester.

a transesterification reaction

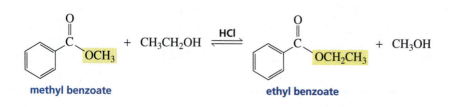

Both the hydrolysis and the alcoholysis of an ester are very slow reactions because water and alcohols are poor nucleophiles and esters have very basic (poor) leaving groups. These reactions are therefore always catalyzed when carried out in the laboratory. Both hydrolysis and alcoholysis of an ester can be catalyzed by an acid (Section 11.9).

Esters also react with amines to form amides. A reaction with an amine that converts one compound into two compounds is called **aminolysis**.

an aminolysis reaction

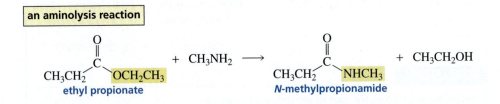

ethyl propionate   *N*-methylpropionamide

The reaction of an ester with an amine is not as slow as the reaction of an ester with water or an alcohol, because an amine is a better nucleophile. This is fortunate, because the reaction of an ester with an amine cannot be catalyzed by an acid. The acid will protonate the amine and a protonated amine is not a nucleophile, so it cannot react with the ester.

NERVE IMPULSES, PARALYSIS, AND INSECTICIDES

After an impulse is transmitted between two nerve cells, an ester called acetylcholine must be rapidly hydrolyzed to enable the recipient cell to receive another impulse.

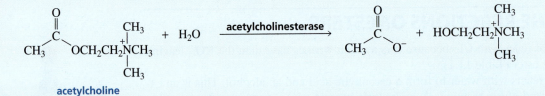

acetylcholine

Acetylcholinesterase, the enzyme that catalyzes this hydrolysis, has a CH$_2$OH group that is necessary for its catalytic activity. Diisopropyl fluorophosphate (DFP), a military nerve gas used during World War II, inactivates acetylcholinesterase by reacting with the CH$_2$OH group. When the enzyme is inactivated, the nerve impulses cannot be transmitted properly, and paralysis occurs. DFP is extremely toxic. Its LD$_{50}$ (the lethal dose for 50% of the test animals) is only 0.5 mg/kg of body weight.

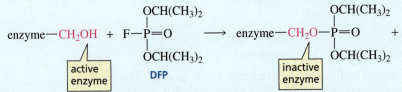

active enzyme   DFP   inactive enzyme

Malathion and parathion, widely used as insecticides, are compounds related to DFP. The LD$_{50}$ of malathion is 2800 mg/kg.

Parathion is more toxic, with an LD$_{50}$ of 2 mg/kg.

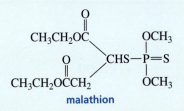

malathion

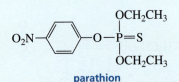

parathion

BIODEGRADABLE POLYMERS

Biodegradable polymers are polymers that can be broken into small segments by enzyme-catalyzed reactions. The enzymes are produced by microorganisms. The carbon–carbon bonds of chain-growth polymers are inert to enzyme-catalyzed reactions, so the polymers are nonbiodegradable unless bonds that can be broken by enzymes are inserted into the polymer. Then when the polymer is buried as waste, microorganisms in the ground can degrade the polymer. One method used to make a polymer biodegradable inserts ester groups into it. For example, if the acetal (Section 12.9) shown below is added to an alkene that is undergoing radical polymerization (Section 5.16), ester groups will be inserted into the polymer, forming "weak links" that are susceptible to enzyme-catalyzed ester hydrolysis.

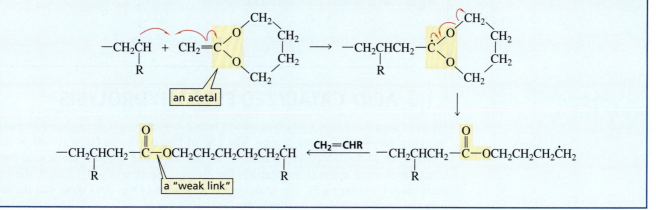

PROBLEM 11

Write the mechanism for each of the following reactions:
a. the noncatalyzed hydrolysis of methyl propionate
b. the aminolysis of phenyl formate, using methylamine

PROBLEM 12   *SOLVED*

a. List the following esters in order of decreasing reactivity toward hydrolysis:

$$CH_3\overset{O}{\overset{\|}{C}}{-}O{-}\langle \text{phenyl} \rangle \qquad CH_3\overset{O}{\overset{\|}{C}}{-}O{-}\langle \text{phenyl} \rangle{-}NO_2 \qquad CH_3\overset{O}{\overset{\|}{C}}{-}O{-}\langle \text{phenyl} \rangle{-}OCH_3$$

b. How would the rate of hydrolysis of the *para*-methylphenyl ester compare with the rates of hydrolysis of these three esters?

Solution to 12a   Both *formation of the tetrahedral intermediate* and *collapse of the tetrahedral intermediate* are fastest for the ester with the electron-withdrawing nitro substituent and slowest for the ester with the electron-donating methoxy substituent.

Formation of the tetrahedral intermediate: An electron-withdrawing substituent increases the ester's susceptibility to nucleophilic attack, and an electron-donating substituent decreases its susceptibility.

Collapse of the tetrahedral intermediate: Electron withdrawal increases acidity and electron donation decreases acidity (Section 8.17). Therefore, *para*-nitrophenol with a strong electron-withdrawing group is a stronger acid than phenol, which in turn is a stronger acid than *para*-methoxyphenol with a strong electron-donating group. Therefore, the *para*-nitrophenoxide ion is the weakest base and the best leaving group of the three, whereas the *para*-methoxyphenoxide ion is the strongest base and the worst leaving group. Thus,

$$CH_3\overset{O}{\overset{\|}{C}}{-}O{-}\langle \text{phenyl} \rangle{-}NO_2 \;>\; CH_3\overset{O}{\overset{\|}{C}}{-}O{-}\langle \text{phenyl} \rangle \;>\; CH_3\overset{O}{\overset{\|}{C}}{-}O{-}\langle \text{phenyl} \rangle{-}OCH_3$$

Solution to 12b   The methyl substituent donates electrons inductively to the benzene ring, but donates electrons to a lesser extent than does the methoxy substituent (Section 8.14).

Therefore, the rate of hydrolysis of the methyl-substituted ester is slower than the rate of hydrolysis of the unsubstituted ester, but faster than the rate of hydrolysis of the methoxy-substituted ester.

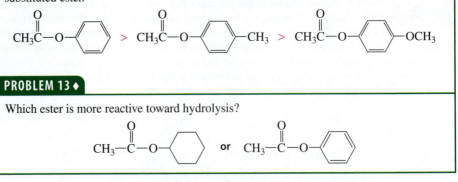

PROBLEM 13 ◆

Which ester is more reactive toward hydrolysis?

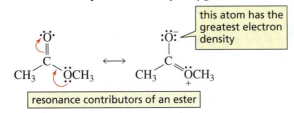

11.9 ACID-CATALYZED ESTER HYDROLYSIS

We have seen that esters hydrolyze slowly because water is a poor nucleophile and esters have very basic leaving groups. The rate of hydrolysis can be increased by acid.

When an acid is added to a reaction, the first thing that happens is the acid protonates the atom in the reactant that has the greatest electron density, that is, the most basic atom (Section 10.2). The resonance contributors of the ester show that the atom with the greatest electron density is the carbonyl oxygen.

The mechanism for acid-catalyzed ester hydrolysis is shown below.

mechanism for acid-catalyzed ester hydrolysis

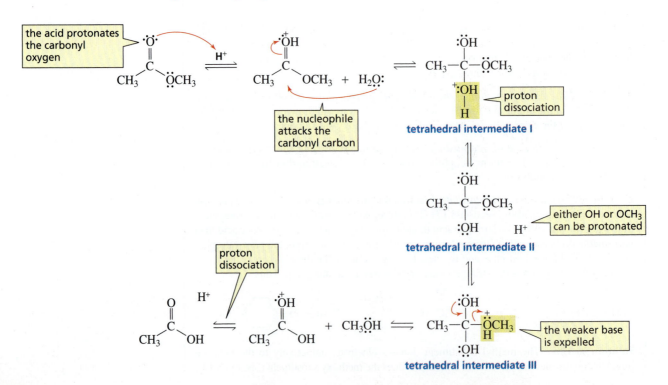

- The acid protonates the carbonyl oxygen.

- The nucleophile (H_2O) attacks the carbonyl carbon of the protonated carbonyl group, forming a protonated tetrahedral intermediate (tetrahedral intermediate I).

- Tetrahedral intermediate I is in equilibrium with its nonprotonated form (tetrahedral intermediate II).

- Once tetrahedral intermediate II has been formed, either the OH or the OCH_3 group of this intermediate I can be protonated, in one case reforming tetrahedral intermediate I (OH is protonated), and in the other case forming tetrahedral intermediate III (OCH_3 is protonated).

- When tetrahedral intermediate I collapses, it expels H_2O in preference to CH_3O^- (because H_2O is a weaker base) and reforms the ester. When tetrahedral intermediate III collapses, it expels CH_3OH rather than HO^- (because CH_3OH is a weaker base) and forms the carboxylic acid.

Because H_2O and CH_3OH have approximately the same basicity, it will be as likely for tetrahedral intermediate I to form and then collapse to reform the ester as it will for tetrahedral intermediate III to form and then collapse to form the carboxylic acid. Consequently, when the reaction has reached equilibrium, both ester and carboxylic acid will be present in approximately equal amounts.

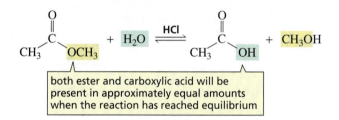

both ester and carboxylic acid will be present in approximately equal amounts when the reaction has reached equilibrium

Now let's see how the acid increases the rate of ester hydrolysis. The acid is a catalyst. Recall that **catalyst** is a substance that increases the rate of a reaction without being consumed or changed in the overall reaction (Section 4.8). For a catalyst to increase the rate of a reaction, it must increase the rate of the slow step of the reaction, because changing the rate of a fast step will not affect the rate of the overall reaction. There are two relatively slow steps in the mechanism: formation of the tetrahedral intermediate and collapse of the tetrahedral intermediate. The acid increases the rates of both slow steps.

The acid increases *the rate of formation of the tetrahedral intermediate* by protonating the carbonyl oxygen. Protonated carbonyl groups are more susceptible than nonprotonated carbonyl groups to nucleophilic attack because a positively charged oxygen is more electron withdrawing than a neutral oxygen. Increased electron withdrawal by the oxygen makes the carbonyl carbon more electron deficient, which increases its attractiveness to nucleophiles.

protonation of the carbonyl oxygen increases the susceptibility of the carbonyl carbon to nucleophilic attack

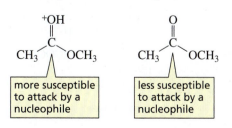

more susceptible to attack by a nucleophile

less susceptible to attack by a nucleophile

An acid catalyst increases the reactivity of a carbonyl group.

An acid catalyst can make a group a better leaving group.

The acid increases *the rate of collapse of the tetrahedral intermediate* by decreasing the basicity of the leaving group, so the group is more easily eliminated. In the acid-catalyzed hydrolysis of an ester, the leaving group is CH_3OH, a weaker base than the leaving group (CH_3O^-) in the uncatalyzed reaction.

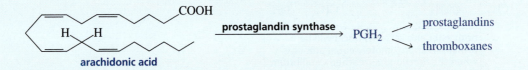

tetrahedral intermediate in acid-catalyzed ester hydrolysis

tetrahedral intermediate in uncatalyzed ester hydrolysis

3-D Molecule:
Aspirin

ASPIRIN

Aspirin, found naturally in willow bark and myrtle leaves, is one of the oldest and most commonly used drugs. As early as the fifth century BC, Hippocrates wrote about the curative powers of willow bark. Even so, aspirin's mode of action was not discovered until 1971, when it was found that the anti-inflammatory and fever-reducing activity of aspirin and related compounds called NSAIDs (nonsteroidal anti-inflammatory agents) were due to a transesterification reaction that blocks prostaglandin synthesis. Prostaglandins have several different biological functions, one being to stimulate inflammation and another to induce fever. The enzyme prostaglandin synthase catalyzes the conversion of arachidonic acid into PGH_2, the precursor of all prostaglandins and the related thromboxanes.

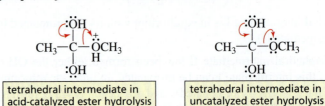

arachidonic acid

Prostaglandin synthase is composed of two enzymes. One of the enzymes—cyclooxygenase—has a CH_2OH group that is necessary for enzymatic activity. The CH_2OH group reacts with aspirin in a transesterification reaction that inactivates the enzyme. When the enzyme is inactivated, prostaglandins cannot be synthesized, and inflammation and fever are suppressed. Because aspirin inhibits the formation of PGH_2, it inhibits the synthesis of thromboxanes, compounds involved in blood clotting. Presumably, this is why low levels of aspirin have been reported to reduce the incidence of strokes and heart attacks that result from blood clot formation. Aspirin's activity as an anticoagulant is why doctors caution patients not to take aspirin for several days before surgery.

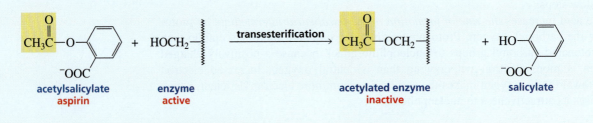

NSAIDs, such as ibuprofen (the active ingredient in Advil, Motrin, and Nuprin) and naproxen (the active ingredient in Aleve), also inhibit the synthesis of prostaglandins.

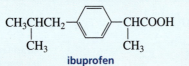

ibuprofen

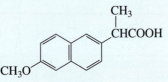

naproxen

(continued...)

Both aspirin and NSAIDs inhibit the synthesis of both the prostaglandins produced under normal physiological conditions and the prostaglandins produced in response to stress. One prostaglandin regulates the production of acid in the stomach, so when prostaglandin synthesis stops, the acidity of the stomach can rise above normal levels. Celebrex, a relatively new drug, inhibits only the enzyme (cyclooxygenase-2) that produces prostaglandin in response to stress. Thus, inflammatory conditions now can be treated without some of the harmful side effects. This drug is known as a COX-2 inhibitor.

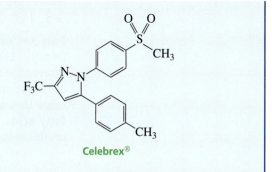

Celebrex®

PROBLEM 14 ♦

What products would be formed from the acid-catalyzed hydrolysis of the following esters?

a. 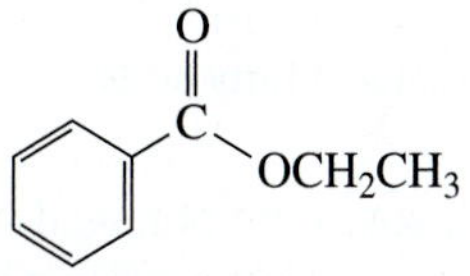 (structure: benzene ring—C(=O)—OCH₂CH₃)

b. 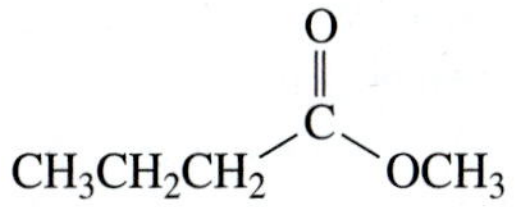 CH₃CH₂CH₂—C(=O)—OCH₃

PROBLEM 15 ♦

What product would be formed from the acid-catalyzed hydrolysis of the following cyclic ester?

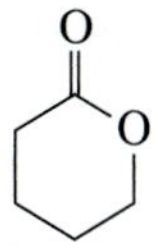

(structure: six-membered lactone ring)

Transesterification

Transesterification—the reaction of an ester with an alcohol—is also catalyzed by acid. The mechanism for transesterification is identical to the mechanism for ester hydrolysis, except that the nucleophile is ROH rather than H₂O. As in the hydrolysis of an ester, the two leaving groups in the tetrahedral intermediate formed in transesterification have approximately the same basicity. Consequently, when the reaction reaches equilibrium, both esters will be present in approximately equal amounts.

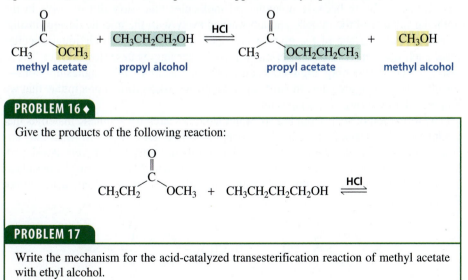

$$CH_3\!-\!\overset{\displaystyle O}{\overset{\|}{C}}\!-\!OCH_3 \;+\; CH_3CH_2CH_2OH \;\underset{}{\overset{HCl}{\rightleftharpoons}}\; CH_3\!-\!\overset{\displaystyle O}{\overset{\|}{C}}\!-\!OCH_2CH_2CH_3 \;+\; CH_3OH$$

methyl acetate　　propyl alcohol　　　　　propyl acetate　　methyl alcohol

PROBLEM 16 ♦

Give the products of the following reaction:

$$CH_3CH_2\!-\!\overset{\displaystyle O}{\overset{\|}{C}}\!-\!OCH_3 \;+\; CH_3CH_2CH_2CH_2OH \;\overset{HCl}{\rightleftharpoons}$$

PROBLEM 17

Write the mechanism for the acid-catalyzed transesterification reaction of methyl acetate with ethyl alcohol.

11.10 SOAPS, DETERGENTS, AND MICELLES

Fats and **oils** are triesters of glycerol. Glycerol contains three alcohol groups and therefore can form three ester groups. When the ester groups of a fat or an oil are hydrolyzed in a basic solution, glycerol and carboxylate ions are formed The carboxylic acids that are bonded to glycerol in fats and oils have long, unbranched R groups. Because they are obtained from fats, unbranched long-chain carboxylic acids are called **fatty acids**. In Section 19.1, we will see that the difference between a fat and an oil resides in the structure of the fatty acid.

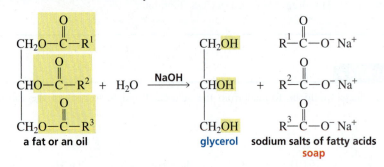

a fat or an oil   **glycerol**   **sodium salts of fatty acids**
                                                       soap

Soaps are sodium or potassium salts of fatty acids. Thus, soaps are obtained when fats or oils are hydrolyzed under basic conditions. The hydrolysis of an ester in a basic solution is called **saponification** (the Latin word for "soap" is *sapo*). Three of the most common soaps are:

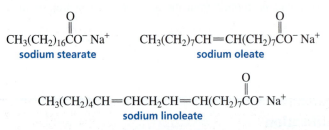

$$CH_3(CH_2)_{16}CO^- Na^+$$
sodium stearate

$$CH_3(CH_2)_7CH{=}CH(CH_2)_7CO^- Na^+$$
sodium oleate

$$CH_3(CH_2)_4CH{=}CHCH_2CH{=}CH(CH_2)_7CO^- Na^+$$
sodium linoleate

Long-chain carboxylate ions do not exist as individual ions in aqueous solution. Instead, they arrange themselves in spherical clusters called **micelles**, as shown in Figure 11.2. Each micelle contains 50 to 100 long-chain carboxylate ions and resembles a large ball: the polar heads of the carboxylate ions, each accompanied by a counterion, are on the outside of the ball because of their attraction for water, whereas the nonpolar tails are buried in the interior of the ball to minimize their contact with water. Soap has cleansing ability because nonpolar oil molecules that carry dirt dissolve in the nonpolar interior of the micelle and are washed away with the micelle during rinsing.

Because the surface of the micelle is negatively charged, the individual micelles repel each other instead of clustering to form larger aggregates. However, in "hard" water—water containing high concentrations of calcium and magnesium ions—micelles do form aggregates. In hard water, therefore, soaps form a precipitate that we know as "bathtub ring" or "soap scum."

The formation of soap scum in hard water led to a search for synthetic materials that would have the cleansing properties of soap, but would not form scum when they encountered calcium and magnesium ions. The synthetic "soaps" that were developed, known as **detergents** (from the Latin *detergere*, which means "to wipe off"), are salts of benzenesulfonic acids. Calcium and magnesium salts of benzenesulfonic acids do not form aggregates.

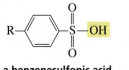 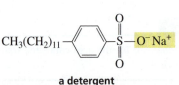

a benzenesulfonic acid   **a detergent**

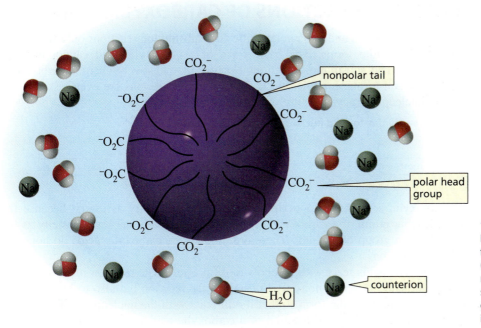

◀ **Figure 11.2**
In aqueous solution, soap molecules form micelles. The polar heads (carboxylate groups) of the soap molecules form the micelle's outer surface; the nonpolar tails (fatty acid R groups) extend into the micelle's interior.

After the initial introduction of detergents into the marketplace, it was discovered that those with straight-chain alkyl groups are biodegradable, whereas those with branched-chain alkyl groups are not. Therefore, to prevent detergents from polluting rivers and lakes, detergents should be made only with straight-chain alkyl groups.

MAKING SOAP

For thousands of years, soap was prepared by heating animal fat with wood ashes. Because the ashes contain potassium carbonate, the solution that results is basic. In the modern commercial method of making soap, fats or oils are boiled in aqueous sodium hydroxide. Sodium chloride is then added to precipitate the soap, which is dried and pressed into bars. Perfume can be added for scented soaps, dyes can be added for colored soaps, sand can be added for scouring soaps, and air can be blown into the soap to make it float in water.

Making soap.

PROBLEM 18 **SOLVED**

An oil obtained from coconuts is unusual in that all three fatty acid components are identical. The molecular formula of the oil is $C_{45}H_{86}O_6$. What is the molecular formula of the carboxylate ion obtained when the oil is saponified?

Solution When the oil is saponified, it forms glycerol and three equivalents of carboxylate ion. In losing glycerol, the fat loses three carbons and five hydrogens. Thus, the three equivalents of carboxylate ion have a combined molecular formula of $C_{42}H_{81}O_6$. Dividing by three gives a molecular formula of $C_{14}H_{27}O_2$ for the carboxylate ion.

acetic acid

11.11 THE REACTIONS OF CARBOXYLIC ACIDS

Carboxylic acids can undergo nucleophilic acyl substitution reactions only when they are in their acidic forms. The basic form of a carboxylic acid cannot undergo nucleophilic acyl substitution reactions because the negatively charged carboxylate ion is resistant to nucleophilic attack. Thus, carboxylate ions are even less reactive toward nucleophilic acyl substitution reactions than are amides.

relative reactivities toward nucleophilic acyl substitution

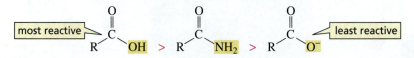

Carboxylic acids have approximately the same reactivity as esters, because the HO^- leaving group of a carboxylic acid has approximately the same basicity as the RO^- leaving group of an ester. Therefore, like esters, carboxylic acids do not react with chloride ions.

Carboxylic acids react with alcohols to form esters. The reaction must be carried out in an acidic solution, not only to catalyze the reaction but also to keep the carboxylic acid in its acidic form so that it will react with the nucleophile. The mechanism for the acid-catalyzed reaction of a carboxylic acid and an alcohol to form an ester and water is the exact reverse of the mechanism for the acid-catalyzed hydrolysis of an ester to form a carboxylic acid and an alcohol (Section 11.9).

Because the tetrahedral intermediate formed in this reaction has two potential leaving groups of approximately the same basicity, both carboxylic acid and ester will be present in approximately equal amounts when the reaction has reached equilibrium. Emil Fischer (Section 15.8) was the first to discover that an ester could be prepared by reacting a carboxylic acid with an alcohol in the presence of an acid catalyst. The reaction, therefore, is called a **Fischer esterification**.

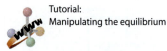

Tutorial:
Manipulating the equilibrium

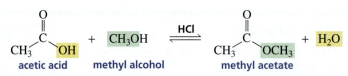

acetic acid   **methyl alcohol**   **methyl acetate**

Carboxylic acids do not undergo nucleophilic acyl substitution reactions with amines. Because a carboxylic acid is an acid and an amine is a base, the carboxylic acid immediately donates a proton to the amine when the two compounds are mixed. The resulting ammonium carboxylate salt is the final product of the reaction; the carboxylate ion is unreactive, and the protonated amine is not a nucleophile.

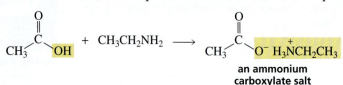

**an ammonium
carboxylate salt**

PROBLEM 19 ◆

Using an acyl chloride and an alcohol, show how the following esters could be synthesized:
a. methyl butyrate (odor of apples)   **b.** octyl acetate (odor of oranges)

PROBLEM 20

Referring to the mechanism for the acid-catalyzed hydrolysis of methyl acetate, write the mechanism—showing all the curved arrows—for the acid-catalyzed reaction of acetic acid and methyl alcohol to form methyl acetate.

PROBLEM 21   *SOLVED*

Loss of water from two molecules of a carboxylic acid forms an **acid anhydride**. Acid anhydrides are also carboxylic acid derivatives—the OH group of the carboxylic acid has been replaced with a carboxylate group. Thus, the carboxylate group is the leaving group of an acid anhydride.

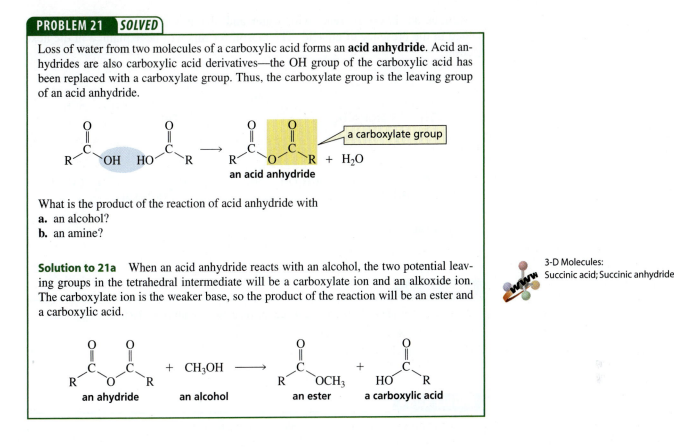

an acid anhydride

a carboxylate group

What is the product of the reaction of acid anhydride with
a. an alcohol?
b. an amine?

Solution to 21a   When an acid anhydride reacts with an alcohol, the two potential leaving groups in the tetrahedral intermediate will be a carboxylate ion and an alkoxide ion. The carboxylate ion is the weaker base, so the product of the reaction will be an ester and a carboxylic acid.

an ahydride   **an alcohol**   **an ester**   **a carboxylic acid**

3-D Molecules:
Succinic acid; Succinic anhydride

11.12 THE REACTIONS OF AMIDES

Amides are very unreactive compounds, which is comforting since proteins, which impart strength to biological structures, are composed of amino acids linked together by amide bonds (Section 16.0). Amides do not react with chloride ions, alcohols, or water because, in each case, the incoming nucleophile is a weaker base than the leaving group of the amide (Table 11.1).

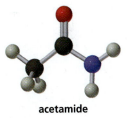

acetamide

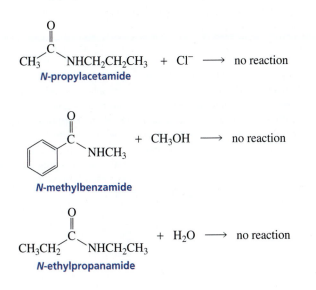

CH_3 $NHCH_2CH_2CH_3$ + Cl^- $\longrightarrow$ no reaction
***N*-propylacetamide**

 + CH_3OH $\longrightarrow$ no reaction
  NHCH_3
***N*-methylbenzamide**

 + H_2O $\longrightarrow$ no reaction
CH_3CH_2 $NHCH_2CH_3$
***N*-ethylpropanamide**

Amides do, however, react with water and alcohols if the reaction mixture is heated in the presence of an acid. The reason for this will be explained in the next section.

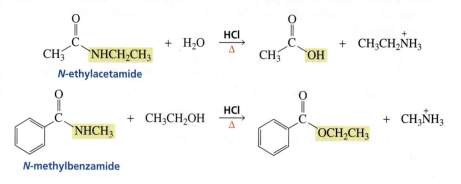

N-ethylacetamide

N-methylbenzamide

PROBLEM 22 ◆

What acyl chloride and what amine would be required to synthesize the following amides?

a. *N*-ethylbutanamide

b. *N,N*-dimethylbenzamide

PROBLEM 23 ◆

Which of the following reactions would lead to the formation of an amide?

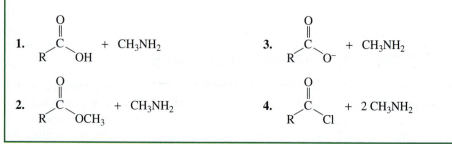

NATURE'S SLEEPING PILL

Melatonin, a naturally occurring amide, is a hormone synthesized by the pineal gland from the amino acid tryptophan. Melatonin regulates the dark–light clock in our brains that governs such things as the sleep–wake cycle, body temperature, and hormone production.

Melatonin levels increase from evening to night and then decrease as morning approaches. People with high levels of melatonin sleep longer and more soundly than those with low levels. The concentration of the hormone in the blood varies with age—6-year-olds have more than five times the concentration that 80-year-olds have—which is one of the reasons young people have less trouble sleeping than older people. Melatonin supplements are used to treat insomnia, jet lag, and seasonal affective disorder.

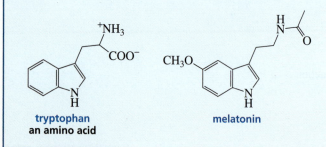

tryptophan
an amino acid

melatonin

11.13 **ACID-CATALYZED AMIDE HYDROLYSIS**

The mechanism for the acid-catalyzed hydrolysis of an amide is similar to the mechanism for the acid-catalyzed hydrolysis of an ester (Section 11.9).

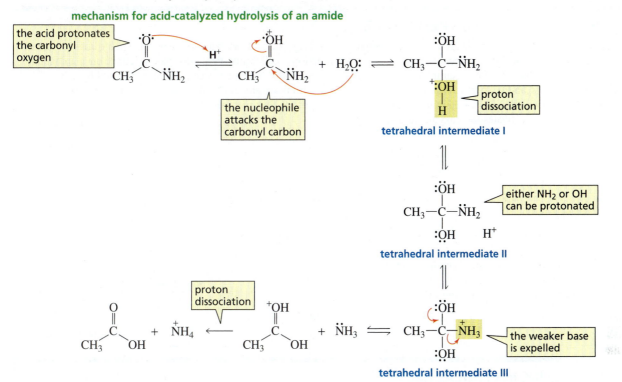

mechanism for acid-catalyzed hydrolysis of an amide

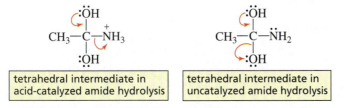

tetrahedral intermediate in acid-catalyzed amide hydrolysis

tetrahedral intermediate in uncatalyzed amide hydrolysis

- The acid protonates the carbonyl oxygen, increasing the susceptibility of the carbonyl carbon to nucleophilic attack.
- Nucleophilic attack by water on the carbonyl carbon leads to tetrahedral intermediate I, which is in equilibrium with its nonprotonated form, tetrahedral intermediate II.
- Reprotonation of tetrahedral intermediate II can occur either on oxygen to reform tetrahedral intermediate I or on nitrogen to form tetrahedral intermediate III. Protonation on nitrogen is favored because the NH_2 group is a stronger base than the OH group.
- Of the two possible leaving groups in tetrahedral intermediate III (HO^- and NH_3), NH_3 is the weaker base, so it is expelled, forming the carboxylic acid as the final product.
- Since the reaction is carried out in an acidic solution, NH_3 will be protonated after it is expelled from the tetrahedral intermediate. This prevents the reverse reaction from occurring, because $^+NH_4$ is not a nucleophile.

Let's take a minute to see why an amide cannot be hydrolyzed without a catalyst. In the uncatalyzed reaction, the amide would not be protonated. Therefore, water, a very poor nucleophile, would have to attack a neutral amide that is much less susceptible to nucleophilic attack than a protonated amide would be. In addition, the NH_2 group of the tetrahedral intermediate would not be protonated in the uncatalyzed reaction. Therefore, HO^- is the group that would be expelled from the tetrahedral intermediate, because HO^- is a weaker base than $^-NH_2$. This would reform the amide.

An amide reacts with an alcohol in the presence of acid (page 314) for the same reason that it reacts with water in the presence of acid. The only difference in the mechanisms of the two reactions is the nucleophile, which is water in one case and an alcohol in the other.

PENICILLIN AND DRUG RESISTANCE

Penicillin contains an amide in a four-membered ring. The strain in this ring increases the amide's reactivity. It is thought that the antibiotic activity of penicillin results from its ability to put an acyl group on a CH_2OH group of an enzyme that has a role in the synthesis of bacterial cell walls. This inactivates the enzyme, and actively

growing bacteria die because they are unable to synthesize functional cell walls. Penicillin has no effect on mammalian cells because mammalian cells are not enclosed by cell walls. Penicillins are stored at cold temperatures to minimize hydrolysis of the four-membered ring.

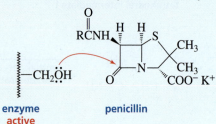

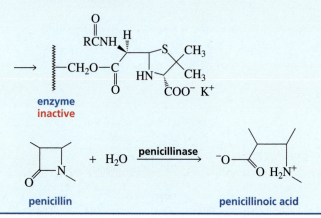

Bacteria that are resistant to penicillin secrete penicillinase, an enzyme that catalyzes the hydrolysis of penicillin's four-membered ring. Penicillinoic acid, the ring-opened product, has no antibacterial activity.

PENICILLINS IN CLINICAL USE

More than 10 different penicillins are currently in clinical use. They differ only in the group (R) attached to the carbonyl group. Some of these penicillins are shown here. In addition to their structural differences, the penicillins differ in the organisms against which

they are most effective. They also differ in their susceptibility to penicillinase. For example, ampicillin, a *synthetic* penicillin, is clinically effective against bacteria that are resistant to penicillin G, the *naturally occurring* penicillin discussed in Section 11.4. Almost 19% of humans are allergic to penicillin G.

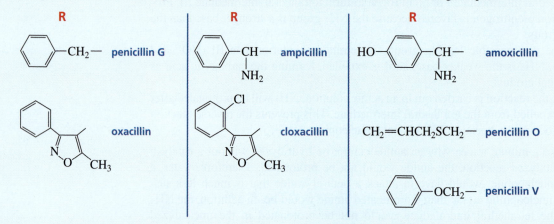

Penicillin V is a *semisynthetic* penicillin in clinical use. It is not a naturally occurring penicillin, but it is also not a true synthetic penicillin because chemists do not synthesize it. The

Penicillium mold synthesizes it after the mold is fed 2-phenoxyethanol, the compound needed for the side chain.

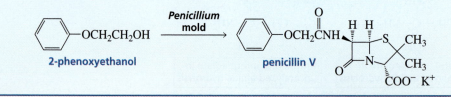

11.14 THE SYNTHESIS OF CARBOXYLIC ACID DERIVATIVES

Of the various classes of carbonyl compounds discussed in this chapter—acyl chlorides, esters, carboxylic acids, and amides—carboxylic acids are the most commonly available. However, we have seen that carboxylic acids are relatively unreactive toward nucleophilic acyl substitution reactions because the OH group of a carboxylic acid is a strong base and thus a poor leaving group. Therefore, chemists need a way to activate carboxylic acids so that they can readily undergo nucleophilic acyl substitution reactions.

Because acyl chlorides are the most reactive of the carboxylic acid derivatives, the easiest way to synthesize any other carboxylic acid derivative is to add the appropriate nucleophile to an acyl chloride. Consequently, organic chemists activate carboxylic acids by converting them into acyl chlorides.

A carboxylic acid can be converted into an acyl chloride by heating it with thionyl chloride ($SOCl_2$). This reagent replaces the OH group with a Cl.

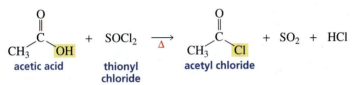

Once the acyl chloride has been prepared, esters and amides can be synthesized simply by adding the appropriate nucleophile (Section 11.7).

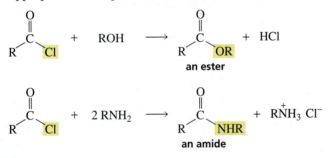

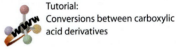

Tutorial:
Conversions between carboxylic acid derivatives

PROBLEM 24 ◆

How would you synthesize the following compounds starting with a carboxylic acid?

a. CH_3CH_2 —C(=O)—O—C₆H₅

b. C₆H₅—C(=O)—NHCH₂CH₃

SYNTHETIC POLYMERS

Synthetic polymers play important roles in our daily lives. Polymers are compounds that are made by linking together many small molecules called monomers. We have seen that **chain-growth polymers** are made by adding monomers to the end of a growing chain (Section 5.16).

A second kind of synthetic polymer, called a **step-growth polymer**, is made with monomers that have reactive functional groups at each end. The functional groups form ester or amide bonds between the monomers. Nylon and Dacron are examles of step-growth polymers. Nylon is a polyamide. Dacron is a polyester.

$$H_3\overset{+}{N}(CH_2)_5CO^- \xrightarrow[-H_2O]{\Delta} -NH(CH_2)_5\overset{O}{\underset{\|}{C}}\!\!\left[NH(CH_2)_5\overset{O}{\underset{\|}{C}}\right]_n\!\!NH(CH_2)_5\overset{O}{\underset{\|}{C}}-$$

6-aminohexanoic acid

nylon 6
a polyamide

(continued...)

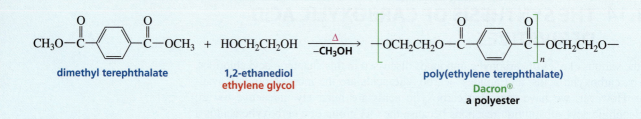

dimethyl terephthalate   1,2-ethanediol / ethylene glycol   poly(ethylene terephthalate) Dacron® a polyester

Synthetic polymers have taken the place of metals, fabrics, glass, ceramics, wood, and paper, allowing us to have a greater variety and larger quantities of materials than nature could have provided. New polymers are continually being designed to fit human needs. For example, Kevlar and Lexan are relatively new step-growth polymers. Kevlar has a tensile strength greater than steel. It is used for high-performance skis and bulletproof vests.

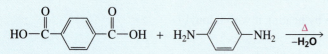

1,4-benzenedicarboxylic acid   1,4-diaminobenzene

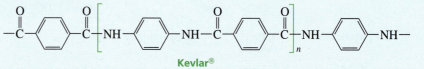

Kevlar®

Lexan is a strong and transparent polymer used for such things as traffic light lenses and compact disks.

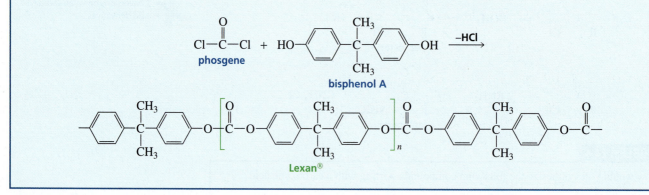

phosgene   bisphenol A

Lexan®

DISSOLVING SUTURES

Dissolving sutures, such as dexon and poly(dioxanone) (PDS), are synthetic polymers that now are routinely used in surgery. The many ester groups they contain are slowly hydrolyzed to small molecules that subsequently are metabolized to compounds easily excreted by the body. Patients no longer have to undergo a second medical procedure that was required to remove the sutures when traditional suture materials were used.

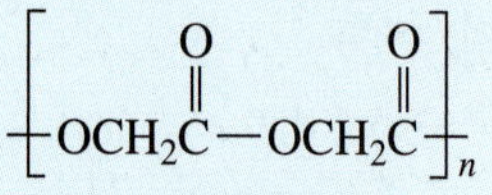

Dexon 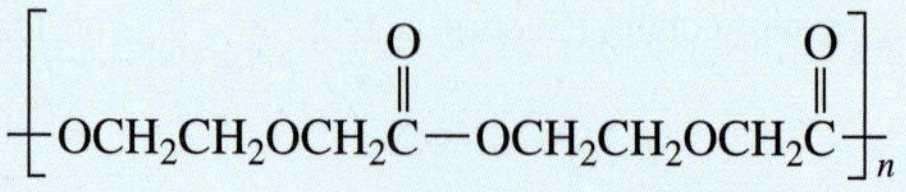   PDS

Depending on their structures, these synthetic sutures lose 50% of their strength after two to three weeks and are completely absorbed within three to six months.

11.15 NITRILES

Nitriles are compounds that contain a C≡N functional group. Nitriles are considered carboxylic acid derivatives because, like all the other carboxylic acid derivatives, they react with water to form carboxylic acids. They are even less reactive than amides, but hydrolyze when heated with water and an acid.

acetonitrile

$$CH_3CH_2C≡N \ + \ H_2O \ \xrightarrow[\Delta]{HCl} \ CH_3CH_2\overset{\overset{\displaystyle O}{\|}}{C}OH \ + \ \overset{+}{N}H_4$$

Nitriles are named by adding "nitrile" to the parent alkane name. Notice that the triple-bonded carbon of the nitrile group is counted in the number of carbons in the longest continuous chain.

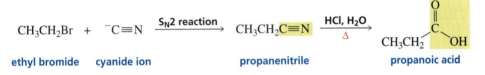

$CH_3C≡N$	$CH_3\overset{\overset{\displaystyle CH_3}{\|}}{C}HCH_2CH_2CH_2C≡N$	$CH_2{=}CHC≡N$

systematic name:	ethanenitrile	5-methylhexanenitrile	propenenitrile
common name:	acetonitrile		acrylonitrile

Nitriles can be prepared from an S_N2 reaction of alkyl halide with cyanide ion. Because a nitrile can be hydrolyzed to carboxylic acid, you now know how to convert an alkyl halide into a carboxylic acid. Notice that the carboxylic acid has one more carbon than the alkyl halide.

$$CH_3CH_2Br \ + \ {}^-C≡N \ \xrightarrow{S_N2 \ reaction} \ CH_3CH_2C≡N \ \xrightarrow[\Delta]{HCl, \ H_2O} \ CH_3CH_2\overset{\overset{\displaystyle O}{\|}}{C}OH$$

ethyl bromide   **cyanide ion**   **propanenitrile**   **propanoic acid**

A nitrile can be reduced to a primary amine by the same reagents that reduce an alkyne to an alkane (Section 5.12).

$$CH_3CH_2CH_2CH_2C≡N \ \xrightarrow[Pt/C]{H_2} \ CH_3CH_2CH_2CH_2CH_2NH_2$$

pentanenitrile   **pentylamine**

PROBLEM 25 ◆

Which alkyl halides form the carboxylic acids listed below after reaction with sodium cyanide followed by heating the product in an acidic aqueous solution?

a. butyric acid   **b.** 4-methylpentanoic acid

Tutorial:
Common terms pertaining to carboxylic acids and their derivatives

---

SUMMARY

A **carbonyl group** is a carbon double bonded to an oxygen; an **acyl group** is a carbonyl group attached to an alkyl or aryl group. **Acyl chlorides**, **esters**, and **amides** are called **carboxylic acid derivatives** because they differ from a carboxylic acid only in the nature of the group that has replaced the OH group of the carboxylic acid.

Carbonyl compounds can be placed in one of two classes. Class I carbonyl compounds contain a group that can be replaced by another group; carboxylic acids and carboxylic acid derivatives belong to this class. Class II

carbonyl compounds do not contain a group that can be replaced by another group; aldehydes and ketones belong to this class.

The reactivity of carbonyl compounds resides in the polarity of the carbonyl group; the carbonyl carbon has a partial positive charge that is attractive to nucleophiles. Class I carbonyl compounds undergo nucleophilic acyl substitution reactions, in which a nucleophile replaces the substituent that was attached to the acyl group in the reactant. All Class I carbonyl compounds react with nucleophiles in the same

way: the nucleophile attacks the carbonyl carbon, forming an unstable tetrahedral intermediate, which reforms a carbonyl compound by eliminating the weakest base.

A carboxylic acid derivative will undergo a nucleophilic acyl substitution reaction provided that the newly added group in the tetrahedral intermediate is not a much weaker base than the group that is attached to the acyl group in the reactant. The weaker the base attached to the acyl group, the more easily both steps of the nucleophilic acyl substitution reaction can take place. The relative reactivities toward nucleophilic acyl substitution: acyl chlorides > esters and carboxylic acids > amides > carboxylate ions.

Hydrolysis, **alcoholysis**, and **aminolysis** are reactions in which water, alcohols, and amines, respectively, convert one compound into two compounds. A **transesterification reaction** converts one ester to another ester. Reacting a carboxylic acid with an alcohol and an acid catalyst is called a **Fischer esterification**.

The rate of hydrolysis or alcoholysis can be increased by an acid. An acid increases the rate of formation of the tetrahedral intermediate by protonating the carbonyl oxygen, which increases the electrophilicity of the carbonyl group, and by decreasing the basicity of the leaving group, which makes it easier to eliminate.

Amides are unreactive compounds but do react with water and alcohols if the reaction mixture is heated in the presence of an acid. Nitriles are harder to hydrolyze than amides. Carboxylic acids are activated by being converted to acyl chlorides.

SUMMARY OF REACTIONS

1. Reactions of acyl halides (Section 11.7)

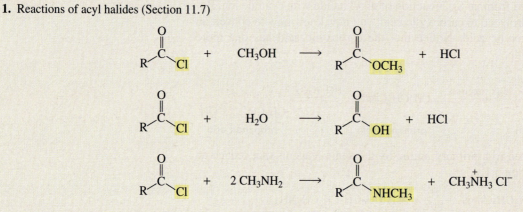

2. Reactions of esters (Sections 11.8–11.9)

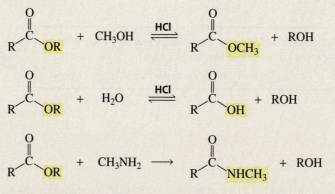

3. Reactions of carboxylic acids (Section 11.11)

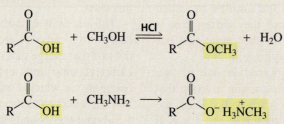

4. Reactions of amides (Sections 11.12 and 11.13)

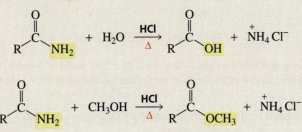

5. Activation of carboxylic acids (Section 11.14)

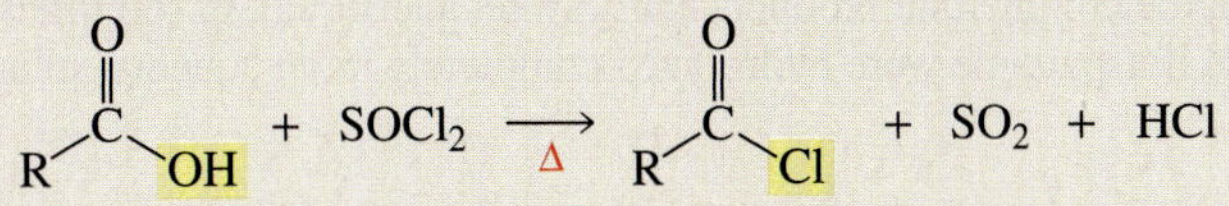

6. Hydrolysis of nitriles (Section 11.15)

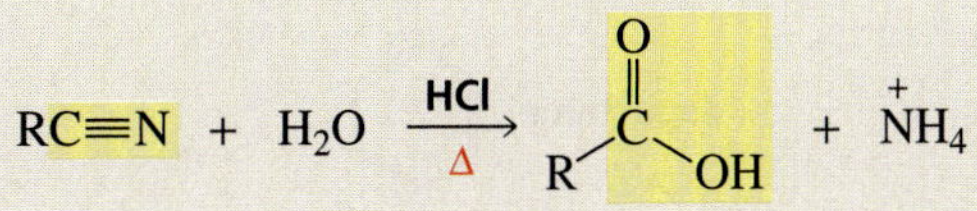

PROBLEMS

26. Write a structure for each of the following:
    a. *N*,*N*-dimethylhexanamide   **c.** 3-methylpentanoyl chloride   **e.** sodium acetate
    b. 3,3-dimethylhexanamide   **d.** 2-bromohexanoic acid   **f.** propionyl chloride

27. Name the following:

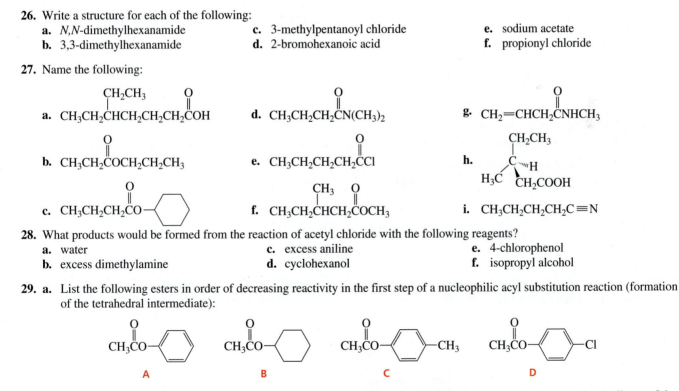

28. What products would be formed from the reaction of acetyl chloride with the following reagents?
    a. water   **c.** excess aniline   **e.** 4-chlorophenol
    b. excess dimethylamine   **d.** cyclohexanol   **f.** isopropyl alcohol

29. a. List the following esters in order of decreasing reactivity in the first step of a nucleophilic acyl substitution reaction (formation of the tetrahedral intermediate):

    b. List the same esters in order of decreasing reactivity in the last step of a nucleophilic acyl substitution reaction (collapse of the tetrahedral intermediate).

30. Using an alcohol for one method and an alkyl halide for the other, show two ways to make each of the following esters:
    a. propyl acetate (odor of pears)   **c.** isopentyl acetate (odor of bananas)
    b. ethyl butyrate (odor of pineapple)   **d.** methyl phenylethanoate (odor of honey)

31. Which compound would you expect to have a higher boiling point, the ester or the carboxylic acid?

$$CH_3\overset{\overset{O}{\|}}{C}OCH_3 \qquad CH_3CH_2\overset{\overset{O}{\|}}{C}OH$$

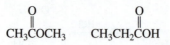

32. If propionyl chloride is added to one equivalent of methylamine, only a 50% yield of *N*-methylpropanamide is obtained. If, however, the acyl chloride is added to two equivalents of methylamine, the yield of *N*-methylpropanamide is almost 100%. Explain these observations.

33. What reagents would you use to convert methyl propanoate into the following compounds?

a. isopropyl propanoate
b. sodium propanoate
c. *N*-ethylpropanamide
d. propanoic acid

34. Aspartame, the sweetener used in the commercial products NutraSweet and Equal, is 160 times sweeter than sucrose. What products would be obtained if aspartame were hydrolyzed completely in an aqueous solution of HCl?

aspartame

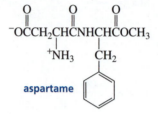

35. a. Which of the following reactions will not give the carbonyl product shown?

1. $CH_3\overset{\overset{O}{\|}}{C}NH_2 + Cl^- \longrightarrow CH_3\overset{\overset{O}{\|}}{C}Cl$

2. $CH_3\overset{\overset{O}{\|}}{C}OH + CH_3NH_2 \longrightarrow CH_3\overset{\overset{O}{\|}}{C}NHCH_3$

3. $CH_3\overset{\overset{O}{\|}}{C}OCH_3 + CH_3NH_2 \longrightarrow CH_3\overset{\overset{O}{\|}}{C}NHCH_3$

4. $CH_3\overset{\overset{O}{\|}}{C}OCH_3 + Cl^- \longrightarrow CH_3\overset{\overset{O}{\|}}{C}Cl$

5. $CH_3\overset{\overset{O}{\|}}{C}Cl + H_2O \longrightarrow CH_3\overset{\overset{O}{\|}}{C}OH$

6. $CH_3\overset{\overset{O}{\|}}{C}NHCH_3 + H_2O \longrightarrow CH_3\overset{\overset{O}{\|}}{C}OH$

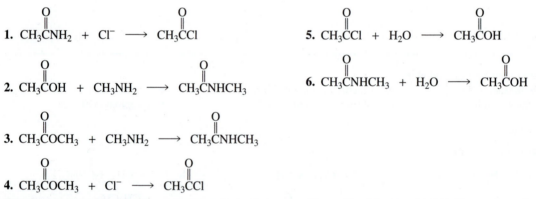

b. Which of the reactions that do not occur can be made to occur if an acid catalyst is added to the reaction mixture?

36. Identify the major and minor products of the following reaction:

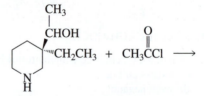

37. D. N. Kursanov, a Russian chemist, studied the hydrolysis of the following ester in a 1.0 M solution of sodium hydroxide, and he was able to prove that the bond that is broken in the reaction is the acyl C—O bond rather than the alkyl C—O bond:

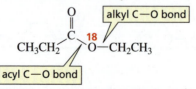

a. Which of the products contained the ^{18}O label?
b. What product would have contained the ^{18}O label if the alkyl C—O bond had broken?

38. Give the products of the following reactions:

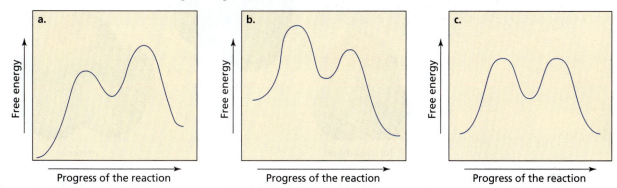

a. 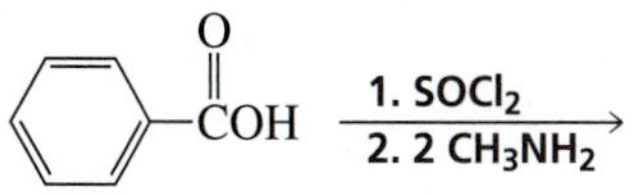 [benzene ring]–COH $\xrightarrow[\text{2. 2 CH}_3\text{NH}_2]{\text{1. SOCl}_2}$

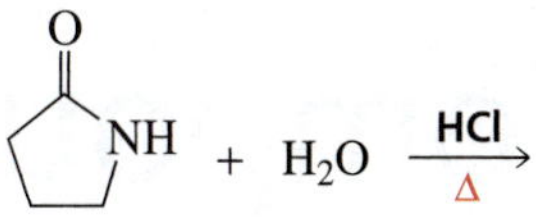

b. 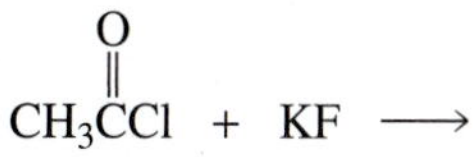 $CH_3\overset{O}{\overset{\|}{C}}Cl + KF \longrightarrow$

c. [pyrrolidinone]NH $+ H_2O \xrightarrow[\Delta]{\text{HCl}}$

d. 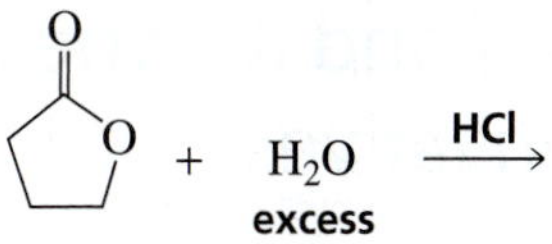 [butyrolactone] $+ H_2O \xrightarrow{\text{HCl}}$ **excess**

39. Which of the reaction coordinate diagrams represents the reaction of an ester with chloride ion?

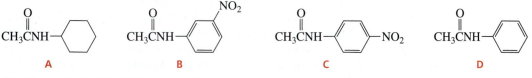

a. Free energy vs. Progress of the reaction

b. Free energy vs. Progress of the reaction

c. Free energy vs. Progress of the reaction

40. Which ester is more reactive, methyl acetate or phenyl acetate?

41. List the following amides in order of decreasing reactivity toward acid-catalyzed hydrolysis:

$CH_3\overset{O}{\overset{\|}{C}}NH$–[cyclohexyl]

A

$CH_3\overset{O}{\overset{\|}{C}}NH$–[phenyl with NO_2 meta]

B

$CH_3\overset{O}{\overset{\|}{C}}NH$–[phenyl]–$NO_2$

C

$CH_3\overset{O}{\overset{\|}{C}}NH$–[phenyl]

D

42. Give the products of the following reactions:

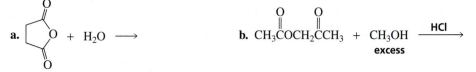

a. [succinic anhydride] $+ H_2O \longrightarrow$

b. $CH_3\overset{O}{\overset{\|}{C}}OCH_2\overset{O}{\overset{\|}{C}}CH_3 + CH_3OH \xrightarrow{\text{HCl}}$ **excess**

43. An aqueous solution of a primary or secondary amine reacts with an acyl chloride to form an amide as the major product. However, if the amine is tertiary, an amide is not formed. What product *is* formed? Explain.

44. Is the acid-catalyzed hydrolysis of acetamide a reversible or an irreversible reaction? Explain.

45. What product would you expect to obtain from each of the following reactions?

a.  $CH_3CH_2\overset{OH}{\overset{\|}{C}H}CH_2CH_2CH_2\overset{O}{\overset{\|}{C}}OH \xrightarrow{\text{HCl}}$

b. 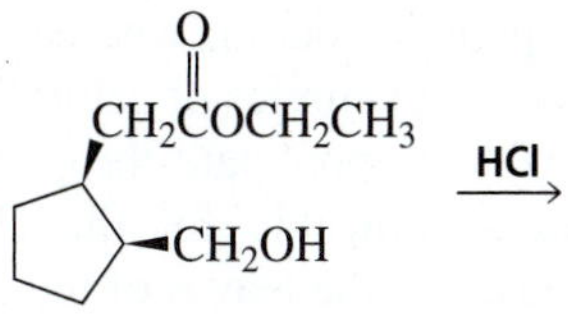 [cyclopentane with $CH_2COCH_2CH_3$ and CH_2OH substituents] $\xrightarrow{\text{HCl}}$

46. a. When a carboxylic acid is dissolved in isotopically labeled water (H_2O^{18}) in the presence of an acid catalyst, the label is incorporated into both oxygens of the acid. Propose a mechanism to account for this.

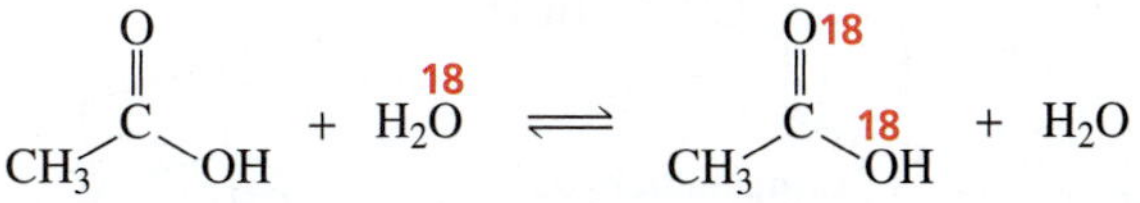

$$CH_3\overset{O}{\overset{\|}{C}}OH + H_2\overset{18}{O} \rightleftharpoons CH_3\overset{\overset{18}{O}}{\overset{\|}{C}}\overset{18}{O}H + H_2O$$

b. If a carboxylic acid is dissolved in isotopically labeled methanol ($CH_3{}^{18}OH$) and an acid catalyst is added, where will the label reside in the product?

CHAPTER 12

Carbonyl Compounds II

Reactions of Aldehydes and Ketones • More Reactions of Carboxylic Acid Derivatives

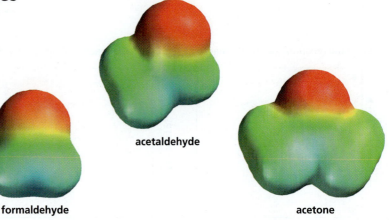

acetaldehyde

formaldehyde

acetone

3-D Molecules:
Formaldehyde; Acetaldehyde;
Acetone

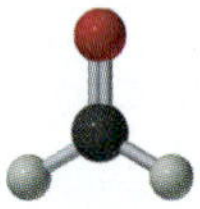

formaldehyde

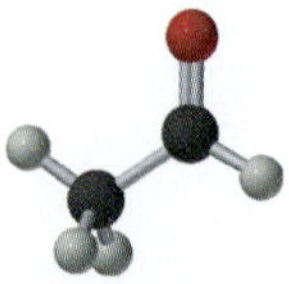

acetaldehyde

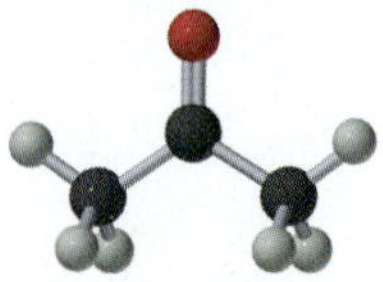

acetone

At the beginning of Chapter 11, we saw that carbonyl compounds—compounds that possess a carbonyl group (C=O)—can be divided into two classes: Class I carbonyl compounds, which have a group that can be replaced by a nucleophile, and Class II carbonyl compounds, which do not have a group that can be replaced by a nucleophile. Class II carbonyl compounds comprise aldehydes and ketones.

The carbonyl carbon of the simplest aldehyde, formaldehyde, is bonded to two hydrogens. The carbonyl carbon in all other **aldehydes** is bonded to a hydrogen and to an alkyl (or an aryl) group. The carbonyl carbon of a **ketone** is bonded to two alkyl (or aryl) groups. Aldehydes and ketones *do not have* a group that can be replaced by another group because hydride ions (H⁻) and carbanions (R⁻) are too basic to be displaced by nucleophiles under normal conditions.

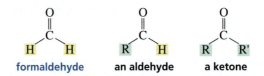

formaldehyde   **an aldehyde**   **a ketone**

The physical properties of aldehydes and ketones are discussed in Section 11.3 (see also Appendix I).

Many compounds found in nature have aldehyde or ketone functional groups. Aldehydes have pungent odors, whereas ketones tend to smell sweet. Vanillin and cinnamaldehyde are examples of naturally occurring aldehydes. A whiff of vanilla extract will allow you to appreciate the pungent odor of vanilla. The ketones carvone and camphor are responsible for the characteristic sweet odors of spearmint leaves, caraway seeds, and the leaves of the camphor tree.

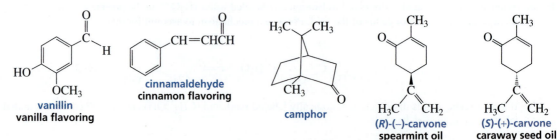

vanillin
vanilla flavoring

cinnamaldehyde
cinnamon flavoring

camphor

(R)-(−)-carvone
spearmint oil

(S)-(+)-carvone
caraway seed oil

In ketosis, a pathological condition that can occur in people with diabetes, the body produces more acetoacetate than can be metabolized. The excess acetoacetate breaks down to acetone (a ketone) and CO_2. Ketosis can be recognized by the smell of acetone on a person's breath.

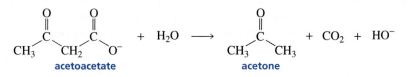

Two ketones that are of biological importance illustrate how a small difference in structure can be responsible for a large difference in biological activity: progesterone is a sex hormone synthesized primarily in the ovaries, whereas testosterone is a sex hormone synthesized primarily in the testes.

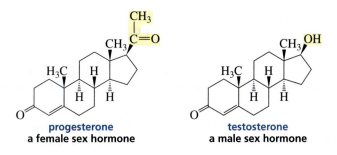

progesterone
a female sex hormone

testosterone
a male sex hormone

12.1 THE NOMENCLATURE OF ALDEHYDES AND KETONES

Naming Aldehydes

The systematic name of an aldehyde is obtained by replacing the final "e" on the name of the parent hydrocarbon with "al." For example, a one-carbon aldehyde is methan*al*; a two-carbon aldehyde is ethan*al*. The position of the carbonyl carbon does not have to be designated because it is always at the end of the parent hydrocarbon and therefore always has the 1-position.

systematic name:	**methanal**	**ethanal**
common name:	**formaldehyde**	**acetaldehyde**

2-bromopropanal
α-bromopropionaldehyde

The common name of an aldehyde is the same as the common name of the corresponding carboxylic acid (Section 11.1), except that "aldehyde' is substituted for "ic acid" (or "oic acid"). When common names are used, the position of a substituent is designated by a lowercase Greek letter. The carbonyl carbon is not given a letter; the carbon adjacent to the carbonyl carbon is the α-carbon.

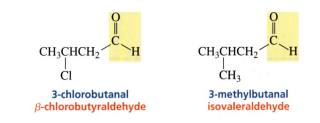

systematic name:	**3-chlorobutanal**	**3-methylbutanal**
common name:	**β-chlorobutyraldehyde**	**isovaleraldehyde**

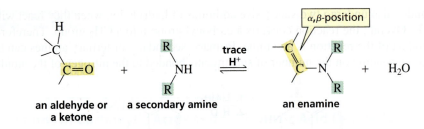

an aldehyde or a ketone   a secondary amine   an enamine

When you first look at the products of imine and enamine formation, they appear to be quite different. However, when you look at the mechanisms for the reactions, you will see that the mechanisms are exactly the same except for the site from which a proton is lost in the last step.

Primary Amines Form Imines

Aldehydes and ketones react with primary amines to form imines.

Aldehyde and ketones react with primary amines to form imines. The reaction requires a trace amount of acid.

benzaldehyde   **ethylamine**   **an imine**
an aldehyde   a primary amine

3-D Molecules:
The *N*-methylimine of *acetone*

3-pentanone   **benzylamine**   **an imine**
a ketone   a primary amine

The mechanism for imine formation is shown below.

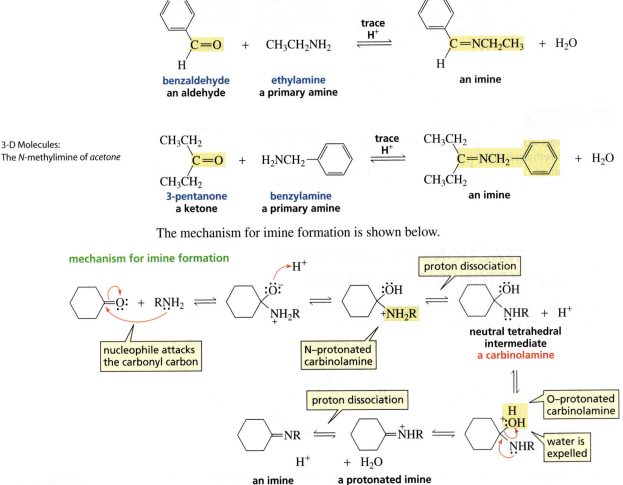

mechanism for imine formation

nucleophile attacks the carbonyl carbon

proton dissociation

N–protonated carbinolamine

neutral tetrahedral intermediate
a carbinolamine

O–protonated carbinolamine

water is expelled

proton dissociation

an imine   a protonated imine

Tutorial:
Imine formation

- The amine attacks the carbonyl carbon.
- Gain of a proton by the alkoxide ion and loss of a proton by the ammonium ion form a neutral tetrahedral intermediate.
- The neutral tetrahedral intermediate, called a *carbinolamine*, is in equilibrium with two protonated forms because either its nitrogen atom or its oxygen atom can be protonated.
- Elimination of water from the oxygen-protonated intermediate forms a protonated imine.
- The protonated imine loses a proton to yield the imine.

Unlike the stable tetrahedral compounds that are formed when a Grignard reagent or a hydride ion adds to an aldehyde or a ketone, the tetrahedral compound formed when an amine adds to an aldehyde or a ketone is unstable because the newly formed sp^3 carbon is bonded to an oxygen and to another electronegative atom (a nitrogen).

stable tetrahedral compounds   unstable tetrahedral compounds

In an acidic aqueous solution, an imine is hydrolyzed back to the carbonyl compound and amine. Notice that the amine is protonated because the solution is acidic.

An imine undergoes acid-catalyzed hydrolysis to form a carbonyl compound and a primary amine.

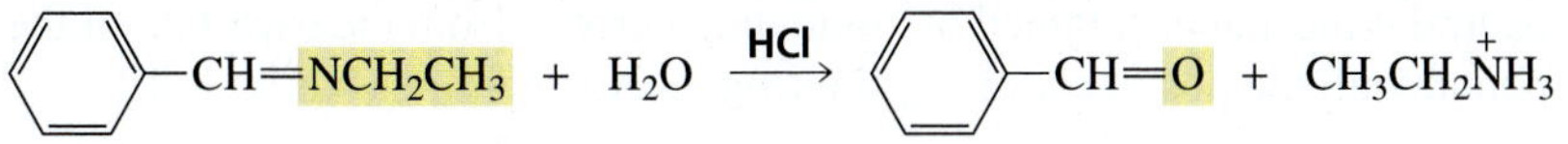

Imine formation and hydrolysis are important reactions in biological systems. For example, we will see that imine hydrolysis is why DNA contains A, G, C, and T nucleotides, whereas RNA contains A, G, C, and U nucleotides (Section 20.9).

Secondary Amines Form Enamines

Aldehydes and ketones react with secondary amines to form enamines. Like imine formation, the reaction requires a trace amount of an acid catalyst.

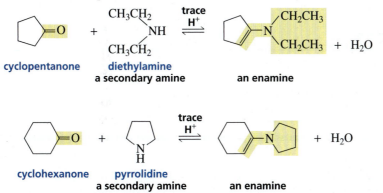

cyclopentanone   diethylamine   an enamine
                        a secondary amine

cyclohexanone   pyrrolidine   an enamine
                       a secondary amine

Aldehydes and ketones react with secondary amines to form enamines.

The mechanism for enamine formation is exactly the same as that for imine formation, except for the last step of the mechanism.

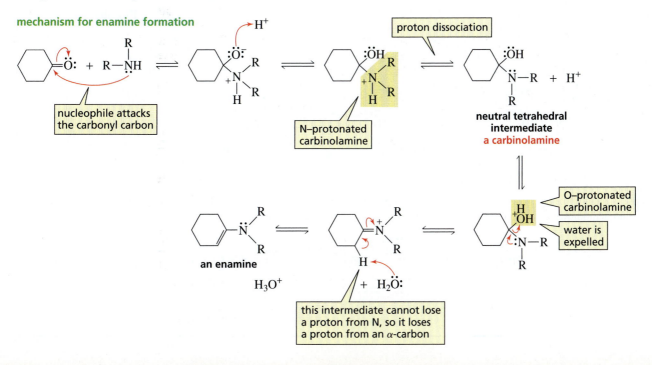

mechanism for enamine formation

nucleophile attacks the carbonyl carbon

proton dissociation

N–protonated carbinolamine

neutral tetrahedral intermediate
a carbinolamine

O–protonated carbinolamine

water is expelled

an enamine

this intermediate cannot lose a proton from N, so it loses a proton from an α-carbon

- The amine attacks the carbonyl carbon.
- Gain of a proton by the alkoxide ion and loss of a proton by the ammonium ion forms a neutral tetrahedral intermediate.
- The neutral tetrahedral intermediate is in equilibrium with two protonated forms because either its nitrogen atom or its oxygen atom can be protonated.
- Elimination of water from the oxygen-protonated intermediate forms a compound with positively charged nitrogen.
- When a primary amine reacts with an aldehyde or a ketone, the compound with a positively charged nitrogen loses a proton from nitrogen in the last step of the mechanism, forming a neutral imine. However, when the amine is secondary, the positively charged nitrogen is not bonded to a hydrogen. In this case, a stable neutral molecule is obtained by removing a proton from the α-carbon of the compound with the positively charged nitrogen. An enamine is the result.

An enamine undergoes acid-catalyzed hydrolysis to form a carbonyl compound and a secondary amine.

In an aqueous acidic solution, an enamine is hydrolyzed back to the carbonyl compound and secondary amine, a reaction that is similar to the acid-catalyzed hydrolysis of an imine back to the carbonyl compound and primary amine. Again, the amine is protonated because the solution is acidic.

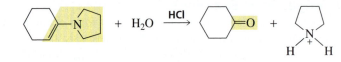

PROBLEM 13 ◆

Give the products of the following reactions. (A trace amount of acid is present in each case.)

a. cyclopentanone + ethylamine

b. cyclopentanone + diethylamine

c. 3-pentanone + hexylamine

d. 3-pentanone + cyclohexylamine

IMINES IN BIOLOGICAL SYSTEMS

Almost all biological reactions are catalyzed by enzymes. Some enzymes need the help of a coenzyme, which is an organic molecule derived from a vitamin, to carry out the catalysis (Sections 17.1 and 17.4). For example, pyridoxal phosphate, a coenzyme derived from vitamin B₆, is required by enzymes that metabolize amino acids (Section 17.9). Pyridoxal phosphate is attached to the enzyme by an imine linkage.

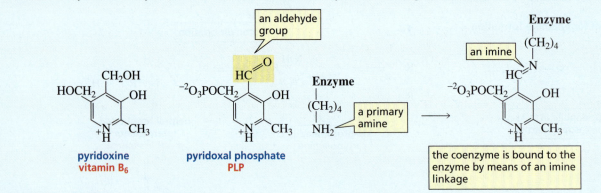

pyridoxine
vitamin B₆

pyridoxal phosphate
PLP

the coenzyme is bound to the enzyme by means of an imine linkage

Glucose is metabolized to pyruvate, which can be converted to the amino acid known as alanine in a two-step process. First, pyruvate reacts with NH_3 to form an imine, which is then reduced by an enzyme to the amino acid.

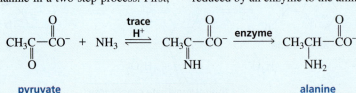

pyruvate

alanine
an amino acid

12.8 THE REACTIONS OF ALDEHYDES AND KETONES WITH WATER

Most hydrates are too unstable to be isolated.

The addition of water to an aldehyde or a ketone forms a *hydrate*. A **hydrate** is a molecule with two OH groups on the same carbon. Hydrates of aldehydes or ketones are generally too unstable to be isolated because the tetrahedral carbon is attached to two electron-withdrawing (oxygen) atoms.

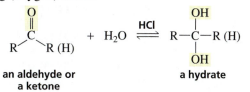

an aldehyde or a ketone   **a hydrate**

Water is a poor nucleophile and therefore adds relatively slowly to a carbonyl group. The rate of the reaction can be increased by an acid catalyst. Keep in mind that the catalyst affects the *rate* at which an aldehyde or a ketone is converted to a hydrate; it has no effect on the *amount* of aldehyde or ketone converted to hydrate (Section 4.8). The mechanism for the reaction is shown below.

mechanism for acid-catalyzed hydrate formation

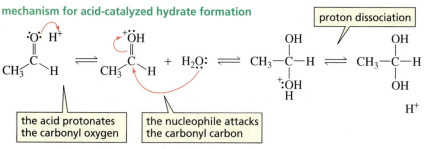

proton dissociation

the acid protonates the carbonyl oxygen

the nucleophile attacks the carbonyl carbon

- The acid protonates the carbonyl oxygen, which makes the carbonyl carbon more susceptible to nucleophilic attack (Figure 12.2).
- Water attacks the carbonyl carbon.
- Loss of a proton from the protonated tetrahedral intermediate gives the hydrate.

The extent to which an aldehyde or a ketone is hydrated in an aqueous solution depends on the substituents attached to the carbonyl compound. For example, only 0.2% of acetone is hydrated at equilibrium, but 99.9% of formaldehyde is hydrated. Bulky substituents and electron-donating substituents (for example, the methyl groups of acetone) *decrease* the percentage of hydrate present at equilibrium, whereas small substituents and electron-withdrawing substituents (the hydrogens of formaldehyde) *increase* it.

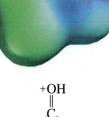

▲ **Figure 12.2**
The electrostatic potential maps show that the carbonyl carbon of the protonated aldehyde is more susceptible to nucleophilic attack (the blue is more intense) than the carbonyl carbon of the unprotonated aldehyde.

PRESERVING BIOLOGICAL SPECIMENS

A 37% solution of formaldehyde in water, known as *formalin*, was commonly used in the past to preserve biological specimens. Formaldehyde is an eye and a skin irritant, however, so formalin has been replaced in many biology laboratories by other preservatives. One preservative frequently used is a solution of 2 to 5% phenol in ethanol with added antimicrobial agents.

PROBLEM 14 ◆

When trichloroacetaldehyde is dissolved in water, almost all of it is converted to the hydrate. Chloral hydrate, the product of the reaction, is a sedative that can be lethal. A cocktail laced with it is commonly known—in detective novels, at least—as a "Mickey Finn." Explain why an aqueous solution of trichloroacetaldehyde is almost all hydrate.

3-D Molecules: Acetone; Acetone hydrate

trichloroacetaldehyde   **chloral hydrate**

Which of the following ketones forms the most hydrate in an aqueous solution?

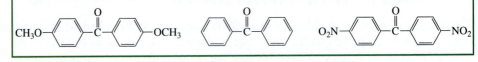

12.9 THE REACTIONS OF ALDEHYDES AND KETONES WITH ALCOHOLS

Aldehydes react with alcohols to form hemiacetals and acetals.

The product formed when one equivalent of an alcohol adds to an *aldehyde* is called a **hemiacetal**. The product formed when a second equivalent of alcohol is added is called an **acetal** (ass-ett-AL). Like water, an alcohol is a poor nucleophile, so an acid catalyst is required for the reaction to take place at a reasonable rate.

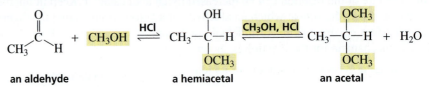

an aldehyde   a hemiacetal   an acetal

Ketones react with alcohols to form hemiketals and ketals.

When the carbonyl compound is a *ketone* instead of an aldehyde, the addition products are called a **hemiketal** and a **ketal**, respectively.

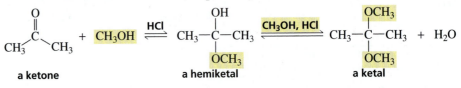

a ketone   a hemiketal   a ketal

Hemi is the Greek word for "half." When one equivalent of alcohol has added to an aldehyde or a ketone, the compound is halfway to the final acetal or ketal, which contains groups from two equivalents of alcohol.

The mechanism for ketal (or acetal) formation is shown below.

mechanism for acid-catalyzed acetal or ketal formation

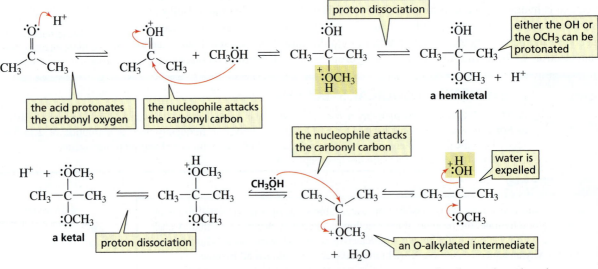

- The acid protonates the carbonyl oxygen to make the carbonyl carbon more susceptible to nucleophilic attack (Figure 12.2).
- The alcohol attacks the carbonyl carbon.
- Loss of a proton from the protonated tetrahedral intermediate gives the hemiacetal (or hemiketal).

- Because the reaction is carried out in an acidic solution, the hemiacetal (or hemiketal) is in equilibrium with its protonated form. The two oxygen atoms of the hemiacetal (or hemiketal) are equally basic, so either one can be protonated.
- Loss of water from the tetrahedral intermediate with a protonated OH group forms an O-alkylated intermediate that is very reactive because of its positively charged oxygen atom. Nucleophilic attack on this compound by a second molecule of alcohol, followed by loss of a proton, forms the acetal (or ketal).

Although the tetrahedral carbon of an acetal (or ketal) is bonded to two oxygen atoms, causing us to predict that it is not stable, the acetal (or ketal) can be isolated if the water that is eliminated from the hemiacetal (or hemiketal) is removed from the reaction mixture. This is because, if water is not available, the only compound the acetal (or ketal) can form is the O-alkylated species, which is less stable than the acetal (or ketal).

The acetal (or ketal) can be hydrolyzed back to the aldehyde (or ketone) in an acidic aqueous solution.

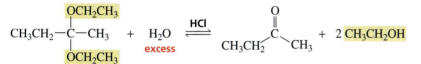

CARBOHYDRATES

When you study carbohydrates in Chapter 15, you will see that the individual sugar units in a carbohydrate are held together by acetal or ketal linkages. For example, the reaction of an alcohol group and an aldehyde group of D-glucose forms a cyclic D-glucose molecule with a hemiacetal linkage. Molecules of cyclic D-glucose are then hooked up as a result of the reaction of the OH group of one molecule with the hemiacetal group of another, resulting in the formation of an acetal linkage.

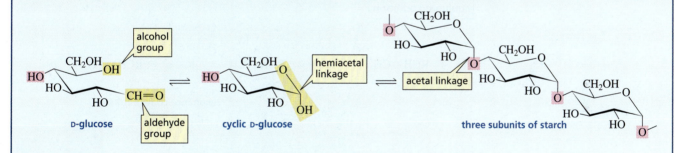

PROBLEM 16

Show the mechanism for the acid-catalyzed hydrolysis of an acetal.

PROBLEM 17 ◆

Which of the following are
a. hemiacetals?   **b.** acetals?   **c.** hemiketals?   **d.** ketals?   **e.** hydrates?

1.
$$CH_3-\underset{\underset{OCH_3}{|}}{\overset{\overset{OH}{|}}{C}}-CH_3$$

3.
$$CH_3-\underset{\underset{OCH_3}{|}}{\overset{\overset{OCH_3}{|}}{C}}-H$$

5.
$$CH_3-\underset{\underset{OCH_3}{|}}{\overset{\overset{OCH_3}{|}}{C}}-CH_3$$

7.
$$CH_3-\underset{\underset{OCH_3}{|}}{\overset{\overset{OH}{|}}{C}}-H$$

2.
$$CH_3-\underset{\underset{OCH_2CH_3}{|}}{\overset{\overset{OCH_2CH_3}{|}}{C}}-H$$

4.
$$CH_3-\underset{\underset{OH}{|}}{\overset{\overset{OH}{|}}{C}}-CH_3$$

6.
$$CH_3-\underset{\underset{OH}{|}}{\overset{\overset{OH}{|}}{C}}-H$$

8.
$$CH_3-\underset{\underset{OCH_3}{|}}{\overset{\overset{OH}{|}}{C}}-CH_2CH_3$$

Tutorial:
Addition to carbonyl compounds

12.10 NUCLEOPHILIC ADDITION TO α,β-UNSATURATED CARBONYL COMPOUNDS

The resonance contributors for an α,β-unsaturated carbonyl compound show that the molecule has two electrophilic sites: the carbonyl carbon and the β-carbon.

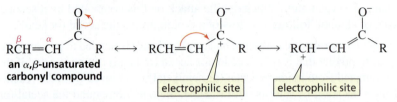

This means that if an aldehyde or a ketone has a double bond in the α,β-position, a nucleophile can add either to the carbonyl carbon or to the β-carbon.

Nucleophilic addition to the carbonyl carbon is called **direct addition** or **1,2-addition**.

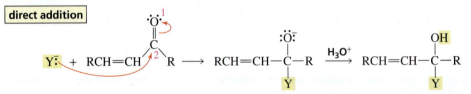

Nucleophilic addition to the β-carbon is called **conjugate addition** or **1,4-addition**, because addition occurs at the 1- and 4-positions (that is, across the conjugated system). After 1,4-addition has occurred, the product—an enol—tautomerizes to a ketone (or to an aldehyde) because the keto tautomer is more stable than the enol tautomer (Section 5.11).

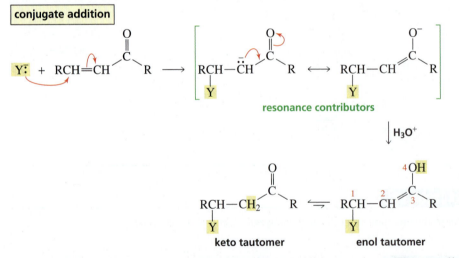

Nucleophiles that are weak bases form conjugate addition products. The overall reaction amounts to addition to the carbon–carbon double bond, with the nucleophile adding to the β-carbon of the double bond and a proton from the reaction mixture adding to the α-carbon.

> **Nucleophiles that are weak bases form conjugate addition products.**

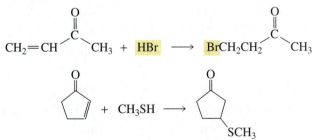

In general, nucleophiles that are strong bases form direct addition products. Ethyl alcohol is used in the second step to protonate the alkoxide ion.

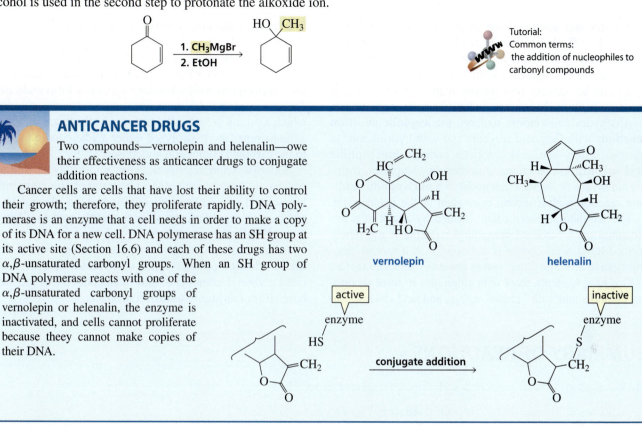

Tutorial:
Common terms:
the addition of nucleophiles to carbonyl compounds

ANTICANCER DRUGS

Two compounds—vernolepin and helenalin—owe their effectiveness as anticancer drugs to conjugate addition reactions.

Cancer cells are cells that have lost their ability to control their growth; therefore, they proliferate rapidly. DNA polymerase is an enzyme that a cell needs in order to make a copy of its DNA for a new cell. DNA polymerase has an SH group at its active site (Section 16.6) and each of these drugs has two α,β-unsaturated carbonyl groups. When an SH group of DNA polymerase reacts with one of the α,β-unsaturated carbonyl groups of vernolepin or helenalin, the enzyme is inactivated, and cells cannot proliferate because theey cannot make copies of their DNA.

vernolepin   **helenalin**

active enzyme   →**conjugate addition**→   inactive enzyme

PROBLEM 18 ◆

Give the major product of each of the following reactions.

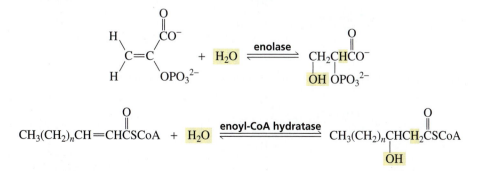

a. ⟶ HBr

b. 1. CH_3MgBr 2. EtOH

12.11 CONJUGATE ADDITION REACTIONS IN BIOLOGICAL SYSTEMS

Several reactions in biological systems involve conjugate addition to α,β-unsaturated carbonyl compounds. Below are examples of two of them. The first occurs in gluconeogenesis—the synthesis of glucose from pyruvate (Section 18.4). The second occurs in the oxidation of fatty acids (Section 18.3).

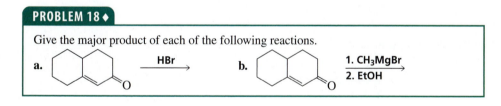

$$CH_3(CH_2)_nCH=CHCSCoA \ + \ H_2O \ \underset{}{\overset{\text{enoyl-CoA hydratase}}{\rightleftharpoons}} \ CH_3(CH_2)_nCHCH_2CSCoA$$
$$\qquad\qquad\qquad\qquad\qquad\qquad\qquad\qquad\qquad OH$$

SUMMARY

Aldehydes and **ketones** have an acyl group attached to a group (H or R) that cannot be readily replaced by another group. Steric and electronic factors cause aldehydes to be more reactive than ketones toward nucleophilic attack. Aldehydes and ketones are less reactive than acyl chlorides and are more reactive than esters, carboxylic acids, and amides.

Aldehydes and ketones undergo **nucleophilic addition reactions** with Grignard reagents and with hydride ion. In contrast, esters and acyl chlorides undergo **nucleophilic acyl substitution** reactions with these nucleophiles, forming aldehydes or ketones that undergo a **nucleophilic addition** reaction with a second equivalent of the nucleophile.

The tetrahedral intermediate formed by attack of a nucleophile on a carbonyl compound is stable if the newly formed tetrahedral carbon is not bonded to a second electronegative atom or group and is generally unstable if it is.

Grignard reagents react with aldehydes to form secondary alcohols and with ketones, esters, and acyl chlorides to form tertiary alcohols. Hydride ion reduces aldehydes, acyl chlorides, and carboxylic acids to primary alcohols, ketones to secondary alcohols, and amides to amines.

Aldehydes and ketones react with primary amines to form **imines** and with secondary amines to form **enamines**. The mechanisms are the same, except for the site from which a proton is lost in the last step of the reaction. Imines and enamines are hydrolyzed under acidic conditions back to the carbonyl compound and amine.

Aldehydes and ketones undergo acid-catalyzed addition of water to form hydrates. Most hydrates are too unstable to be isolated. Acid-catalyzed addition of an alcohol to an aldehyde forms **hemiacetals** and **acetals**, and to a ketone forms **hemiketals** and **ketals**.

In general, nucleophiles that are strong bases form **direct addition** products when they react with α,β-unsaturated carbonyl compound, and nucleophiles that are weak bases form **conjugate addition** products.

SUMMARY OF REACTIONS

1. Reactions of *carbonyl compounds* with Grignard reagents (Section 12.5)

   a. Reaction of *formaldehyde* with a Grignard reagent forms a *primary alcohol*:

$$ \underset{H \quad\quad H}{\overset{O}{C}} \quad \xrightarrow[\text{2. } H_3O^+]{\text{1. } CH_3MgBr} \quad CH_3CH_2OH $$

   b. Reaction of an *aldehyde* (other than formaldehyde) with a Grignard reagent forms a *secondary alcohol*:

$$ \underset{R \quad\quad H}{\overset{O}{C}} \quad \xrightarrow[\text{2. } H_3O^+]{\text{1. } CH_3MgBr} \quad R-\underset{CH_3}{\overset{OH}{C}}-H $$

   c. Reaction of a *ketone* with a Grignard reagent forms a *tertiary alcohol*:

$$ \underset{R \quad\quad R'}{\overset{O}{C}} \quad \xrightarrow[\text{2. } H_3O^+]{\text{1. } CH_3MgBr} \quad R-\underset{CH_3}{\overset{OH}{C}}-R' $$

   d. Reaction of an *ester* with a Grignard reagent forms a *tertiary alcohol* with two identical substituents:

$$ \underset{R \quad\quad OR'}{\overset{O}{C}} \quad \xrightarrow[\text{2. } H_3O^+]{\text{1. 2 } CH_3MgBr} \quad R-\underset{CH_3}{\overset{OH}{C}}-CH_3 $$

   e. Reaction of an *acyl chloride* with a Grignard reagent forms a *tertiary alcohol* with two identical substituents:

$$ \underset{R \quad\quad Cl}{\overset{O}{C}} \quad \xrightarrow[\text{2. } H_3O^+]{\text{1. 2 } CH_3MgBr} \quad R-\underset{CH_3}{\overset{OH}{C}}-CH_3 $$

2. Reactions of *carbonyl compounds* with hydride ion donors (Section 12.6)

   a. Reaction of an *aldehyde* with sodium borohydride forms a *primary alcohol*:

$$\underset{R}{\overset{O}{\underset{\parallel}{C}}}\diagdown H \quad \xrightarrow[\text{2. }H_3O^+]{\text{1. NaBH}_4} \quad RCH_2OH$$

   b. Reaction of a *ketone* with sodium borohydride forms a *secondary alcohol*:

$$\underset{R}{\overset{O}{\underset{\parallel}{C}}}\diagdown R \quad \xrightarrow[\text{2. }H_3O^+]{\text{1. NaBH}_4} \quad R-\overset{OH}{\underset{|}{C}H}-R$$

   c. Reaction of an *acyl chloride* with sodium borohydride forms a *primary alcohol*:

$$\underset{R}{\overset{O}{\underset{\parallel}{C}}}\diagdown Cl \quad \xrightarrow[\text{2. }H_3O^+]{\text{1. NaBH}_4} \quad R-CH_2-OH$$

   d. Reaction of an *ester* with lithium aluminum hydride forms *two alcohols*:

$$\underset{R}{\overset{O}{\underset{\parallel}{C}}}\diagdown OR' \quad \xrightarrow[\text{2. }H_3O^+]{\text{1. LiAlH}_4} \quad RCH_2OH \; + \; R'OH$$

   e. Reaction of a *carboxylic acid* with lithium aluminum hydride forms a *primary alcohol*:

$$\underset{R}{\overset{O}{\underset{\parallel}{C}}}\diagdown OH \quad \xrightarrow[\text{2. }H_3O^+]{\text{1. LiAlH}_4} \quad R-CH_2-OH$$

   f. Reaction of an *amide* with lithium aluminum hydride forms an *amine*:

$$\underset{R}{\overset{O}{\underset{\parallel}{C}}}\diagdown NH_2 \quad \xrightarrow[\text{2. }H_2O]{\text{1. LiAlH}_4} \quad R-CH_2-NH_2$$

$$\underset{R}{\overset{O}{\underset{\parallel}{C}}}\diagdown NHR' \quad \xrightarrow[\text{2. }H_2O]{\text{1. LiAlH}_4} \quad R-CH_2-NHR'$$

$$\underset{R}{\overset{O}{\underset{\parallel}{C}}}\diagdown \underset{\underset{R''}{|}}{N}R' \quad \xrightarrow[\text{2. }H_2O]{\text{1. LiAlH}_4} \quad R-CH_2-\underset{\underset{R''}{|}}{N}-R'$$

3. Reactions of *aldehydes* and *ketones* with amines (Section 12.7)

   a. Reaction with a *primary amine* forms an *imine*:

$$\underset{R}{\overset{R}{\diagdown}}C=O \; + \; H_2NR \quad \underset{\text{trace}}{\overset{H^+}{\rightleftharpoons}} \quad \underset{R}{\overset{R}{\diagdown}}C=NR \; + \; H_2O$$

   b. Reaction with a *secondary amine* forms an *enamine*:

$$\underset{-CH}{\overset{R}{\diagdown}}C=O \; + \; RNHR \quad \underset{\text{trace}}{\overset{H^+}{\rightleftharpoons}} \quad \underset{-C}{\overset{R}{\diagdown}}C-\underset{R}{\overset{R}{\diagdown}}N \; + \; H_2O$$

4. Reaction of an *aldehyde* or a *ketone* with water forms a *hydrate* (Section 12.8).

$$\underset{R}{\overset{O}{\underset{\parallel}{C}}}\diagdown R' \; + \; H_2O \quad \overset{\text{HCl}}{\rightleftharpoons} \quad R-\overset{OH}{\underset{\underset{OH}{|}}{\overset{|}{C}}}-R'$$

5. Reaction of an *aldehyde* or a *ketone* with excess alcohol forms an *acetal* or a *ketal* (Section 12.9).

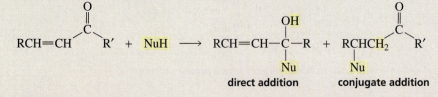

6. Reaction of α,β-unsaturated carbonyl compounds with a nucleophile (Section 12.10):

Nucleophiles that are strong bases (H^-, RMgBr) form direct addition products; nucleophiles that are weak bases (RSH, RNH_2, Br^-) form conjugate addition products.

PROBLEMS

19. Draw the structure for each of the following:
   a. isobutyraldehyde
   b. 4-octanone
   c. 4-bromohexanal
   d. 4-bromo-3-heptanone
   e. 3-methylcyclohexanone
   f. 2,4-pentanedione

20. Give the products of each of the following reactions:

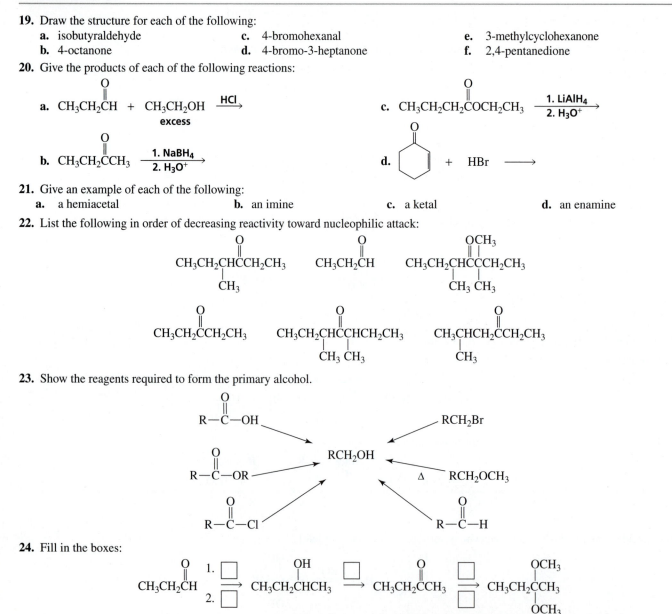

21. Give an example of each of the following:
   a. a hemiacetal
   b. an imine
   c. a ketal
   d. an enamine

22. List the following in order of decreasing reactivity toward nucleophilic attack:

23. Show the reagents required to form the primary alcohol.

24. Fill in the boxes:

25. Using cyclohexanone as the starting material, describe how each of the following could be synthesized:

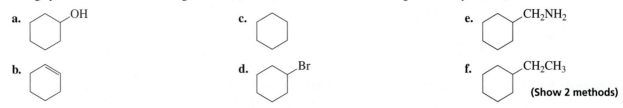

a. (cyclohexane with OH)   c. (cyclohexane)   e. (cyclohexane with CH₂NH₂)

b. (cyclohexene)   d. (cyclohexane with Br)   f. (cyclohexane with CH₂CH₃)

(Show 2 methods)

26. a. How many isomers are obtained from the reaction of 2-pentanone with ethylmagnesium bromide followed by addition of dilute acid?
   b. How many isomers are obtained from the reaction of 2-pentanone with methylmagnesium bromide followed by addition of dilute aqueous acid?

27. How would you convert *N*-methylbenzamide into the following compounds?
   a. *N*-methylbenzylamine   **b.** benzoic acid   **c.** methyl benzoate   **d.** benzyl alcohol

28. Propose a mechanism for the following reaction:

$$ HOCH_2CH_2CH_2CH_2 - \overset{\overset{\textstyle O}{\|}}{C} - H \xrightarrow[CH_3OH]{HCl} $$

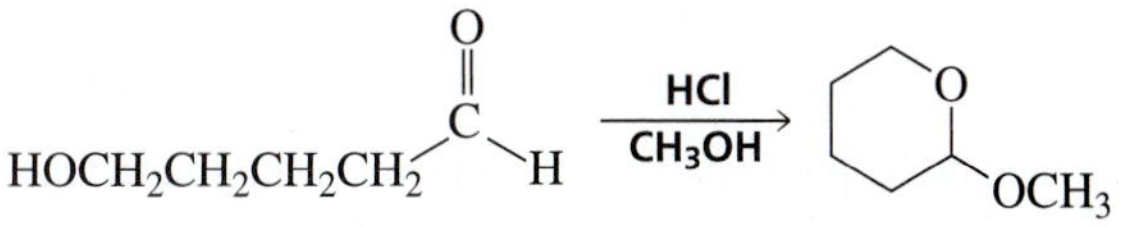

29. List the following in order of decreasing amount of hydrate formed in an acidic aqueous solution:

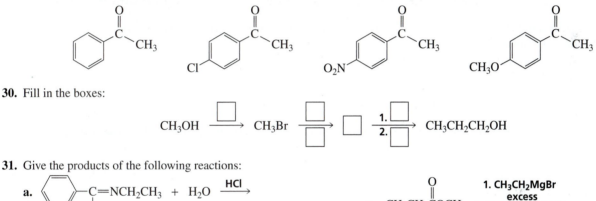

30. Fill in the boxes:

$$ CH_3OH \xrightarrow{\square} CH_3Br \quad \square \atop\square \quad \square \xrightarrow[2.]{1.\square} CH_3CH_2CH_2OH $$

31. Give the products of the following reactions:

a. (phenyl)—C(=NCH₂CH₃)(CH₂CH₃) + H₂O $\xrightarrow{HCl}$

b. CH₃CH₂CCH₃ (O) $\xrightarrow[2.\ H_3O^+]{1.\ CH_3CH_2MgBr}$

c. CH₃CH₂COCH₃ (O) $\xrightarrow[2.\ H_3O^+]{1.\ CH_3CH_2MgBr \atop excess}$

d. (phenyl ketone) + CH₃NH₂ $\xrightarrow[]{trace \atop H^+}$

32. List three different sets of reagents (each set consisting of a carbonyl compound and a Grignard reagent) that could be used to prepare each of the following tertiary alcohols:

a. 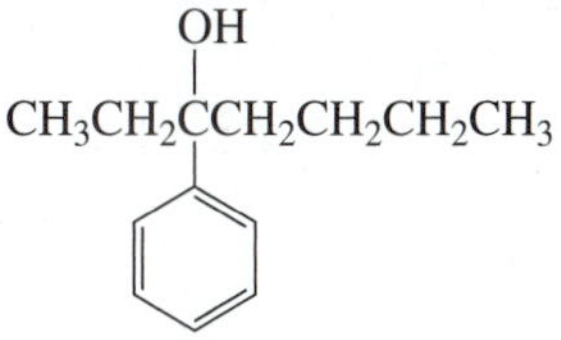 CH₃CH₂C(OH)(phenyl)CH₂CH₂CH₂CH₃

b. CH₃CH₂C(OH)(CH₂CH₃)CH₂CH₂CH₃

33. Give the product of the reaction of 3-methyl-2-cyclohexenone with each of the following reagents:
   a. CH₃CH₂SH   **b.** HBr   **c.** H₂, Pd/C

34. Give the product of each of the following reactions:

a. 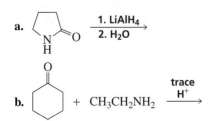 $\xrightarrow[2.\ H_2O]{1.\ LiAlH_4}$

b. (cyclohexanone) + CH₃CH₂NH₂ $\xrightarrow[]{trace \atop H^+}$

c. 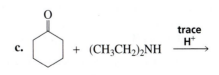 (cyclohexanone) + (CH₃CH₂)₂NH $\xrightarrow[]{trace \atop H^+}$

d.  CH₃C=CHCCH₃ + HBr ⟶

35. Indicate how the following compounds could be prepared from the given starting materials:

a. 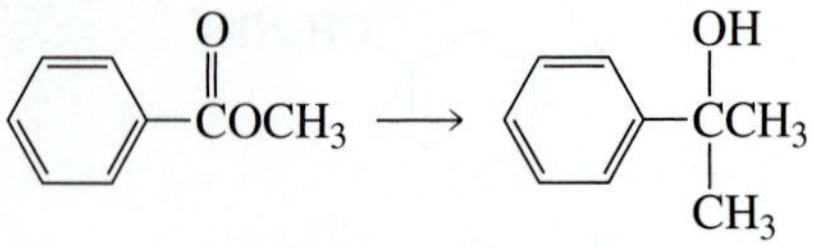 (aromatic ring)–COCH₃ ⟶ (aromatic ring)–C(OH)(CH₃)(CH₃)

b. 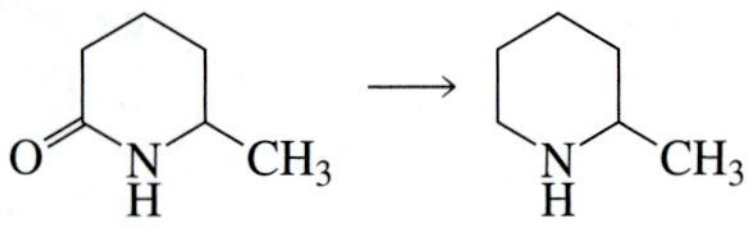 (piperidinone with N–H, O=, CH₃) ⟶ (piperidine with N–H, CH₃)

36. Give the products of the following reactions. Show all stereoisomers that are formed.

a. $CH_3CH_2CCH_2CH_2CH_2CH_3$ $\xrightarrow{\text{1. NaBH}_4}{\text{2. H}_3O^+}$

b. 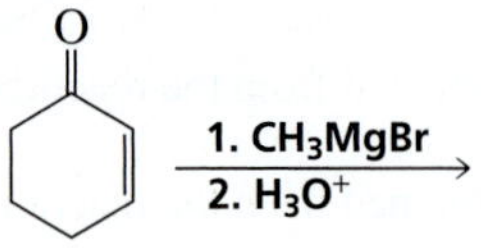 (cyclohexenone) $\xrightarrow{\text{1. CH}_3\text{MgBr}}{\text{2. H}_3O^+}$

37. Propose a mechanism for the following reaction:

(dihydropyran) + CH_3CH_2OH $\xrightarrow{\text{HCl}}$ (tetrahydropyran)–OCH₂CH₃

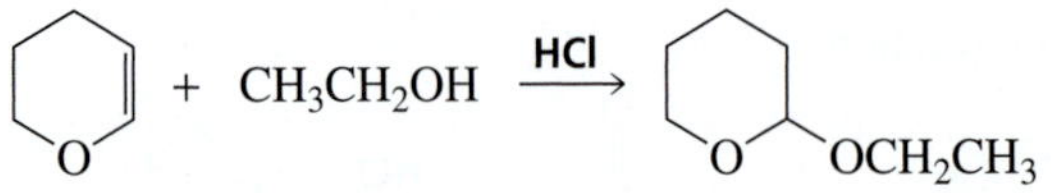

38. Indicate how the following compounds could be prepared from the given starting materials:

a. $CH_3CH_2CH_2CH_2Br$ ⟶ $CH_3CH_2CH_2CH_2COH$ b. $CH_3CH_2CH_2CH_2Br$ ⟶ $CH_3CH_2CH_2CH_2CH_2NH_2$

39. What class of alcohol (primary, secondary, or tertiary) is formed from the reaction of methyl formate with excess Grignard reagent followed by addition of dilute acid?

40. Put the appropriate compound in each box:

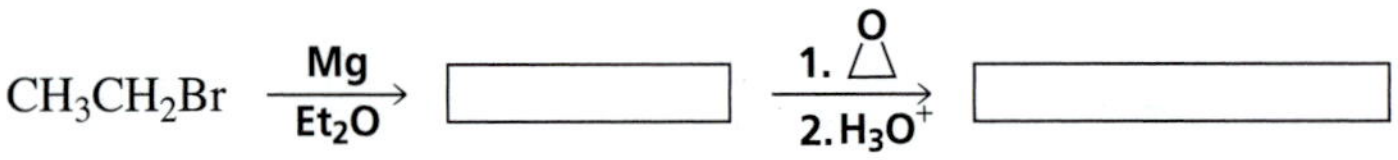

CH_3CH_2Br $\xrightarrow{\text{Mg}}{\text{Et}_2O}$ [   ] $\xrightarrow{\text{1. } \triangle \text{O}}{\text{2. H}_3O^+}$ [   ]

41. What alcohol would be formed from the reaction of the following Grignard reagent with ethylene oxide followed by the addition of acid?

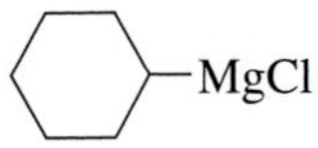

(cyclohexyl)–MgCl

42. a. Write the mechanism for the following reactions:
   1. The acid-catalyzed hydrolysis of an imine to a carbonyl compound and a protonated primary amine.
   2. The acid-catalyzed hydrolysis of an enamine to a carbonyl compound and a protonated secondary amine.
   b. How do these mechanisms differ?

43. Which of the following alkyl halides could be successfully used to prepare a Grignard reagent?

   a. $HOCH_2CH_2CH_2CH_2Br$ **b.** $BrCH_2CH_2CH_2COH$ **c.** $CH_3NCH_2CH_2CH_2Br$
                                                                                              $|$
                                                                                             CH_3

44. The pK_a values of oxaloacetic acid are 2.22 and 3.98.

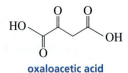

   oxaloacetic acid

   a. Which carboxyl group is more acidic?
   b. The amount of hydrate present in an aqueous solution of oxaloacetic acid depends on the pH of the solution: 95% at pH = 0, 81% at pH = 1.3, 35% at pH = 3.1, 13% at pH = 4.7, 6% at pH = 6.7, and 6% at pH = 12.7. Explain this pH dependence.

45. Propose a reasonable mechanism for the following reaction:

$CH_3CCH_2CH_2COCH_2CH_3$ $\xrightarrow{\text{1. CH}_3\text{MgBr}}{\text{2. H}_3O^+}$ (lactone with H₃C, H₃C, O, =O) + CH_3CH_2OH

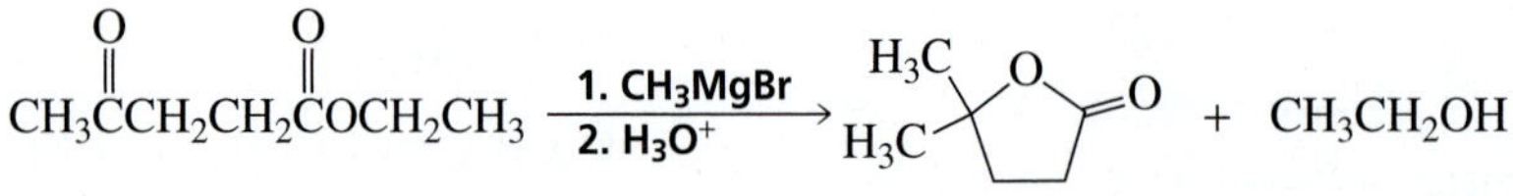

CHAPTER 13

Carbonyl Compounds III

Reactions at the α-Carbon

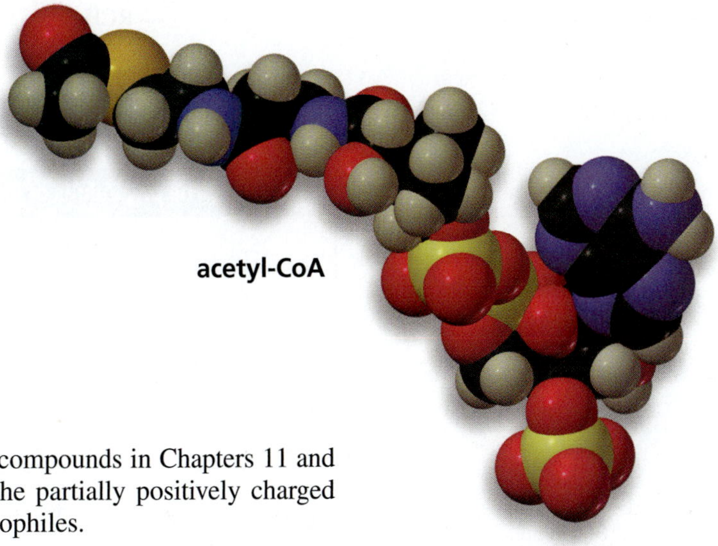

acetyl-CoA

W hen we looked at the reactions of carbonyl compounds in Chapters 11 and 12, we saw that their site of reactivity is the partially positively charged carbonyl carbon, which is attacked by nucleophiles.

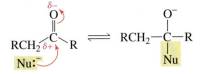

Aldehydes, ketones, and esters have a second site of reactivity. A hydrogen bonded to a carbon *adjacent* to a carbonyl carbon is sufficiently acidic to be removed by a strong base. The carbon adjacent to a carbonyl carbon is called an **α-carbon**. A hydrogen bonded to an α-carbon is called an **α-hydrogen**.

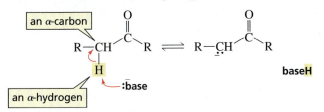

13.1 THE ACIDITY OF AN α-HYDROGEN

Hydrogen and carbon have similar electronegativities, which means that the electrons binding them together are shared almost equally by the two atoms. Consequently, a hydrogen bonded to a carbon is usually not acidic. This is particularly true for hydrogens bonded to sp^3 carbons because these carbons are the most similar to hydrogen in electronegativity (Section 5.13). The high pK_a of ethane provides evidence for the low acidity of a hydrogen bonded to an sp^3 carbon.

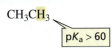

A hydrogen bonded to an sp^3 carbon adjacent to a carbonyl carbon, however, is much more acidic than hydrogens bonded to other sp^3 carbons. For example, the pK_a value for dissociation of a proton from an α-carbon of an aldehyde or a ketone ranges from 16 to

20, and the pK_a value for dissociation of a proton from the α-carbon of an ester is about 25 (Table 13.1). Notice that although an α-hydrogen is more acidic than most other carbon-bound hydrogens, it is less acidic than a hydrogen of water (p$K_a = 15.7$).

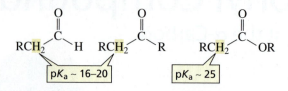

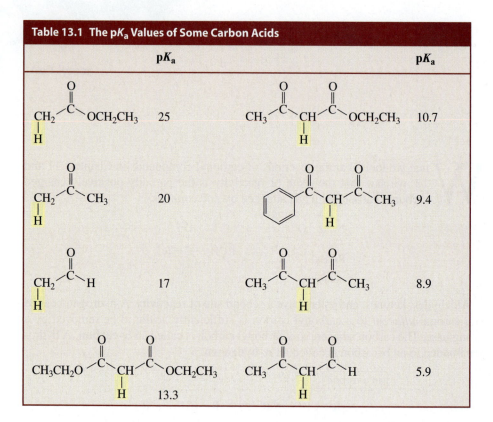

Table 13.1 The pK_a Values of Some Carbon Acids			
	pK_a		pK_a
CH_2—C(=O)—OCH_2CH_3 (with H)	25	CH_3—C(=O)—CH—C(=O)—OCH_2CH_3 (with H)	10.7
CH_2—C(=O)—CH_3 (with H)	20	phenyl—C(=O)—CH—C(=O)—CH_3 (with H)	9.4
CH_2—C(=O)—H (with H)	17	CH_3—C(=O)—CH—C(=O)—CH_3 (with H)	8.9
CH_3CH_2O—C(=O)—CH—C(=O)—OCH_2CH_3 (with H)	13.3	CH_3—C(=O)—CH—C(=O)—H (with H)	5.9

Why is a hydrogen bonded to an sp^3 carbon adjacent to a carbonyl carbon so much more acidic than hydrogens bonded to other sp^3 carbons? An α-hydrogen is more acidic because the base formed when a proton is removed from an α-carbon is more stable than the base formed when a proton is removed from other sp^3 carbons. As we have seen, the more stable the base, the stronger is its conjugate acid (Section 2.6).

Why is the base formed when a proton is removed from α-carbon more stable? When a proton is removed from ethane, the electrons left behind reside solely on a carbon atom. Because carbon is not very electronegative, a carbanion is unstable. As a result, the pK_a of its conjugate acid is very high.

localized electrons

$$CH_3CH_3 \ \rightleftharpoons \ CH_3\ddot{C}H_2 \ + \ H^+$$

In contrast, when a proton is removed from a carbon adjacent to a carbonyl carbon, two factors combine to increase the stability of the base that is formed. First, the electrons left behind, when the proton is removed, are delocalized, and we have seen that electron delocalization increases stability (Section 7.6). More importantly, the electrons

are delocalized onto an oxygen, an atom that is better able to accommodate them because it is more electronegative than carbon.

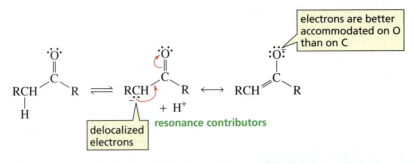

Now we can understand why aldehydes and ketones ($pK_a = 16 - 20$) are more acidic than esters ($pK_a = 25$). The electrons left behind when a proton is removed from the α-carbon of an ester are not as readily delocalized onto the carbonyl oxygen as they would be in an aldehyde or a ketone. This is because the oxygen of the OR group of the ester also has a lone pair that can be delocalized onto the carbonyl oxygen. Thus, the two pairs of electrons compete for delocalization onto the same oxygen.

The α-hydrogen of a ketone or an aldehyde is more acidic than the α-hydrogen of an ester.

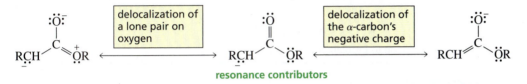

If the α-carbon is *between* two carbonyl groups, the acidity of an α-hydrogen is even greater (Table 13.1). For example, the pK_a value for dissociation of a proton from the α-carbon of 2,4-pentanedione, a compound with an α-carbon between two ketone carbonyl groups, is 8.9; the pK_a value for dissociation of a proton from the α-carbon of ethyl 3-oxobutyrate, a compound with an α-carbon between a ketone carbonyl group and an ester carbonyl group, is 10.7.

2,4-pentanedione
acetylacetone
a β-diketone

ethyl 3-oxobutyrate
ethyl acetoacetate
a β-keto ester

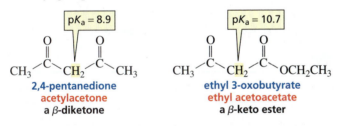

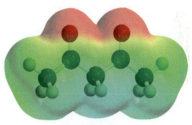

2,4-pentanedione

The acidity of α-hydrogens bonded to carbons flanked by two carbonyl groups increases because the electrons left behind when the proton is removed can be delocalized onto *two* oxygen atoms.

resonance contributors for the 2,4-pentanedione anion

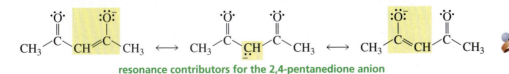

3-D Molecules: Enol of acetone; Enol of a β-diketone

PROBLEM 1

Identify the most acidic hydrogen in each compound.

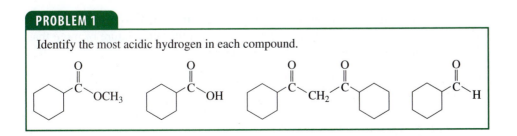

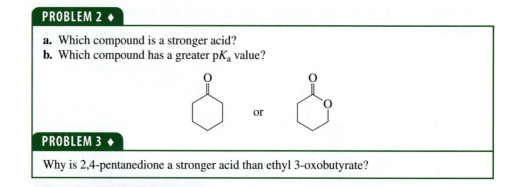
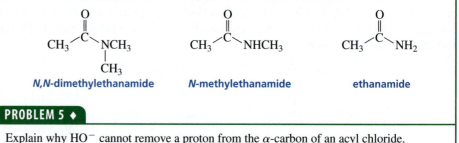
13.2 KETO–ENOL TAUTOMERS

A ketone exists in equilibrium with its enol tautomer. Recall that **tautomers** are isomers that are in rapid equilibrium (Section 5.11). Keto–enol tautomers differ in the location of a double bond and a hydrogen.

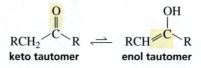

For most ketones, the enol tautomer is much less stable than the keto tautomer. For example, an aqueous solution of acetone exists as an equilibrium mixture of more than 99.9% keto tautomer and less than 0.1% enol tautomer.

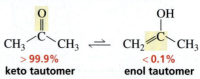

Phenol is unusual in that its enol tautomer is *more* stable than its keto tautomer because the enol tautomer is aromatic, but the keto tautomer is not.

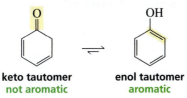

keto tautomer
not aromatic

enol tautomer
aromatic

13.3 KETO–ENOL INTERCONVERSION

Now that we know that a hydrogen bonded to a carbon adjacent to a carbonyl carbon is somewhat acidic, we can understand why keto and enol tautomers interconvert as we first saw in Chapter 5. The interconversion of keto and enol tautomers is called **keto–enol interconversion** or **tautomerization**. The interconversion can be catalyzed by either a base or an acid. The mechanism for base-catalyzed keto–enol interconversion is shown below.

mechanism for base-catalyzed keto–enol interconversion

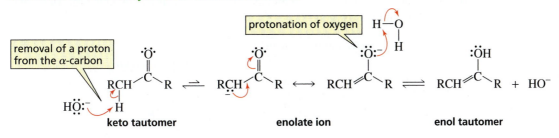

keto tautomer

enolate ion

enol tautomer

- Hydroxide ion removes a proton from the α-carbon of the keto tautomer, forming an anion called an **enolate ion**. The enolate ion has two resonance contributors.
- Protonation on oxygen forms the enol tautomer, whereas protonation on the α-carbon reforms the keto tautomer.

The mechanism for acid-catalyzed keto–enol interconversion follows.

mechanism for acid-catalyzed keto–enol interconversion

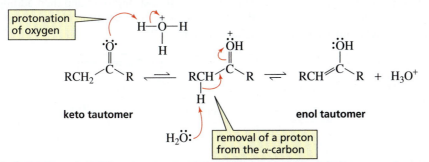

keto tautomer

enol tautomer

- The acid protonates the carbonyl oxygen of the keto tautomer.
- Water removes a proton from the α-carbon, forming the enol.

Notice that the steps are reversed in the base- and acid-catalyzed reactions. In the base-catalyzed reaction, the base removes a proton from the α-carbon in the first step and the oxygen is protonated in the second step. In the acid-catalyzed reaction, the oxygen is protonated in the first step and the proton is removed from the α-carbon in the second step.

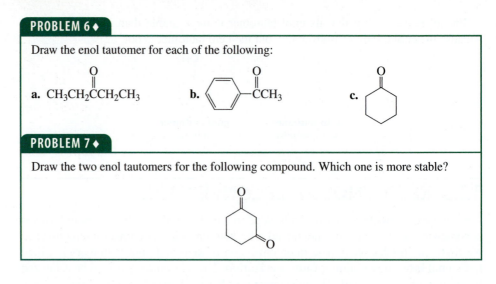

PROBLEM 6 ♦

Draw the enol tautomer for each of the following:

a. $CH_3CH_2CCH_2CH_3$ (with O double-bonded above the third carbon)

b. (benzene ring)—CCH_3 (with O double-bonded above)

c. (cyclohexanone ring with O)

PROBLEM 7 ♦

Draw the two enol tautomers for the following compound. Which one is more stable?

(1,3-cyclohexanedione structure with two O atoms)

13.4 ALKYLATION OF ENOLATE IONS

The resonance contributors of an enolate ion show that it has two electron-rich sites: the α-carbon and the oxygen.

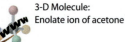
3-D Molecule:
Enolate ion of acetone

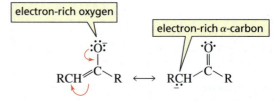

resonance contributors of an enolate ion

Which nucleophilic site (C or O) reacts with an electrophile depends on the electrophile. Protonation occurs preferentially on oxygen because of the greater concentration of negative charge on the more electronegative oxygen atom. However, when the electrophile is something other than a proton, carbon is more likely to be the nucleophile because carbon is a better nucleophile than oxygen.

Alkylation of the α-carbon of a carbonyl compound is an important reaction because it gives us another way to form a carbon–carbon bond. Alkylation is carried out by first removing a proton from the α-carbon with a strong base and then adding the appropriate alkyl halide. Because the alkylation is an S_N2 reaction, it works best with primary alkyl halides and methyl halides (Section 9.2).

Enolate ions can be alkylated on the α-carbon.

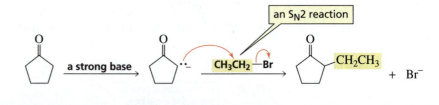

THE SYNTHESIS OF ASPIRIN

In the first step in the industrial synthesis of aspirin, a phenolate ion reacts with carbon dioxide under pressure to form *o*-hydroxybenzoic acid, also known as salicylic acid. Salicylic acid reacts with acetic anhydride to form acetylsalicylic acid (aspirin).

During World War I, the Bayer Company bought as much phenol as it could on the international market, knowing that eventually all the phenol could be used to manufacture aspirin. This left little phenol available for other countries to purchase for the synthesis of 2,4,6-trinitrophenol, a common explosive at that time.

(continued...)

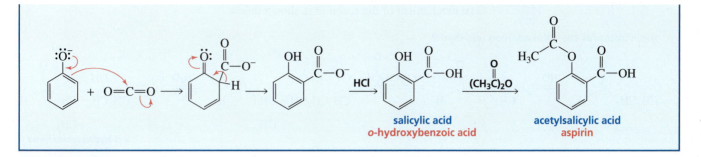

salicylic acid
o-hydroxybenzoic acid

acetylsalicylic acid
aspirin

PROBLEM 8

Draw the contributing resonance structures for the enolate ion of
a. 3-pentanone   **b.** cyclohexanone

PROBLEM 9 ♦

Give the product that would be formed if the enolate ion of each compound in Problem 8 were treated with ethyl bromide.

13.5 AN ALDOL ADDITION FORMS β-HYDROXYALDEHYDES AND β-HYDROXYKETONES

We saw in Chapter 12 that the carbonyl carbon of aldehydes and ketones is an electrophile. We have just seen that a proton can be removed from the α-carbon of an aldehyde or a ketone, converting the α-carbon into a nucleophile. An **aldol addition** is a reaction in which *both* of these activities are observed: one molecule of a carbonyl compound—after a proton is removed from an α-carbon—reacts as a *nucleophile* and attacks the *electrophilic* carbonyl carbon of a second molecule of the carbonyl compound.

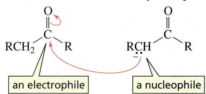

Thus, an aldol addition is a reaction between two molecules of an *aldehyde* or two molecules of a *ketone*. When the reactant is an aldehyde, the addition product is a β-hydroxyaldehyde, which is why the reaction is called an aldol addition ("ald" for aldehyde, "ol" for alcohol). When the reactant is a ketone, the addition product is a β-hydroxyketone. Notice that the reaction forms a new C—C bond between the α-carbon of one molecule and the carbon that formerly was the carbonyl carbon of the other molecule.

aldol additions

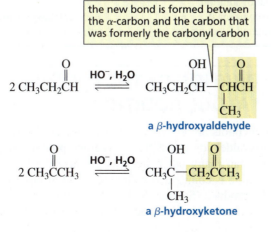

The mechanism of the reaction is shown below.

mechanism for the aldol addition (aldehyde)

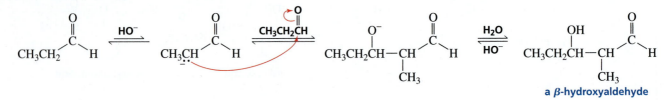

a β-hydroxyaldehyde

The new C—C bond formed in an aldol addition connects the α-carbon of one molecule and the carbon that formerly was the carbonyl carbon of the other molecule.

- A base removes a proton from an α-carbon, creating an enolate ion (a nucleophile).
- The enolate ion adds to the carbonyl carbon of a second molecule of the carbonyl compound.
- The negatively charged oxygen is protonated by the solvent.

Because an aldol addition occurs between two molecules of the same carbonyl compound, the product has twice as many carbons as the reacting aldehyde or ketone.

mechanism for the aldol addition (ketone)

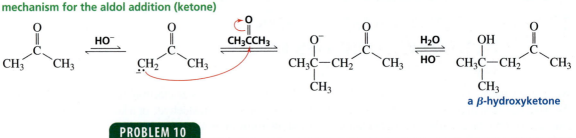

a β-hydroxyketone

3-D Molecules:
β-Hydroxyaldehyde;
β-Hydroxyketone

PROBLEM 10

Show the aldol addition product that would be formed from each of the following compounds:

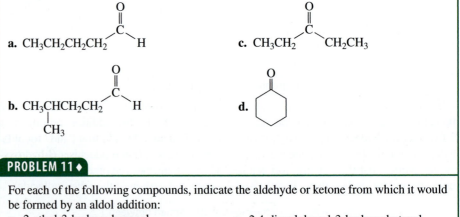

a. CH₃CH₂CH₂CH₂—C(=O)—H

b. CH₃CHCH₂CH₂—C(=O)—H
   |
   CH₃

c. CH₃CH₂—C(=O)—CH₂CH₃

d.

PROBLEM 11 ♦

For each of the following compounds, indicate the aldehyde or ketone from which it would be formed by an aldol addition:
a. 2-ethyl-3-hydroxyhexanal
b. 4-hydroxy-4-methyl-2-pentanone
c. 2,4-dicyclohexyl-3-hydroxybutanal
d. 5-ethyl-5-hydroxy-4-methyl-3-heptanone

13.6 DEHYDRATION OF THE PRODUCT OF AN ALDOL ADDITION

We have seen that alcohols are dehydrated when they are heated with acid (Section 10.3). The β-hydroxyaldehyde and β-hydroxyketone products of aldol additions are easier to dehydrate than many other alcohols because the double bond formed as the result of dehydration is conjugated with a carbonyl group. Conjugation increases the stability of the product (Section 6.7) and therefore makes it easier to form. If the product of an aldol addition is dehydrated, the overall reaction is called an

aldol condensation. A **condensation reaction** is a reaction that combines two molecules by forming a new C—C bond while removing a small molecule (usually water or an alcohol). Notice that an aldol condensation forms an α,β-unsaturated aldehyde or an α,β-unsaturated ketone.

An aldol addition product loses water to form an aldol condensation product.

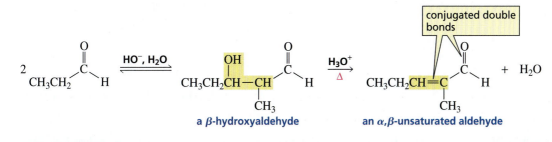

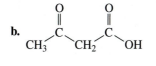

PROBLEM 12 ◆

Give the product obtained from the aldol condensation of cyclohexanone.

PROBLEM 13   *SOLVED*

How could you prepare the following compounds using a starting material containing no more than three carbons?

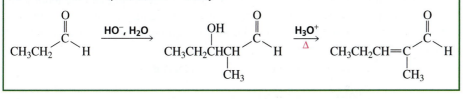

Tutorial:
Aldol reactions—synthesis

Solution to 13a   A compound with the correct six-carbon skeleton can be obtained if a three-carbon aldehyde undergoes an aldol addition. Dehydration of the addition product forms the desired α,β-unsaturated aldehyde.

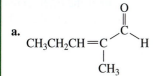

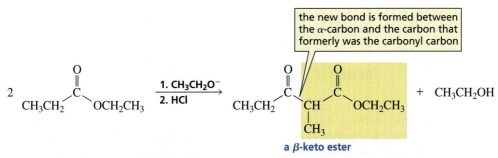

13.7 A CLAISEN CONDENSATION FORMS A β-KETO ESTER

When two molecules of an *ester* undergo a condensation reaction, the reaction is called a **Claisen condensation**. The product of a Claisen condensation is a β-keto ester.

In a Claisen condensation, as in an aldol addition, one molecule of carbonyl compound is converted into an enolate ion that reacts with a second molecule of the carbonyl compound.

The new C—C bond formed in a Claisen condensation connects the α-carbon of one molecule and the carbon that formerly was the carbonyl carbon of the other molecule.

mechanism for the Claisen condensation

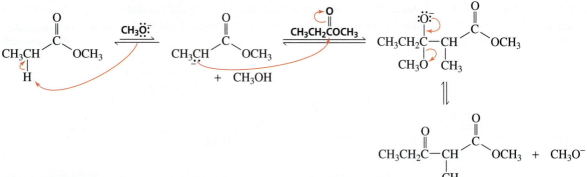

Ludwig Claisen (1851–1930) *was born in Germany and received a Ph.D. from the University of Bonn, studying under Kekulé (Section 7.1). Claisen was a professor of chemistry at the University of Bonn, Owens College (Manchester, England), the University of Munich, the University of Aachen, the University of Kiel, and the University of Berlin.*

3-D Molecule:
β-Keto ester

- A strong base removes a proton from an α-carbon, forming an enolate ion.
- The enolate ion attacks the carbonyl carbon of a second molecule of ester. The new C—C bond is formed between the α-carbon of one molecule and the carbon that formerly was the carbonyl carbon of the other molecule.
- The negatively charged oxygen reforms the carbon–oxygen π bond and expels the ⁻OR group.

When the reaction is over, HCl is added to protonate the ⁻OR group. The base employed in a Claisen condensation corresponds to the leaving group of the ester so that the reactant will not change if the base were to act as a nucleophile and attack the carbonyl group.

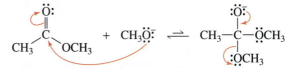

Notice that after nucleophilic attack, the Claisen condensation and the aldol addition differ. In the Claisen condensation, *the negatively charged oxygen* reforms the carbon–oxygen π bond. In the aldol addition, *the negatively charged oxygen* obtains a proton from the solvent.

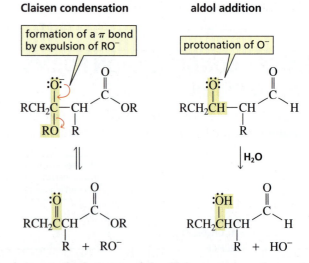

The difference between the last step of the Claisen condensation and the last step of the aldol addition arises from the difference between esters and aldehydes or ketones. With esters, the carbon bonded to the negatively charged oxygen is also bonded to a group that can be expelled. With aldehydes or ketones, the carbon bonded to the negatively charged oxygen is not bonded to a group that can be expelled. Thus, the Claisen condensation is a *nucleophilic substitution reaction*, whereas the aldol addition is a *nucleophilic addition reaction*.

PROBLEM 14 ◆

Give the products of the following reactions:

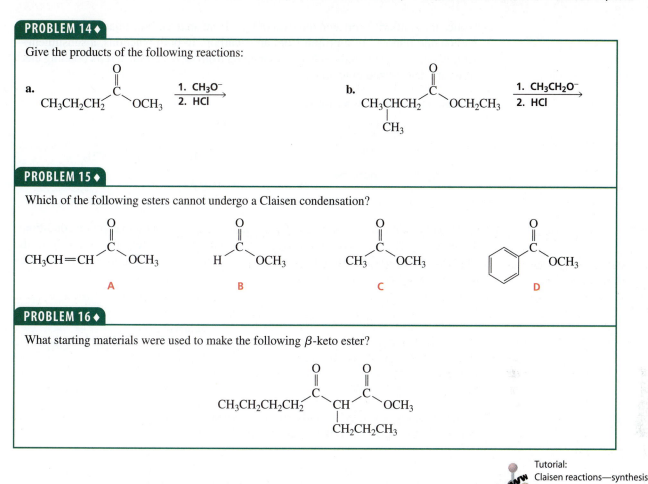

a.

$$CH_3CH_2CH_2 - C(=O) - OCH_3 \xrightarrow[\text{2. HCl}]{\text{1. } CH_3O^-}$$

b.

$$CH_3CHCH_2 - C(=O) - OCH_2CH_3 \xrightarrow[\text{2. HCl}]{\text{1. } CH_3CH_2O^-}$$
$$\quad\quad\quad |$$
$$\quad\quad CH_3$$

PROBLEM 15 ◆

Which of the following esters cannot undergo a Claisen condensation?

$$CH_3CH=CH - C(=O) - OCH_3$$
A

$$H - C(=O) - OCH_3$$
B

$$CH_3 - C(=O) - OCH_3$$
C

$$C_6H_5 - C(=O) - OCH_3$$
D

PROBLEM 16 ◆

What starting materials were used to make the following β-keto ester?

$$CH_3CH_2CH_2CH_2 - C(=O) - CH - C(=O) - OCH_3$$
$$\quad\quad\quad\quad\quad\quad |$$
$$\quad\quad\quad\quad\quad CH_2CH_2CH_3$$

Tutorial:
Claisen reactions—synthesis

13.8 CARBOXYLIC ACIDS WITH A CARBONYL GROUP AT THE 3-POSITION CAN BE DECARBOXYLATED

Carboxylate ions do not lose CO_2, for the same reason that alkanes such as ethane do not lose a proton—because the leaving group would be a carbanion. Carbanions are very strong bases and therefore are very poor leaving groups.

$$CH_3CH_2 - H \quad\quad CH_3CH_2 - C(=O) - \ddot{O}\!:^-$$

If, however, the CO_2 group is bonded to a carbon adjacent to a carbonyl carbon, the CO_2 group can be removed, because the electrons left behind can be delocalized onto the carbonyl oxygen. Consequently, carboxylate ions with a carbonyl group at the 3-position lose CO_2 when they are heated. Loss of CO_2 from a molecule is called **decarboxylation**.

removing CO₂ from an α-carbon

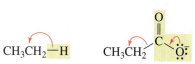

3-oxobutanoate ion
acetoacetate ion

$$+ \; CO_2$$

A carboxylic acid with a carbonyl group in the 3-position loses CO_2 when heated.

Notice the similarity between removal of CO_2 from an α-carbon adjacent to a carbonyl carbon and removal of a proton from an α-carbon. In both reactions, a substituent—CO_2 in one case, H^+ in the other—is removed from an α-carbon and its bonding electrons are delocalized onto an oxygen.

removing a proton from an α-carbon

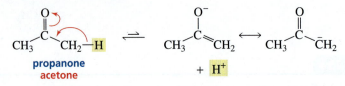

Decarboxylation is even easier if the reaction is carried out under acidic conditions, because the reaction is catalyzed by an intramolecular transfer of a proton from the carboxyl group to the carbonyl oxygen. The enol that is formed immediately tautomerizes to a ketone.

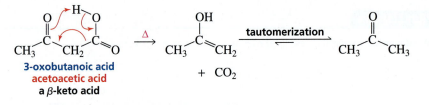

In summary, carboxylic acids and carboxylate ions with a carbonyl group at the 3-position lose CO_2 when they are heated.

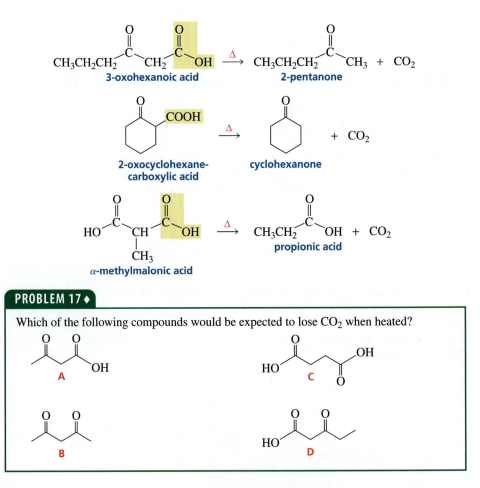

PROBLEM 17 ◆

Which of the following compounds would be expected to lose CO_2 when heated?

13.9 THE MALONIC ESTER SYNTHESIS: A METHOD TO SYNTHESIZE A CARBOXYLIC ACID

A combination of two of the reactions discussed in this chapter—alkylation of an α-carbon and decarboxylation of a carboxylic acid that has a carbonyl group at the 3-position—can be used to prepare carboxylic acids of any desired chain length. The procedure is called the **malonic ester synthesis** because the starting material for the synthesis is the diethyl ester of malonic acid.

The first two carbons of the carboxylic acid being synthesized come from malonic ester, and the rest of the carboxylic acid comes from the alkyl halide used in the second step of the synthesis.

malonic ester synthesis

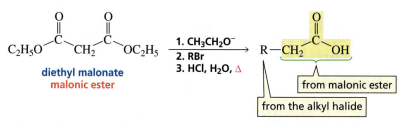

A malonic ester synthesis forms a carboxylic acid with two more carbons than the alkyl halide.

The steps in the malonic ester synthesis are shown below.

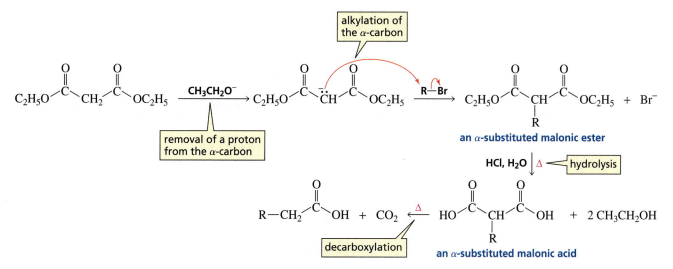

- A base easily removes a proton from the α-carbon because it is flanked by two ester groups ($pK_a = 13$).
- The resulting α-carbanion reacts with an alkyl halide, forming an α-substituted malonic ester. Because alkylation is an S_N2 reaction, it works best with primary alkyl halides and methyl halides (Section 9.2).
- Heating the α-substituted malonic ester in an acidic aqueous solution hydrolyzes both ester groups.
- Further heating decarboxylates the α-substituted malonic acid.

Tutorial:
Malonic ester synthesis

PROBLEM 18 ◆

What alkyl bromide should be used in the malonic ester synthesis of each of the following carboxylic acids?

a. propanoic acid   **b.** 3-phenylpropanoic acid   **c.** 4-methylpentanoic acid

> **PROBLEM 21**
>
> Propose a mechanism for the formation of fructose-1,6-diphosphate from dihydroxyace-tone phosphate and glyceraldehyde-3-phosphate, using HO^- as the catalyst.

A Biological Aldol Condensation

Collagen is the most abundant protein in mammals, amounting to about one-fourth of the total protein. It is the major fibrous component of bone, teeth, skin, cartilage, and tendons. Individual collagen molecules can be isolated only from tissues of young animals. As animals age, the individual molecules become cross-linked, which is why meat from older animals is tougher than meat from younger ones. Collagen cross-linking is an example of an aldol condensation.

Before collagen molecules can cross-link, their ammonium groups must be converted to aldehyde groups. The enzyme that catalyzes this reaction is called lysyl oxidase. An aldol condensation between two aldehyde groups results in a cross-linked protein.

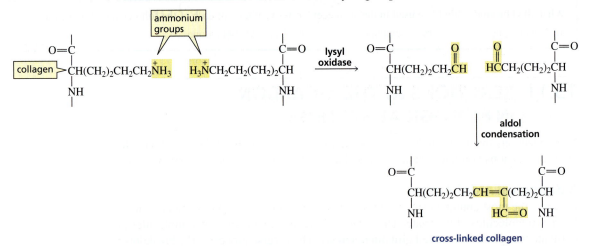

cross-linked collagen

A Biological Claisen Condensation

Fatty acids are long-chain, unbranched carboxylic acids (Sections 11.10 and 19.1). Most naturally occurring fatty acids contain an even number of carbons because they are synthesized from acetic acid, which has two carbons.

Because the biological reaction occurs at physiological pH (7.3), the reactant is acetate—acetic acid without its proton. We have seen that carboxylate ions are very unreactive toward nucleophilic attack (Section 11.11). Biological systems can activate carboxylate ions by converting them into *thioesters*. A **thioester** is an ester with sulfur in place of the carboxylate oxygen. This reaction requires ATP; ATP puts a leaving group on the carboxylate ion that can be replaced by the thiol (Section 18.2). A **thiol** is an alcohol with sulfur in place of the oxygen. The particular *thiol* used to make the thioester shown below is called coenzyme A.

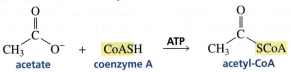

One of the necessary reactants for fatty acid synthesis is malonyl-CoA, which is obtained by carboxylation of acetyl-CoA (Section 17.8).

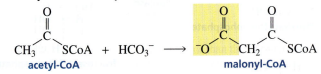

Before fatty acid synthesis can occur, the acyl groups of acetyl-CoA and malonyl-CoA are transferred to other thiols by means of a transesterification reaction (Section 11.8).

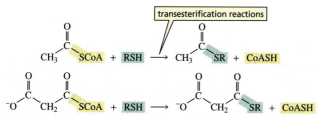

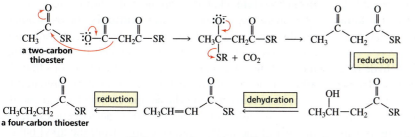

A molecule of acetyl thioester and a molecule of malonyl thioester are the reactants for the first round in the biosynthesis of a fatty acid.

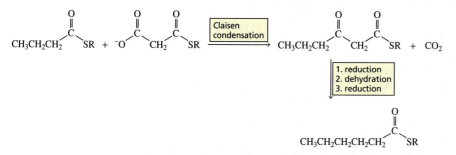

- The first step is a Claisen condensation. We have seen that the nucleophile need- ed for a Claisen condensation is obtained by using a strong base to remove an α-hydrogen. Strong bases are not available in living cells, however, because bio- logical reactions take place at neutral pH. Thus, the required nucleophile is gen- erated by removing CO_2—rather than a proton—from the α-carbon of malonyl thioester. (Recall that a carboxylic acid with a carbonyl group in the 3-position is easily decarboxylated; see Section 13.8.)

- The product of the condensation reaction undergoes a reduction, a dehydration, and a second reduction to give a four-carbon thioester.

- The four-carbon thioester and a molecule of malonyl thioester are the reactants for the second round. Again, the product of the condensation reaction undergoes a reduction, a dehydration, and a second reduction, this time to form a six-carbon thioester.

The sequence of reactions is repeated, and each time two more carbons are added to the chain. Now we can understand why naturally occurring fatty acids are unbranched and generally contain an even number of carbons.

Once a thioester with the appropriate number of carbons is obtained, it undergoes a transesterification reaction with glycerol in order to form fats, oils, and phospholipids (see Sections 11.10, 19.3, and 19.5).

PROBLEM 22 ◆

Palmitic acid is a straight-chain saturated 16-carbon fatty acid. How many moles of malonyl-CoA are required for the synthesis of one mole of palmitic acid?

PROBLEM 23 ◆

a. If the biosynthesis of palmitic acid were carried out with CD_3COSR and nondeuterated malonyl thioester, how many deuteriums would be incorporated into palmitic acid?

b. If the biosynthesis of palmitic acid were carried out with $^-OOCCD_2COSR$ and nondeute- rated acetyl thioester, how many deuteriums would be incorporated into palmitic acid?

Tutorial:
Common terms:
  reactions at the α-carbon

33. Both 2,6-heptanedione and 2,8-nonanedione form a product with a six-membered ring when treated with sodium hydroxide. Give the structure of the six-membered ring products.

34. a. Explain why a racemic mixture of 2-methyl-1-phenyl-1-butanone is formed when (R)-2-methyl-1-phenyl-1-butanone is dissolved in a basic aqueous solution.
   b. Give an example of another ketone that would form a racemic mixture in a basic aqueous solution.

35. An intramolecular Claisen condensation is called a Dieckmann condensation. Give the mechanism for the following Dieckmann condensation.

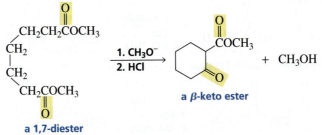

36. What product is formed when a 1,6-diester undergoes a Dieckmann condensation?

37. Show how the following compounds could be synthesized from the given starting materials:

38. Give the products of the following reactions:
   a. diethyl heptanedioate: (1) sodium ethoxide; (2) HCl
   b. diethyl 2-ethylhexanedioate: (1) sodium ethoxide; (2) HCl
   c. diethyl malonate: (1) sodium ethoxide; (2) isobutyl bromide; (3) HCl, H_2O + Δ
   d. 2,7-octanedione + aqueous sodium hydroxide

39. Which would require a higher temperature, decarboxylation of a β-dicarboxylic acid or decarboxylation of a β-keto acid?

40. What compound is formed when a dilute solution of cyclohexanone is shaken with NaOD in D_2O for several hours?

41. Explain why the following carboxylic acid cannot be prepared by the malonic ester synthesis:

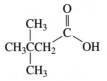

42. Give the structures of the four β-hydroxyaldehydes that would be obtained from a mixture of butanal and pentanal in a basic aqueous solution.

43. a. Draw the enol tautomer of 2,4-pentanedione.
   b. Most ketones form less than 1% enol in an aqueous solution. Explain why the enol tautomer of 2,4-pentanedione is much more prevalent (15%).

44. Give the major product of the following reaction:

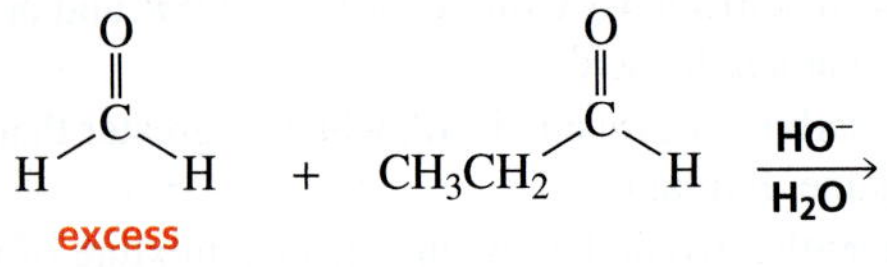

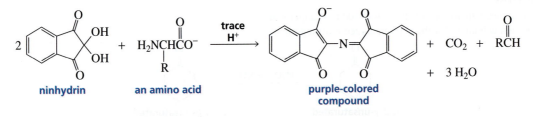

45. Ninhydrin reacts with an amino acid to form a purple-colored compound. Propose a mechanism to account for the formation of the colored compound.

CHAPTER 14

Determining the Structures of Organic Compounds

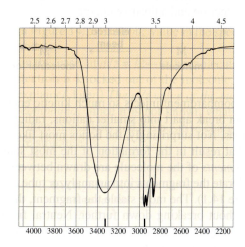

E
wav
give
the

T
of th

whe
Ger
freq
divi

A
mos
in 1
tion
equ;

S
higl

PF
O
th

a

Determining the structures of organic compounds is an important part of organic chemistry. Whenever a chemist synthesizes a compound, its structure must be confirmed. For example, you were told that a ketone is formed when an alkyne undergoes the acid-catalyzed addition of water (Section 5.11). But how was it determined that the product of that reaction is actually a ketone?

Scientists search the world for new compounds with physiological activity. If a promising compound is found, its structure needs to be determined. Without knowing its structure, chemists cannot design ways to synthesize the compound, nor can they undertake studies to provide insights into its biological behavior.

At one time, determining the structure of an organic compound was a daunting task and required a relatively large amount of the compound—a real problem for the analysis of compounds that were difficult to obtain. Today, a number of different instrumental techniques are used to determine the structures of organic compounds. These techniques can be performed quickly on small amounts of a compound. We have already discussed one such technique: ultraviolet/visible (UV/Vis) spectroscopy, which provides information about organic compounds with conjugated double bonds (Section 7.10). In this chapter, we will look at three more instrumental techniques.

- **mass spectrometry**, which allows us to determine the *molecular mass* and the *molecular formula* of a compound, as well as some of its *structural features*,

- **infrared (IR) spectroscopy**, which allows us to determine the *kinds of functional groups* in a compound, and

- **nuclear magnetic resonance (NMR) spectroscopy**, which provides information about the carbon–hydrogen framework of a compound.

We will be referring to different classes of organic compounds as we discuss various instrumental techniques; these classes are listed inside the back cover of the book for easy reference.

14.10 C—H ABSORPTION BANDS

The strength of a C—H bond depends on the hybridization of the carbon: a C—H bond is stronger when the carbon is sp hybridized than when it is sp^2 hybridized, which in turn is stronger than when the carbon is sp^3 hybridized. More energy is needed to stretch a stronger bond, and this is reflected in the C—H stretch absorption bands, which occur at ~3300 cm^{-1} when the carbon is sp hybridized, at ~3100 cm^{-1} when the carbon is sp^2 hybridized, and at ~2900 cm^{-1} when the carbon is sp^3 hybridized (Table 14.3).

Table 14.3 IR Absorptions of Carbon–Hydrogen Bonds	
Carbon–Hydrogen Stretching Vibrations	**Wavenumber (cm^{-1})**
C≡C—H	~3300
C=C—H	3100–3020
C—C—H	2960–2850
R—C(=O)—H	~2820 and ~2720

A useful step in the analysis of a spectrum is to look at the absorption bands in the vicinity of 3000 cm^{-1}. Figures 14.10 and 14.11 show the IR spectra for methylcyclohexane and cyclohexene, respectively. The only absorption band in the vicinity of 3000 cm^{-1} in Figure 14.10 is slightly to the right of that value. This tells us that the compound has hydrogens bonded to sp^3 carbons, but none bonded to sp^2 or to sp carbons. Figure 14.11 shows absorption bands slightly to the left and slightly to the right of 3000 cm^{-1}, indicating that the compound that produced that spectrum contains hydrogens bonded to sp^2 and sp^3 carbons.

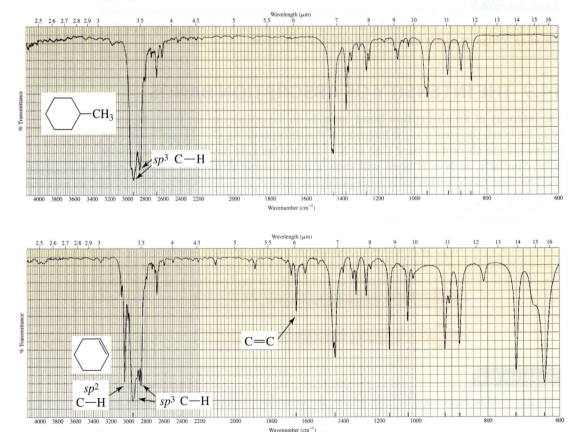

▶ **Figure 14.10**
The IR spectrum of methylcyclohexane.

▶ **Figure 14.11**
The IR spectrum of cyclohexene.

The stretch of the C—H bond in an aldehyde group shows two absorption bands, one at ~2820 cm^{-1} and the other at ~2720 cm^{-1} (Figure 14.12). This makes aldehydes relatively easy to identify because essentially no other absorption occurs at these wavenumbers.

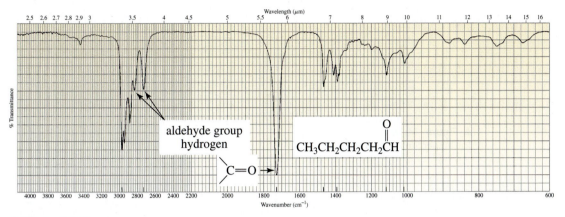

▲ **Figure 14.12**
The IR spectrum of pentanal. The absorptions at ~2820 and ~2720 cm^{-1} readily identify an aldehyde group. Note also the intense absorption band at ~1730 cm^{-1} indicating a C=O bond.

14.11 THE SHAPE OF ABSORPTION BANDS

The **shape of an absorption band** can be helpful in identifying the compound responsible for an IR spectrum. For example, both O—H and N—H bonds stretch at wavenumbers above 3100 cm^{-1}, but the shapes of their stretching absorption bands are distinctive. Notice the difference in the shape of these absorption bands in the IR spectra of 1-hexanol (Figure 14.13), pentanoic acid (Figure 14.14), and isopentylamine (Figure 14.15). An N—H absorption band (~3300 cm^{-1}) is narrower and less intense than an O—H absorption band (~3300 cm^{-1}), and the O—H absorption band of a carboxylic acid (~3300–2500 cm^{-1}) is broader than the O—H absorption band of an alcohol. Notice that two absorption bands are detectable in Figure 14.15 for the N—H stretch because there are two N—H bonds in the compound.

The position, intensity, and shape of an absorption band are helpful in identifying functional groups.

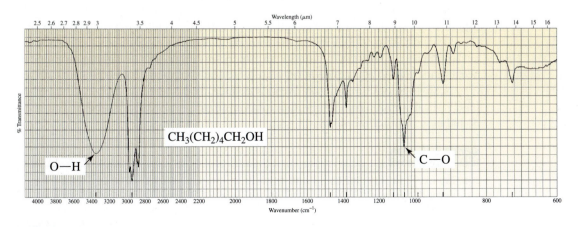

▲ **Figure 14.13**
The IR spectrum of 1-hexanol.

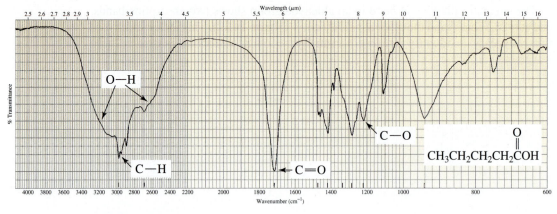

▲ **Figure 14.14**
The IR spectrum of pentanoic acid.

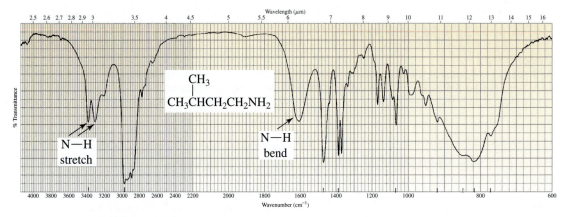

▲ **Figure 14.15**
The IR spectrum of isopentylamine. The double peak at ~3300 cm^{-1} indicates the presence of two N—H bonds.

PROBLEM 16

Why is the O—H stretch of a carboxylic acid broader than the O—H stretch of an alcohol?

14.12 THE ABSENCE OF ABSORPTION BANDS

The absence of an absorption band can be as useful as the presence of an absorption band in identifying a compound by IR spectroscopy. For example, the IR spectrum in Figure 14.16 shows a strong absorption at ~1100 cm^{-1}, indicating the presence of a C—O bond (Table 14.3). Clearly, the compound is not an alcohol because there is no absorption above 3100 cm^{-1}. Nor is it an ester or any other kind of carbonyl compound because there is no absorption at ~1700 cm^{-1}. The compound has no C≡C, C=C, C≡N, C=N, or C—N bonds. We may deduce, then, that the compound is an ether. Its C—H absorption bands show that it has hydrogens only on sp^3 carbons. The compound is in fact diethyl ether.

PROBLEM 17 ♦

a. An oxygen-containing compound shows an absorption band at ~1700 cm^{-1} and no absorption bands at ~3300 cm^{-1}, ~2700 cm^{-1}, or ~1100 cm^{-1}. What class of compound is it?

b. A nitrogen-containing compound shows no absorption band at ~3400 cm^{-1} and no absorption bands between 1700 cm^{-1} and 1600 cm^{-1}. What class of compound is it?

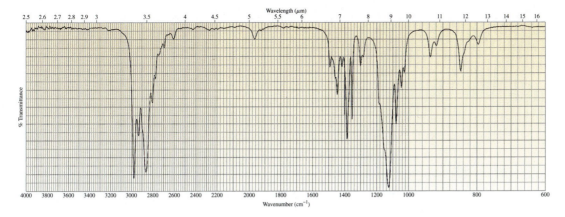

▲ **Figure 14.16**
The IR spectrum of diethyl ether.

14.13 **IDENTIFYING INFRARED SPECTRA**

We will now look at a few IR spectra and see what we can deduce about the structure of the compounds that give rise to the spectra. We might not be able to identify the compound precisely, but when we are told what it is, its structure should fit our observations.

Compound 1.   The absorption in the 3000 cm^{-1} region in Figure 14.17 indicates that hydrogens are attached to sp^2 carbons (3050 cm^{-1}) but not to sp^3 carbons. The absorptions at 1600 cm^{-1} and 1460 cm^{-1} indicate that the compound has a benzene ring. The absorptions at 2810 cm^{-1} and 2730 cm^{-1} show that the compound is an aldehyde. The absorption band for the carbonyl group (C=O) is lower (1700 cm^{-1}) than normal (1720 cm^{-1}), so the carbonyl group has partial single-bond character. Thus, it must be attached directly to the benzene ring. The compound is benzaldehyde.

Tutorial:
IR spectra

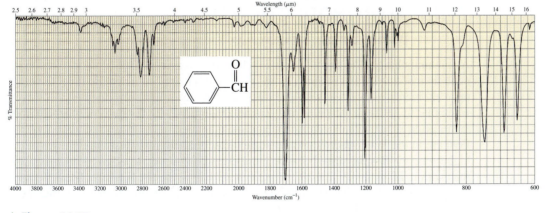

▲ **Figure 14.17**
The IR spectrum of Compound 1.

Compound 2.   The absorptions in the 3000 cm^{-1} region in Figure 14.18 indicate that hydrogens are attached to sp^3 carbons (2950 cm^{-1}) but not to sp^2 carbons. The shape of the strong absorption band at 3300 cm^{-1} is characteristic of an O—H group of an alcohol. The absorption at ~ 2100 cm^{-1} indicates that the compound has a triple bond. The sharp absorption band at 3300 cm^{-1} indicates that the compound has a hydrogen attached to an sp carbon, so we know it is a terminal alkyne. The compound is 2-propyn-1-ol.

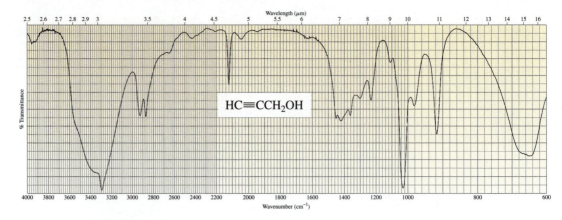

▲ **Figure 14.18**
The IR spectrum of Compound 2.

PROBLEM 18

How could IR spectroscopy distinguish between the following compounds?
a. a ketone and an aldehyde **c.** cyclohexene and cyclohexane
b. benzene and toluene **d.** a primary amine and a tertiary amine

---

PROBLEM 19

For each of the following pairs of compounds, give one absorption band that could be used to distinguish between them:

a. $CH_3CH_2COCH_3$ and CH_3CH_2COH **c.** $CH_3CH_2C{\equiv}CCH_3$ and $CH_3CH_2C{\equiv}CH$

b. CH_3CH_2COH and $CH_3CH_2CH_2OH$ **d.** 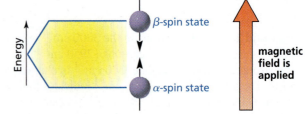 and

---

14.14 AN INTRODUCTION TO NMR SPECTROSCOPY

A spinning charged ^1H or ^{13}C nucleus generates a magnetic field similar to the magnetic field of a small bar magnet. In the absence of an applied magnetic field, the magnetic moments associated with the nuclear spins are randomly oriented. However, when placed between the poles of a magnet (Figure 14.19), the nuclear magnetic moments align either *with* or *against* the applied magnetic field. Those that align with the field are in the lower-energy **α-spin state**; those that align against the field are in the higher-energy **β-spin state** because more energy is needed to align against the field than with it. More nuclei are in the α-spin state than in the β-spin state. The difference in the populations is very small (about 20 out of 1 million nuclei) but is sufficient to form the basis of NMR spectroscopy.

▶ **Figure 14.19**
In the absence of an applied magnetic field, the magnetic moments of the nuclei are randomly oriented.
In the presence of an applied magnetic field, the magnetic moments of the nuclei line up with or against the field.

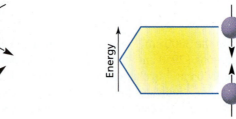

The energy difference (ΔE) between the α- and β-spin states depends on the strength of the **applied magnetic field** (B_0), measured in tesla (T). The greater the strength of the magnetic field, the greater is the difference in energy between the α- and β-spin states (Figure 14.20).

Earth's magnetic field is 5 $\times$ 10^{-5} T, measured at the equator. Its maximum surface magnetic field is 7 $\times$ 10^{-5} T, measured at the south magnetic pole.

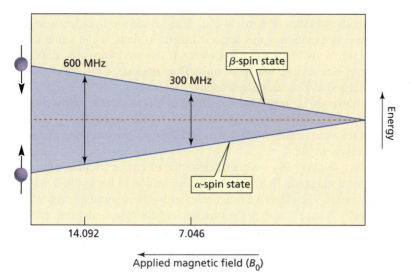

◀ **Figure 14.20**
The greater the strength of the applied magnetic field, the greater is the difference in energy between the α- and β-spin states.

When a sample is subjected to a pulse of radiation whose energy corresponds to the difference in energy (ΔE) between the α- and β-spin states, nuclei in the α-spin state are promoted to the β-spin state. When the nuclei absorb the radiation, they flip their spin, causing them to generate signals whose frequency depends on the difference in energy (ΔE) between the α- and β-spin states. The NMR spectrometer detects these signals and displays them as a plot of signal frequency versus intensity; this plot is the NMR spectrum. The nuclei are said to be *in resonance* with the radiation, hence the term **nuclear magnetic resonance (NMR)**. In this context, "resonance" refers to the flipping back and forth of nuclei between the α- and β-spin states in response to the radiation.

Today's NMR spectrometers operate at frequencies between 60 and 950 MHz. The greater the operating frequency of the instrument—and the stronger the magnet— the better is the resolution of the NMR spectrum.

NIKOLA TESLA (1856–1943)

Nikola Tesla was born in Croatia, the son of a clergyman. He immigrated to the United States in 1884 and became a citizen in 1891. He was a proponent of alternating current to distribute electricity and bitterly fought Thomas Edison, who promoted direct current. Tesla was granted a patent for developing the radio in 1900, but Guglielmo Marconi was also given a patent for its development in 1904. Not until shortly after his death was Tesla's patent upheld by the U.S. Supreme Court. Tesla also is given credit for developing neon and fluorescent lighting, the electron microscope, the refrigerator motor, and the Tesla coil, a type of transformer for changing the voltage of alternating current.

Nikola Tesla in his laboratory

14.15 SHIELDING CAUSES DIFFERENT HYDROGENS TO SHOW SIGNALS AT DIFFERENT FREQUENCIES

The frequency of an NMR signal depends on the strength of the applied magnetic field experienced by the nucleus (Figure 14.20). Thus, if all the hydrogens in an organic compound were to experience the applied magnetic field to the same degree, they would all give signals with the same frequency. If this were the case, all NMR spectra would consist of one signal, which would tell us nothing about the structure of the compound, except that it contains hydrogens.

A nucleus, however, is embedded in a cloud of electrons that partly *shields* it from the applied magnetic field. Fortunately for chemists, the **shielding** varies for different hydrogens within a molecule. In other words, all the hydrogens do not experience the same applied magnetic field.

What causes shielding? In a magnetic field, the electrons circulate about the nuclei and induce a local magnetic field that acts in opposition to the applied magnetic field and subtracts from it. The **effective magnetic field**—the amount of magnetic field that the nuclei actually "sense" through the surrounding electronic environment—is, therefore, somewhat smaller than the applied field:

$$B_{\text{effective}} = B_{\text{applied}} - B_{\text{local}}$$

This means that the greater the electron density of the environment in which the proton* is located, the greater B_{local} is and the more the proton is shielded from the applied magnetic field. Thus, protons in electron-rich environments sense a *smaller effective magnetic field*. They, therefore, will require a *lower frequency* to come into resonance—that is, flip their spins—because ΔE is smaller (Figure 14.20). Protons in electron-poor environments sense a *larger effective magnetic field* and, therefore, will require a *higher frequency* to come into resonance because ΔE is larger.

An NMR spectrum gives a signal for each proton in a different environment. Protons in electron-rich environments are more shielded from the applied magnetic field and appear at lower frequencies (on the right-hand side of the spectrum) (Figure 14.21). Protons in electron-poor environments are less shielded from the applied magnetic field and appear at higher frequencies (on the left-hand side of the spectrum). For example, the signal for the methyl protons of CH_3F occurs at a higher frequency than the signal for the methyl protons of CH_3Br because fluorine is more electron withdrawing than bromine; thus, the protons in CH_3F are in a less electron-rich environment.

The larger the magnetic field sensed by the proton, the higher is the frequency of the signal.

deshielded = less shielded

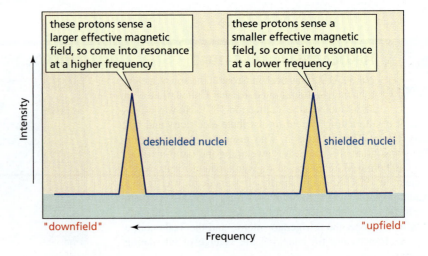

▶ **Figure 14.21**
Shielded nuclei come into resonance at lower frequencies than deshielded nuclei.

---

* The terms *proton* and *hydrogen* are both used to describe covalently bonded hydrogens in discussions of NMR spectroscopy.

The terms *upfield* and *downfield* are used to describe the position of a signal: **upfield** means farther to the right-hand side of the spectrum, and **downfield** means farther to the left-hand side of the spectrum.

14.16 THE NUMBER OF SIGNALS IN AN ^{1}H NMR SPECTRUM

Protons in the same environment are called **chemically equivalent protons**. For example, 1-bromopropane has three different sets of chemically equivalent protons: (1) the three methyl protons are chemically equivalent because of rotation about the C—C bonds; (2) the two methylene (CH_2) protons on the middle carbon are chemically equivalent; and (3) the two methylene protons on the carbon bonded to the bromine atom make up the third set of chemically equivalent protons.

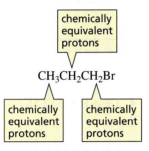

Each set of chemically equivalent protons in a compound gives rise to a signal in the ^{1}H NMR spectrum of that compound. Because 1-bromopropane has three sets of chemically equivalent protons, it has three signals in its ^{1}H NMR spectrum.

Each set of chemically equivalent protons gives rise to a signal.

2-Bromopropane has two sets of chemically equivalent protons and, therefore, it has two signals in its ^{1}H NMR spectrum; the six methyl protons in 2-bromopropane are equivalent, so they give rise to only one signal; the hydrogen bonded to the middle carbon gives the second signal. Ethyl methyl ether has three sets of chemically equivalent protons: the methyl protons on the carbon adjacent to the oxygen, the methylene (CH_2) protons on the carbon adjacent to the oxygen, and the methyl protons on the carbon that is one carbon removed from the oxygen. The chemically equivalent protons in the following compounds are designated by the same letter:

You can tell how many sets of chemically equivalent protons a compound has from the number of signals in its ^{1}H NMR spectrum.

PROBLEM-SOLVING STRATEGY

How many signals would you expect to see in the ^{1}H NMR spectrum of ethylbenzene?

$$CH_3CH_2-\hspace{-0.5em}\bigcirc$$

To determine the number of signals you would expect to see in the spectrum, replace each hydrogen in turn by another atom (here we use Br) and name the resulting compound. The number of different names corresponds to the number of signals in the ^{1}H NMR spectrum. We get five different names for the bromosubstituted benzenes, so we expect to see five signals in the ^{1}H NMR spectrum of ethylbenzene.

14.18 THE RELATIVE POSITIONS OF ^{1}H NMR SIGNALS

The ^{1}H NMR spectrum of 1-bromo-2,2-dimethylpropane in Figure 14.22 has two signals because the compound has two different kinds of protons. The methylene protons are in a less electron-rich environment than are the methyl protons because the methylene protons are closer to the electron-withdrawing bromine. Because the methylene protons are in a less electron-rich environment, they are less shielded from the applied magnetic field. The signal for these protons, therefore, occurs at a higher frequency than the signal for the more shielded methyl protons. *Remember that the right-hand side of an NMR spectrum is the low-frequency side, where protons in electron-rich environments (more shielded) show a signal. The left-hand side is the high-frequency side, where protons in electron-poor environments (less shielded) show a signal* (Figure 14.21).

Protons in electron-poor environments show signals at high frequencies.

We would expect the ^{1}H NMR spectrum of 1-nitropropane to have three signals because the compound has three different kinds of protons. The closer the protons are to the electron-withdrawing nitro group, the less they are shielded from the applied magnetic field, so the higher the frequency at which their signal will appear. Thus, the protons closest to the nitro group show a signal at the highest frequency (4.37 ppm), and the ones farthest from the nitro group show a signal at the lowest frequency (1.04 ppm).

Electron withdrawal causes NMR signals to appear at higher frequencies (at larger δ values).

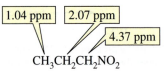

$$CH_3CH_2CH_2NO_2$$

Compare the chemical shifts of the methylene protons immediately adjacent to the halogen in each of the following alkyl halides. The position of the signal depends on the electronegativity of the halogen—the more electronegative the halogen, the higher is the frequency of the signal. Thus, the signal for the methylene protons adjacent to fluorine (the most electronegative of the halogens) occurs at the highest frequency, whereas the signal for the methylene protons adjacent to iodine (the least electronegative of the halogens) occurs at the lowest frequency.

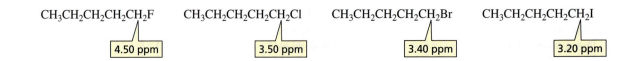

PROBLEM 25

a. Which set of protons in each of the following compounds is the least shielded?

$$\text{O}$$
$$\|$$
1. $CH_3CH_2CH_2Cl$ **2.** $CH_3CH_2COCH_3$ **3.** $CH_3CHCHBr$
$$\qquad\qquad\qquad\qquad\qquad\qquad\qquad\qquad\qquad\qquad\qquad | \; |$$
$$\qquad\qquad\qquad\qquad\qquad\qquad\qquad\qquad\qquad\qquad\text{Br Br}$$

b. Which set of protons in each compound is the most shielded?

PROBLEM 26 ♦

One of the spectra in Figure 14.23 is due to 1-chloropropane, and the other to 1-iodopropane. Which is which?

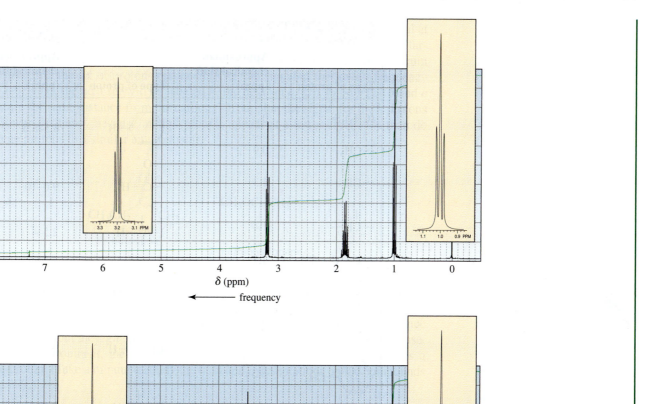

14.19 THE CHARACTERISTIC VALUES OF CHEMICAL SHIFTS

Approximate values of chemical shifts for different kinds of protons are shown in Table 14.4 on page 394. (A more extensive compilation is given in Appendix III.) An ^{1}H NMR spectrum can be divided into seven regions, one of which is empty. If you can remember the kinds of protons that appear in each region, you will be able to tell what kinds of protons a molecule has from a quick look at its NMR spectrum.

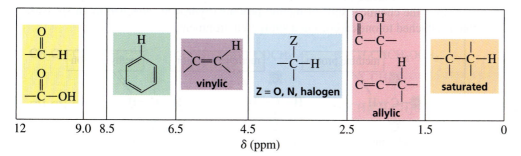

14.20 THE INTEGRATION OF NMR SIGNALS REVEALS THE RELATIVE NUMBER OF PROTONS CAUSING THE SIGNAL

The two signals in the ^{1}H NMR spectrum of 1-bromo-2,2-dimethylpropane in Figure 14.22 are not the same size because *the area under each signal is proportional to the number of protons giving rise to the signal.* The spectrum is shown again in Figure 14.24. The area under the signal occurring at the lower frequency is larger because the signal is caused by *nine* methyl protons, whereas the smaller, higher-frequency signal results from *two* methylene protons.

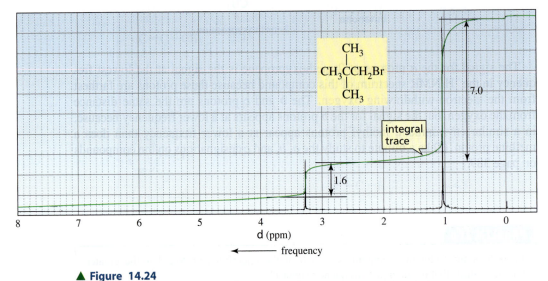

▲ **Figure 14.24**
Analysis of the integral trace in the ^{1}H NMR spectrum of 1-bromo-2,2-dimethylpropane.

The area under each signal can be determined by integration. An ^{1}H NMR spectrometer is equipped with a computer that calculates the integrals electronically and then displays them as an integral trace superimposed on the original spectrum (Figure 14.24). The height of each step in the integral trace is proportional to the number of protons giving rise to the signal. The heights of the integration steps in Figure 14.24, for example, tell us that the ratio of the integrals is approximately 1.6:7.0 = 1:4.4. We multiply the ratio by a number that will cause all the numbers making up the ratio to be close to whole numbers—in this case, we multiply by 2. That means that the ratio of protons in the compound is 2:8.8, which is rounded to 2:9, as there can be only whole numbers of protons. (The measured integrals are approximate because of experimental error.) Modern spectrometers print the integrals as numbers on the spectrum; see Figure 14.26 on page 398.

The **integration** tells us the *relative* number of protons that give rise to each signal, not the *absolute* number. In other words, integration cannot distinguish between 1,1-dichloroethane and 1,2-dichloro-2-methylpropane because both compounds will show an integral ratio of 1:3.

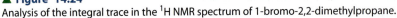

1,1-dichloroethane
ratio of protons = 1:3

1,2-dichloro-2-methylpropane
ratio of protons 2:6 = 1:3

PROBLEM 30

How can integration distinguish the ^{1}H NMR spectra of the following compounds?

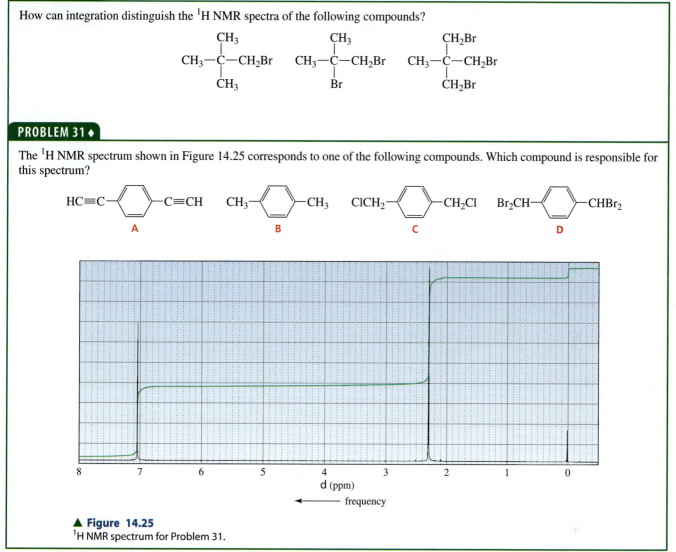

PROBLEM 31 ♦

The ^{1}H NMR spectrum shown in Figure 14.25 corresponds to one of the following compounds. Which compound is responsible for this spectrum?

▲ **Figure 14.25**
^{1}H NMR spectrum for Problem 31.

14.21 THE SPLITTING OF THE SIGNALS IS DESCRIBED BY THE *N* + 1 RULE

Notice that the shapes of the signals in the ^{1}H NMR spectrum of 1,1-dichloroethane (Figure 14.26) are different from the shapes of the signals in the ^{1}H NMR spectrum of 1-bromo-2,2-dimethylpropane (Figure 14.24). Both signals in Figure 14.24 are **singlets** (meaning each is composed of a single peak). In contrast, the signal for the methyl protons of 1,1-dichloroethane (the lower-frequency signal) is split into two peaks (a **doublet**), and the signal for the methine proton is split into four peaks (a **quartet**). (Magnifications of the doublet and quartet are shown as insets in Figure 14.26.)

Splitting is caused by protons bonded to adjacent carbons. The splitting of a signal is described by the ***N* + 1 rule**, where *N* is the number of *equivalent* protons bonded to *adjacent* carbons. By "equivalent protons," we mean that the protons bonded to an adjacent carbon are equivalent to each other, but not equivalent to the proton giving rise to the signal. Both signals in Figure 14.24 are singlets; the three methyl groups of 1-bromo-2,2-dimethylpropane give an unsplit signal because they are attached to a

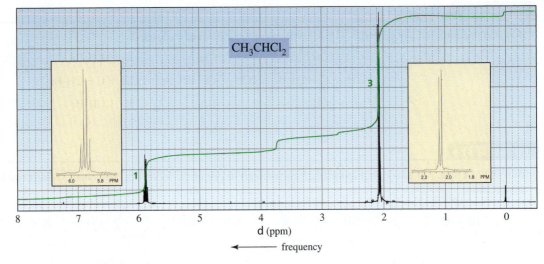

CH_3CHCl_2

▲ **Figure 14.26**
^{1}H NMR spectrum of 1,1-dichloroethane. The higher-frequency signal is an example of a quartet; the lower-frequency signal is a doublet.

An ^{1}H NMR signal is split into N + 1 peaks, where N is the number of equivalent protons bonded to adjacent carbons.

Coupled protons split each other's signal.

Coupled protons are bonded to adjacent carbons.

carbon that is not bonded to a hydrogen; the methylene group also gives an unsplit signal because it too is attached to a carbon that is not bonded to a hydrogen ($N = 0$, so $N + 1 = 1$). In contrast, the carbon adjacent to the methyl group in 1,1-dichloroethane (Figure 14.26) is bonded to one proton, so the signal for the methyl protons is split into a doublet ($N = 1$, so $N + 1 = 2$). The carbon adjacent to the carbon bonded to the methine proton is bonded to three equivalent protons, so the signal for the methine proton is split into a quartet ($N = 3$, so $N + 1 = 4$). The number of peaks in a signal is called the **multiplicity** of the signal. Splitting is always mutual: if the *a* protons split the *b* protons, then the *b* protons must split the *a* protons. The *a* and *b* protons, in this case, are coupled protons. **Coupled protons** split each other's signal. Coupled protons are on adjacent carbons.

   Keep in mind that it is not the number of protons giving rise to a signal that determines the multiplicity of the signal; rather, it is the number of protons bonded to the immediately adjacent carbons that determines the multiplicity. For example, the signal for the *a* protons in the following compound will be split into three peaks (a **triplet**) because the adjacent carbon is bonded to two hydrogens. The signal for the *b* protons will appear as a quartet because the adjacent carbon is bonded to three hydrogens, and the signal for the *c* protons will be a singlet.

 Tutorial:
NMR signal splitting

$$\underset{a \quad\quad b \quad\quad\quad c}{CH_3CH_2\overset{\overset{\textstyle O}{\|}}{C}OCH_3}$$

A signal for a proton is never split by *equivalent* protons. For example, the ^{1}H NMR spectrum of bromomethane shows one singlet. The three methyl protons are chemically equivalent, and chemically equivalent protons do not split each other's signal. The four protons in 1,2-dichloroethane are also chemically equivalent, so its ^{1}H NMR spectrum also shows one singlet.

Equivalent protons do not split each other's signal.

$$CH_3Br \qquad\qquad ClCH_2CH_2Cl$$
bromomethane   **1,2-dichloroethane**

each compound has an NMR spectrum that shows one singlet because equivalent protons do not split each other's signals

PROBLEM 32 ◆

The ^{1}H NMR spectra of two carboxylic acids with molecular formula $C_3H_5O_2Cl$ are shown in Figure 14.27. Identify the carboxylic acids. (The "offset" notation means that the signal has been moved to the right by the indicated amount.)

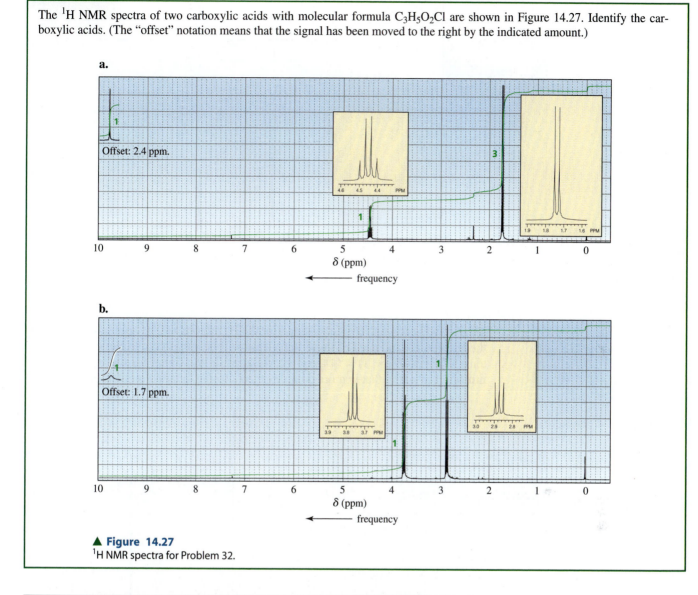

a.

Offset: 2.4 ppm.

b.

Offset: 1.7 ppm.

▲ **Figure 14.27**
^{1}H NMR spectra for Problem 32.

14.22 MORE EXAMPLES OF ^{1}H NMR SPECTRA

We will now look at a few more spectra to give you additional practice in analyzing ^{1}H NMR spectra.

There are two signals in the ^{1}H NMR spectrum of 1,3-dibromopropane (Figure 14.28). The signal for the H_b protons is split into a triplet by the two hydrogens on the adjacent carbon. The carbons adjacent to the carbon bonded to the H_a protons are both bonded to protons. The protons on one of these carbons are equivalent to the protons on the other. Because the two sets of protons are equivalent, the $N + 1$ rule is applied to both sets at the same time. In other words, N is equal to the sum of the equivalent protons on both carbons. So the signal for the H_a protons is split into a quintet ($4 + 1 = 5$).

The ^{1}H NMR spectrum of isopropyl butanoate shows five signals (Figure 14.29). The signal for the H_a protons is split into a triplet by the H_c protons. The signal for the H_b protons is split into a doublet by the H_e proton. The signal for the H_d protons is split into a triplet by the H_c protons, and the signal for the H_e proton is split into a septet by the H_b protons. The signal for the H_c protons is split by both the H_a and H_d protons. Because the H_a and H_d protons are not equivalent, the $N + 1$ rule has to be applied

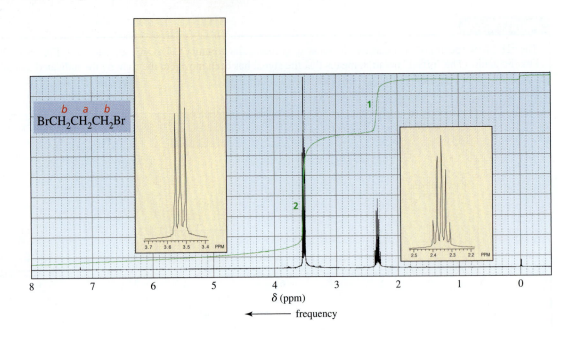

▶ **Figure 14.28**
^{1}H NMR spectrum of 1, 3-dibromopropane.

separately to each set. Thus, the signal for the H_c protons will be split into a quartet by the H_a protons, and each of these four peaks will be split into a triplet by the H_d protons: $(N_a + 1)(N_d + 1) = (4)(3) = 12$. As a result, the signal for the H_c protons is a **multiplet** (a signal that is more complex than a triplet, quartet, quintet, or such).

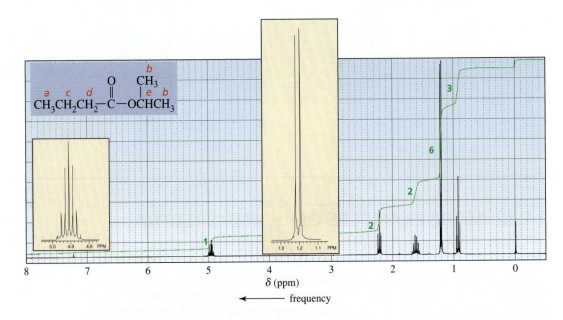

▶ **Figure 14.29**
^{1}H NMR spectrum of isopropyl butanoate.

Ethylbenzene has five sets of chemically equivalent protons (Figure 14.30). We see the expected triplet for the H_a protons and the quartet for the H_b protons. (This is a characteristic pattern for an ethyl group.) The five protons on the benzene ring are not all in the same environment, so we expect to see three signals for them: one for the H_c protons, one for the H_d protons, and one for the H_e protons. However, we do not see three distinct signals because their environments are not sufficiently different to allow them to appear as separate signals.

Notice that the signals for the benzene ring protons in Figure 14.30 occur in the 6.5–8.5 ppm region. Other kinds of protons usually do not resonate in this region, so signals in this region of an ^{1}H NMR spectrum indicate that the compound probably contains an aromatic ring.

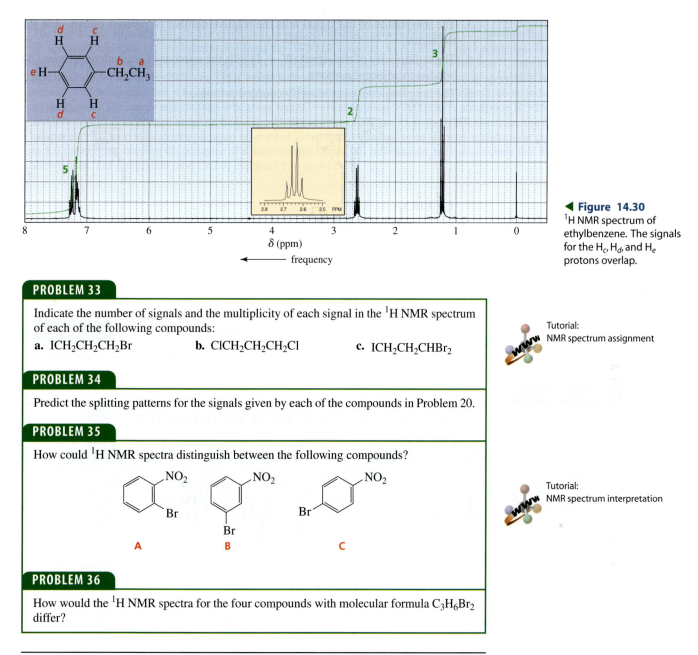

◀ **Figure 14.30**
^{1}H NMR spectrum of ethylbenzene. The signals for the H_c, H_d, and H_e protons overlap.

PROBLEM 33

Indicate the number of signals and the multiplicity of each signal in the ^{1}H NMR spectrum of each of the following compounds:

a. $ICH_2CH_2CH_2Br$ **b.** $ClCH_2CH_2CH_2Cl$ **c.** $ICH_2CH_2CHBr_2$

PROBLEM 34

Predict the splitting patterns for the signals given by each of the compounds in Problem 20.

PROBLEM 35

How could ^{1}H NMR spectra distinguish between the following compounds?

A   B   C

PROBLEM 36

How would the ^{1}H NMR spectra for the four compounds with molecular formula $C_3H_6Br_2$ differ?

Tutorial:
NMR spectrum assignment

Tutorial:
NMR spectrum interpretation

14.23 COUPLING CONSTANTS IDENTIFY COUPLED PROTONS

The distance, in hertz, between two adjacent peaks of a split NMR signal is called the **coupling constant** (denoted by J). The coupling constant for H_a being split by H_b is denoted by J_{ab}. The signals of coupled protons (protons that split each other's signal) have the same coupling constant; in other words, $J_{ab} = J_{ba}$ (Figure 14.31). Coupling constants are useful in analyzing complex NMR spectra because protons on adjacent carbons can be identified by their identical coupling constants. Coupling constants range from 0 to 15 Hz.

Let's now summarize the kind of information that can be obtained from an ^{1}H NMR spectrum:

1. The number of signals indicates the number of different kinds of protons in the compound.

2. The position of a signal indicates the kind of proton(s) responsible for the signal (methyl, methylene, methine, allylic, vinylic, aromatic, and so on) and the kinds of neighboring substituents.

b.

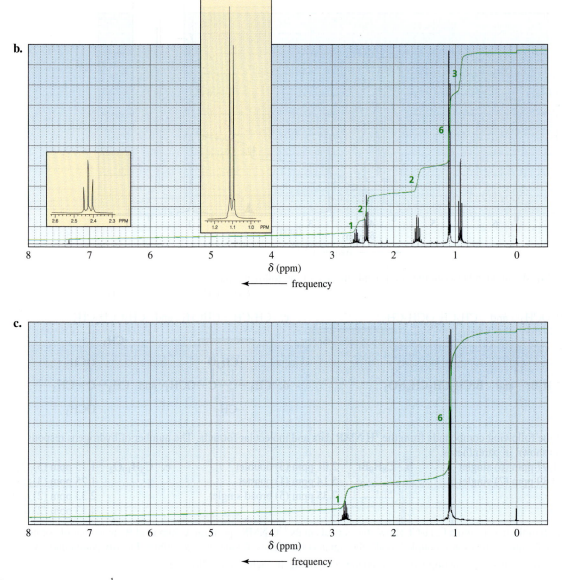

c.

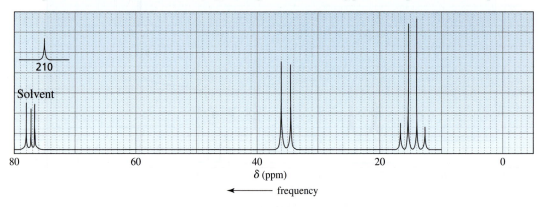

65. How could you use ^{1}H NMR spectroscopy to distinguish the following esters?

66. Identify the compound with molecular formula $C_7H_{14}O$ that gives the following proton-coupled ^{13}C NMR spectrum.

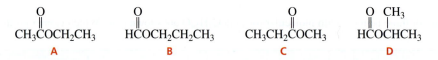

67. An alkyl halide reacts with an alkoxide ion to form a compound whose ^{1}H NMR spectrum is shown here. Identify the alkyl halide and the alkoxide ion.

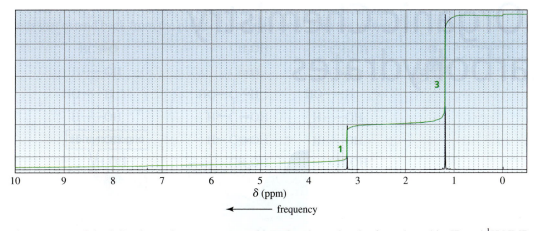

68. Determine the structure of the following unknown compound based on its molecular formula and its IR and ^{1}H NMR spectra.

$C_6H_{12}O_2$

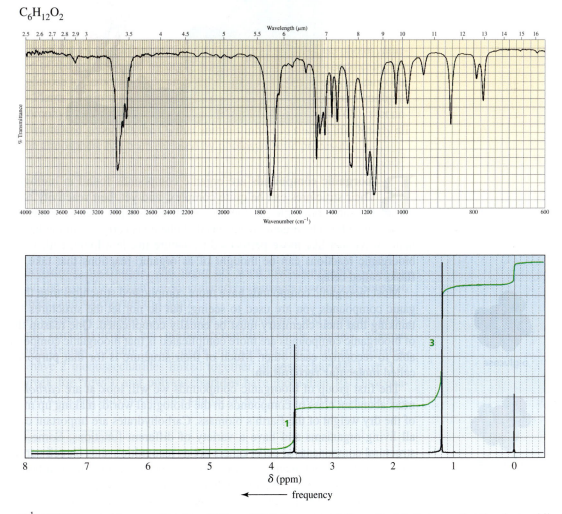

69. How could ^{1}H NMR be used to prove that the addition of HBr to propene follows the rule that says that the electrophile adds to the sp^2 carbon bonded to the greater number of hydrogens (Section 5.3)?

15.5 THE REACTIONS OF MONOSACCHARIDES IN BASIC SOLUTIONS

Monosaccharides cannot undergo reactions with basic reagents because in a basic solution, a monosaccharide is converted to a complex mixture of polyhydroxy aldehydes and polyhydroxy ketones. Let's first look at what happens to D-glucose in a basic solution, beginning with its conversion to its C-2 epimer.

mechanism for the base-catalyzed epimerization of a monosaccharide

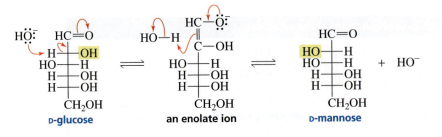

- The base removes a proton from an α-carbon, forming an enolate ion (Section 13.3).
- In the enolate ion, C-2 is no longer an asymmetric center.
- When C-2 is reprotonated, the proton can come from the top or the bottom of the planar sp^2 carbon, forming both D-glucose and D-mannose.

Because the reaction forms a pair of C-2 epimers, it is called an epimerization reaction. **Epimerization** changes the configuration of a carbon by removing a proton and then reprotonating it.

In a basic solution, in addition to forming its C-2 epimer, D-glucose can also undergo a rearrangement, which results in the formation of D-fructose and other ketohexoses.

mechanism for the base-catalyzed rearrangement of a monosaccharide

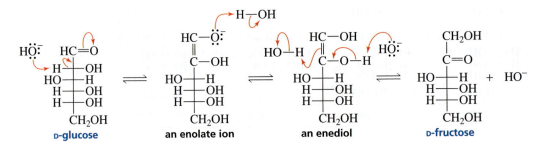

- The base removes a proton from an α-carbon, forming an enolate ion (Section 13.3).
- The enolate ion can be protonated on the oxygen (instead of on the carbon as in the mechanism shown above) to form an enediol.
- The enediol has two OH groups that can form a carbonyl group. Tautomerization of the OH at C-1 reforms D-glucose (Section 5.5); tautomerization of the OH group at C-2 forms D-fructose.

Another rearrangement, initiated by a base removing a proton from C-3 of D-fructose, forms an enediol that can tautomerize to give a ketose with the carbonyl group at C-2 or C-3. Thus, the carbonyl group can be moved up and down the chain.

PROBLEM 6◆

When D-tagatose is added to a basic aqueous solution, an equilibrium mixture of three monosaccharides is obtained. What are these monosaccharides?

PROBLEM 7

Write the mechanism for the base-catalyzed conversion of D-fructose into D-glucose and D-mannose.

15.6 THE OXIDATION–REDUCTION REACTIONS OF MONOSACCHARIDES

Because they contain *alcohol* functional groups and *aldehyde* or *ketone* functional groups, the reactions of monosaccharides are an extension of what you have already learned about the reactions of alcohols, aldehydes, and ketones. For example, an aldehyde group in a monosaccharide can be oxidized or reduced and can react with nucleophiles to form imines, hemiacetals, and acetals. As you read this section and the ones that follow, dealing with the reactions of monosaccharides, you will find cross-references to earlier discussions of simpler organic compounds undergoing the same reactions. Go back and look at these earlier discussions when they are mentioned; they will make learning about carbohydrates a lot easier.

Reduction Reactions

The carbonyl group in aldoses and ketoses can be reduced by sodium borohydride (NaBH$_4$; Section 12.6). The product of the reduction is a polyalcohol, known as an **alditol**. Reduction of an aldose forms one alditol. For example, the reduction of D-mannose forms D-mannitol, the alditol found in mushrooms, olives, and onions. Reduction of a ketose forms two alditols because the reaction creates a new asymmetric center in the product. For example, the reduction of D-fructose forms D-mannitol and D-glucitol, the C-2 epimer of D-mannitol. D-Glucitol—also called sorbitol—is about 60% as sweet as sucrose. It is found in plums, pears, cherries, and berries and is used as a sugar substitute in the manufacture of candy.

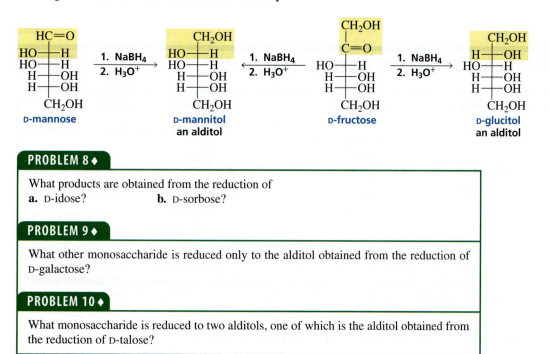

PROBLEM 8◆

What products are obtained from the reduction of
a. D-idose?   **b.** D-sorbose?

PROBLEM 9◆

What other monosaccharide is reduced only to the alditol obtained from the reduction of D-galactose?

PROBLEM 10◆

What monosaccharide is reduced to two alditols, one of which is the alditol obtained from the reduction of D-talose?

CONTROLLING FLEAS

Several different drugs have been developed to help pet owners control fleas. One of these drugs is lufenuron, the active ingredient in Program.

Lufenuron interferes with the flea's production of chitin. The consequence is fatal for the flea because its exoskeleton is composed primarily of chitin.

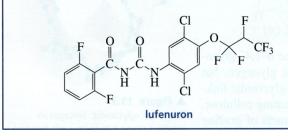

lufenuron

PROBLEM 16

What is the main structural difference between

a. amylose and cellulose?

b. amylose and amylopectin?

c. amylopectin and glycogen?

d. cellulose and chitin?

15.15 SOME NATURALLY OCCURRING PRODUCTS DERIVED FROM CARBOHYDRATES

Deoxy sugars are sugars in which one of the OH groups is replaced by a hydrogen (*deoxy* means "without oxygen"). 2-Deoxyribose is an important example of a deoxy sugar; it is missing the oxygen at the C-2 position. D-Ribose is the sugar component of ribonucleic acid (RNA), whereas 2-deoxyribose is the sugar component of deoxyribonucleic acid (DNA) (Section 20.1).

β-D-ribose   β-D-2-deoxyribose

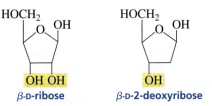

HEPARIN

Heparin is a polysaccharide found principally in cells that line arterial walls. It is released when an injury occurs in order to prevent excessive blood clot formation. Heparin is widely used clinically as an anticoagulant.

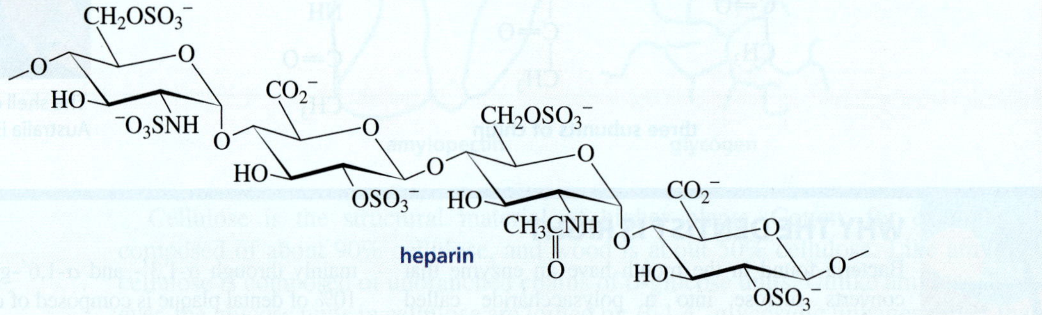

heparin

In **amino sugars**, one of the OH groups is replaced by an amino group. *N*-Acetyl-glucosamine—the subunit of chitin—is an example of an amino sugar (Section 15.14). Some important antibiotics contain amino sugars. For example, the three subunits of the antibiotic gentamicin are deoxyamino sugars. Notice that the middle subunit is missing the ring oxygen, so it is not really a sugar.

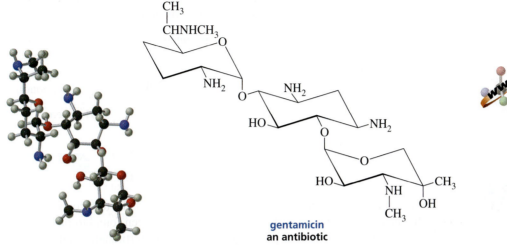

3-D Molecule:
Gentamicin

gentamicin
an antibiotic

L-Ascorbic acid (vitamin C) is synthesized from D-glucose in plants and in the livers of most vertebrates. Humans, monkeys, and guinea pigs do not have the enzymes necessary for the biosynthesis of vitamin C, so they must obtain the vitamin in their diets. The L-configuration of ascorbic acid refers to the configuration at C-5, which was C-2 in D-glucose.

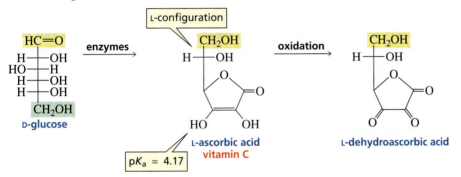

Although L-ascorbic acid does not have a carboxylic acid group, it is an acidic compound because the pK_a of the C-3 OH group is 4.17. L-Ascorbic acid is readily oxidized—notice that when it is oxidized, it loses hydrogens—to L-dehydroascorbic acid, which is also physiologically active. If the five-membered ring is opened by hydrolysis, all vitamin C activity is lost. Therefore, not much intact vitamin C survives in food that has been thoroughly cooked. Worse, if the food is cooked in water and then drained, the water-soluble vitamin is thrown out with the water!

VITAMIN C

Vitamin C is an antioxidant because it prevents oxidation reactions by radicals. It traps radicals formed in aqueous environments (Section 5.17), preventing harmful oxidation reactions the radicals would cause. Not all the physiological functions of vitamin C are known; however, we do know that vitamin C is required for the synthesis of collagen, which is the structural protein of skin, tendons, connective tissue, and bone. Vitamin C is abundant in citrus fruits and tomatoes, but when it is not present in the diet, lesions appear on the skin, bleeding occurs about the gums, in the joints, and under the skin, and any wounds heal slowly. The condition, known as *scurvy*, was the first disease to be treated by adjusting the diet. British sailors who shipped out to sea after the late 1700s were required to eat limes to prevent it (which is how they came to be called "limeys"). *Scorbutus* is Latin for "scurvy"; *ascorbic*, therefore, means "no scurvy."

3. Reduction (Section 15.6)

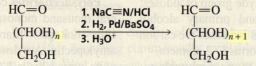

$$
\begin{array}{c}
\text{CH}_2\text{OH} \\
| \\
\text{C}=\text{O} \\
| \\
(\text{CHOH})_n \\
| \\
\text{CH}_2\text{OH}
\end{array}
\xrightarrow[\text{2. H}_3\text{O}^+]{\text{1. NaBH}_4}
\begin{array}{c}
\text{CH}_2\text{OH} \\
| \\
\text{CHOH} \\
| \\
(\text{CHOH})_n \\
| \\
\text{CH}_2\text{OH}
\end{array}
$$

4. Oxidation (Section 15.6)

a.
$$
\begin{array}{c}
\text{HC}=\text{O} \\
| \\
(\text{CHOH})_n \\
| \\
\text{CH}_2\text{OH}
\end{array}
\xrightarrow[\text{H}_2\text{O}]{\text{Br}_2}
\begin{array}{c}
\text{COOH} \\
| \\
(\text{CHOH})_n \\
| \\
\text{CH}_2\text{OH}
\end{array}
+ \ 2\ \text{Br}^-
$$

b.
$$
\begin{array}{c}
\text{HC}=\text{O} \\
| \\
(\text{CHOH})_n \\
| \\
\text{CH}_2\text{OH}
\end{array}
\xrightarrow[\Delta]{\text{HNO}_3}
\begin{array}{c}
\text{COOH} \\
| \\
(\text{CHOH})_n \\
| \\
\text{COOH}
\end{array}
$$

5. Chain elongation: the Kiliani–Fischer synthesis (Section 15.7)

$$
\begin{array}{c}
\text{HC}=\text{O} \\
| \\
(\text{CHOH})_n \\
| \\
\text{CH}_2\text{OH}
\end{array}
\xrightarrow[\substack{\text{2. H}_2\text{, Pd/BaSO}_4 \\ \text{3. H}_3\text{O}^+}]{\text{1. NaC}\equiv\text{N/HCl}}
\begin{array}{c}
\text{HC}=\text{O} \\
| \\
(\text{CHOH})_{n+1} \\
| \\
\text{CH}_2\text{OH}
\end{array}
$$

6. Acetal (and ketal) formation (Section 15.11)

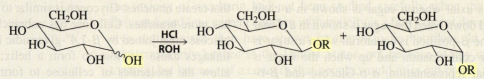

$$
\xrightarrow[\text{ROH}]{\text{HCl}}
$$

PROBLEMS

18. Give the product or products that are obtained when D-galactose reacts with the following substances:

   a. nitric acid   **b.** NaBH$_4$ followed by H$_3$O$^+$ **c.** Br$_2$ in water   **d.** ethanol + HCl

19. What sugar is the C-4 epimer of L-gulose?

20. Identify the sugars in each description:

   a. an aldopentose that is not D-arabinose forms D-arabinitol when it is reduced with NaBH$_4$.

   b. a sugar that is not D-altrose forms D-altraric acid when it reacts with nitric acid.

   c. a ketose, when reduced with NaBH$_4$, forms D-altritol and D-allitol.

21. Answer the following questions about the eight aldopentoses:

   a. Which are enantiomers?   **b.** Which form an optically active compound when oxidized with nitric acid?

22. What other monosaccharide is reduced only to the alditol obtained from the reduction of D-talose?

23. What monosaccharide is reduced to two alditols, one of which is the alditol obtained from the reduction of D-mannose?

24. What monosaccharides would be formed in a Kiliani–Fischer synthesis starting with

   a. D-xylose?   **b.** L-threose?

25. Name the following compounds, and indicate whether each is a reducing sugar or a nonreducing sugar:

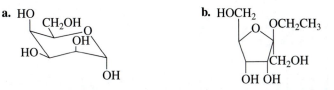

26. The reaction of D-ribose with one equivalent of methanol plus HCl forms four products. Give the structures of the products.

27. Dr. Isent T. Sweet isolated a monosaccharide and determined that it had a molecular weight of 150. Much to his surprise, he found that it was not optically active. What is the structure of the monosaccharide?

28. Indicate whether each of the following is D-glyceraldehyde or L-glyceraldehyde, assuming that the horizontal bonds point toward you and the vertical bonds point away from you:

a.
$$HC=O$$
$$HOCH_2 \!-\!\!\!-\!\!\!- OH$$
$$H$$

b.
$$H$$
$$HO \!-\!\!\!-\!\!\!- CH_2OH$$
$$HC=O$$

c.
$$CH_2OH$$
$$HO \!-\!\!\!-\!\!\!- H$$
$$HC=O$$

29. D-Glucuronic acid is found widely in plants and animals. One of its functions is to detoxify poisonous HO-containing compounds by reacting with them in the liver to form glucuronides. Glucuronides are water soluble and therefore readily excreted. After one ingests a poison such as turpentine, morphine, or phenol, the glucuronides of these compounds are found in the urine. Give the structure of the glucuronide formed by the reaction of β-D-glucuronic acid and phenol.

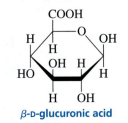

β-D-glucuronic acid

30. In order to synthesize D-galactose, Professor Amy Losse went to the stockroom to get some D-lyxose to use as a starting material. She found that the labels had fallen off the bottles containing D-lyxose and D-xylose. How could she determine which bottle contained D-lyxose?

31. Hyaluronic acid, a component of connective tissue, is the fluid that lubricates the joints. It is an alternating polymer of N-acetyl-D-glucosamine and D-glucuronic acid joined by β-1,3′-glycosidic linkages. Draw a short segment of hyaluronic acid.

32. How many aldaric acids are obtained from the 16 aldohexoses?

33. Give the structure of the cyclic hemiacetal formed by each of the following:

a. 4-hydroxypentanal   **b.** 4-hydroxyheptanal

34. Explain why the C-3 OH group of vitamin C is more acidic than the C-2 OH group.

35. Calculate the percentages of α-D-glucose and β-D-glucose present at equilibrium from the specific rotations of α-D-glucose, β-D-glucose, and the equilibrium mixture. Compare your values with those given in Section 15.9. (*Hint:* The specific rotation of the mixture equals the specific rotation of α-D-glucose times the fraction of glucose present in the α-form plus the specific rotation of β-D-glucose times the fraction of glucose present in the β-form.)

36. Predict whether D-altrose exists preferentially as a pyranose or a furanose. (*Hint:* The most stable arrangement for a five-membered ring is for all the adjacent substituents to be trans.)

CHAPTER 16

The Organic Chemistry of Amino Acids, Peptides, and Proteins

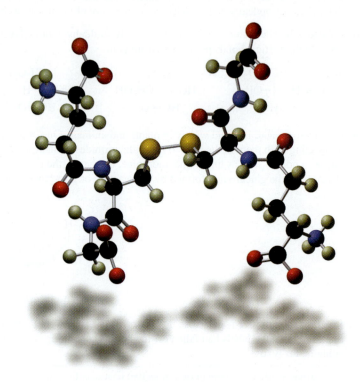

oxidized glutathione

The three kinds of polymers prevalent in nature are polysaccharides, proteins, and nucleic acids. You have already learned about polysaccharides, which are naturally occurring polymers of monosaccharide subunits (Section 15.1). We will now look at proteins and the structurally similar, but shorter, peptides.

Peptides and **proteins** are polymers of **amino acids**. The amino acids are linked together by amide bonds. An **amino acid** is a carboxylic acid with an amino group on the α-carbon. The repeating units are called **amino acid residues**.

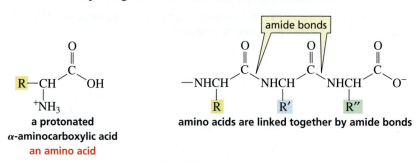

a protonated
α-aminocarboxylic acid
an amino acid

amino acids are linked together by amide bonds

Amino acid polymers can be composed of any number of amino acid residues. A **dipeptide** contains 2 amino acid residues, a **tripeptide** contains 3, an **oligopeptide**

contains 3 to 10, and a **polypeptide** contains many amino acid residues. Proteins are naturally occurring polypeptides that are made up of 40 to 4000 amino acid residues. Proteins and peptides serve many functions in biological systems (Table 16.1).

Proteins can be divided roughly into two classes. **Fibrous proteins** contain long chains of polypeptides arranged in bundles; these proteins are insoluble in water. All structural proteins are fibrous proteins. **Globular proteins** tend to have roughly spherical shapes and are soluble in water. Essentially all enzymes are globular proteins.

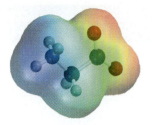

glycine

Table 16.1	Examples of the Many Functions of Proteins in Biological Systems
Structural proteins	These proteins impart strength to biological structures or protect organisms from their environment. For example, collagen is the major component of bones, muscles, and tendons; keratin is the major component of hair, hooves, feathers, fur, and the outer layer of skin.
Protective proteins	Snake venoms and plant toxins protect their owners from predators. Blood-clotting proteins protect the vascular system when it is injured. Antibodies and peptide antibiotics protect us from disease.
Enzymes	Enzymes are proteins that catalyze the reactions that occur in living systems.
Hormones	Some of the hormones, such as insulin, that regulate the reactions that occur in living systems are proteins.
Proteins with physiological functions	These proteins are responsible for physiological functions such as the transport and storage of oxygen in the body, the storage of oxygen in the muscles, and the contraction of muscles.

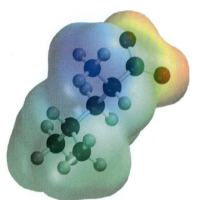

leucine

16.1 THE CLASSIFICATION AND NOMENCLATURE OF AMINO ACIDS

The structures of the 20 most common naturally occurring amino acids and the frequency with which each occurs in proteins are shown in Table 16.2. Other amino acids occur in nature, but only infrequently. Note that the amino acids differ only in the substituent (R) attached to the α-carbon. The wide variation in these substituents (called **side chains**) is what gives proteins their great structural diversity and, as a consequence, their great functional diversity (Table 16.1).

The amino acids are almost always called by their common names. Often, the name tells you something about the amino acid. For example, glycine got its name as a result of its sweet taste (*glykos* is Greek for "sweet"). Asparagine was first found in asparagus, and tyrosine was isolated from cheese (*tyros* is Greek for "cheese"). Table 16.2 shows that each of the amino acids has both a three-letter abbreviation (the first three letters of the name, in most cases) and a single-letter abbreviation.

Notice that, in spite of its name, isoleucine does *not* have an isobutyl substituent; it has a *sec*-butyl substituent. Leucine is the amino acid that has an isobutyl substituent. Proline is the only amino acid that is a secondary amine.

Ten amino acids are *essential amino acids*; these are denoted by an asterisk (*) in Table 16.2. We humans must obtain these 10 **essential amino acids** from our diets because we either cannot synthesize them at all or cannot synthesize them in adequate amounts. For example, we must have a dietary source of phenylalanine because we cannot synthesize benzene rings. However, we do not need tyrosine in our diets because we can synthesize the necessary amounts from phenylalanine. Although humans can synthesize arginine, it is needed for growth in greater amounts than can be synthesized. So arginine is an essential amino acid for children but a nonessential amino acid for adults.

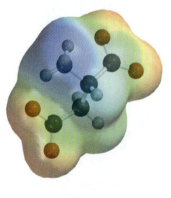

aspartate

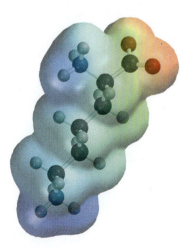

lysine

Table 16.2 The Most Common Naturally Occurring Amino Acids
The amino acids are shown in the form that predominates at physiological pH (7.3).

Formula	Name	Abbreviations		Average relative abundance in proteins
Aliphatic side chain amino acids — Glycine structure	Glycine	Gly	G	7.5%
Alanine structure	Alanine	Ala	A	9.0%
Valine structure	Valine*	Val	V	6.9%
Leucine structure	Leucine*	Leu	L	7.5%
Isoleucine structure	Isoleucine*	Ile	I	4.6%
Hydroxy-containing amino acids — Serine structure	Serine	Ser	S	7.1%
Threonine structure	Threonine*	Thr	T	6.0%
Sulfur-containing amino acids — Cysteine structure	Cysteine	Cys	C	2.8%
Methionine structure	Methionine*	Met	M	1.7%

(*Continued*)

Table 16.2   Continued

Formula	Name	Abbreviations		Average relative abundance in proteins
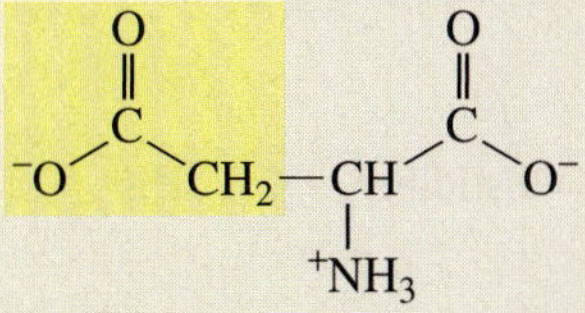 Acidic amino acids — (aspartate structure)	Aspartate (aspartic acid)	Asp	D	5.5%
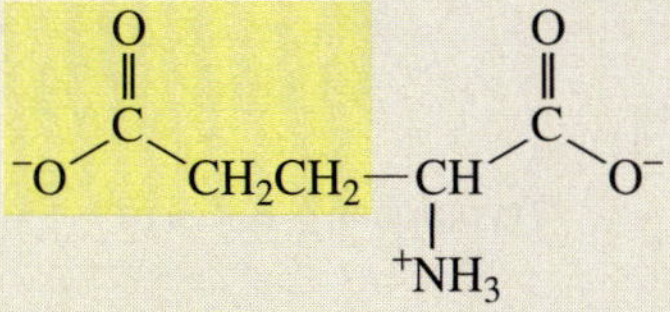 (glutamate structure)	Glutamate (glutamic acid)	Glu	E	6.2%
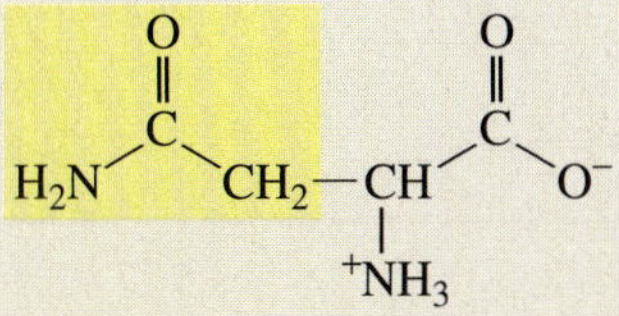 Amides of acidic amino acids — (asparagine structure)	Asparagine	Asn	N	4.4%
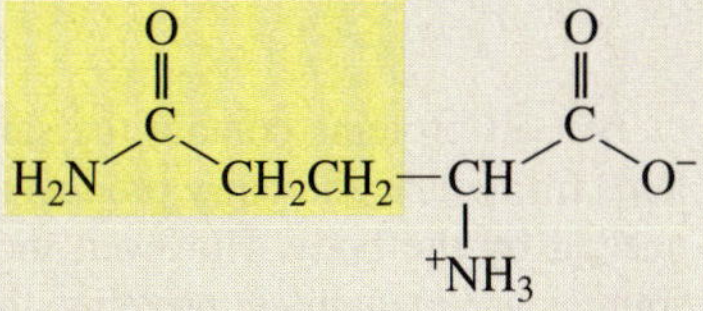 (glutamine structure)	Glutamine	Gln	Q	3.9%
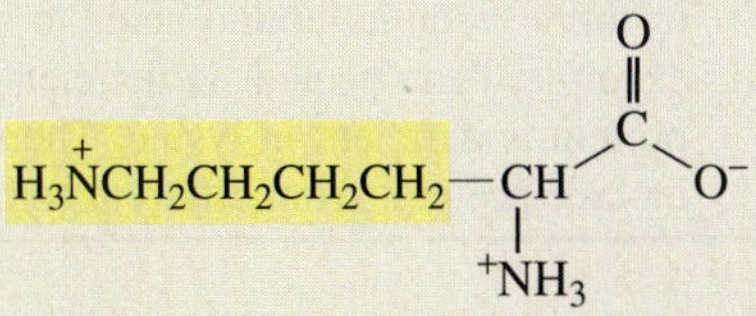 Basic amino acids — (lysine structure)	Lysine*	Lys	K	7.0%
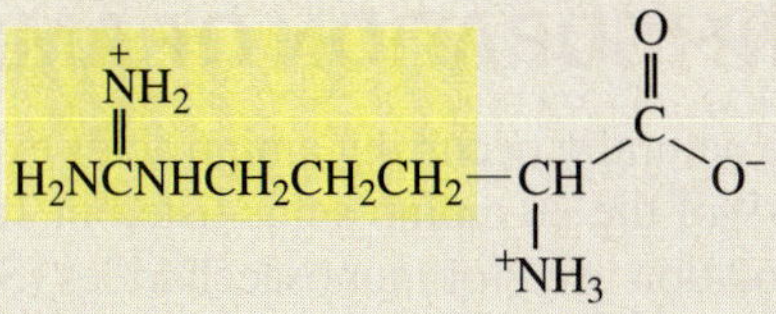 (arginine structure)	Arginine*	Arg	R	4.7%
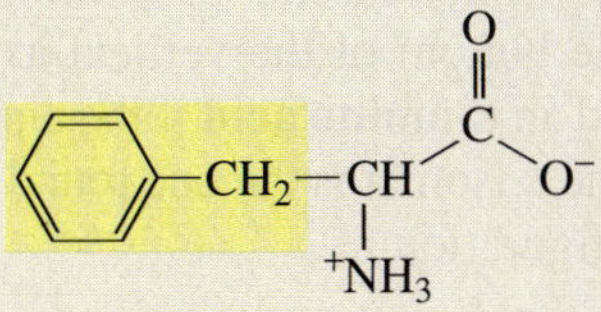 Benzene-containing amino acids — (phenylalanine structure)	Phenylalanine*	Phe	F	3.5%
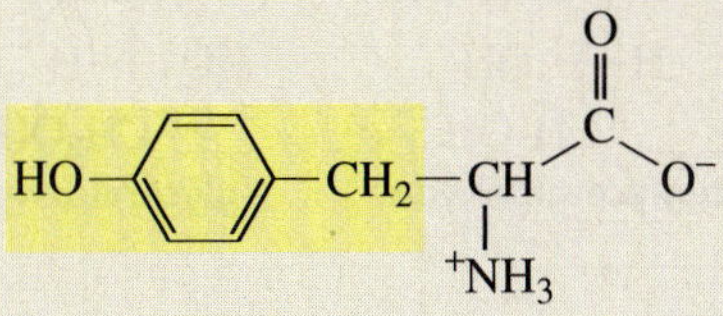 (tyrosine structure)	Tyrosine	Tyr	Y	3.5%
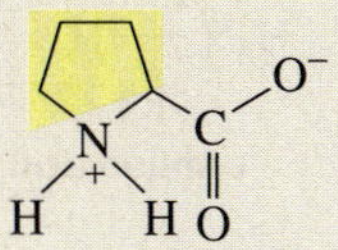 Heterocyclic amino acids — (proline structure)	Proline	Pro	P	4.6%

(Continued)

Table 16.2 Continued

Formula	Name	Abbreviations		Average relative abundance in proteins
Histidine structure	Histidine*	His	H	2.1%
Tryptophan structure	Tryptophan*	Trp	W	1.1%

* An essential amino acid.

PROTEINS AND NUTRITION

Proteins are an important component of our diets. Dietary protein is hydrolyzed in the body to individual amino acids. Some of these amino acids are used to synthesize proteins needed by the body, some are broken down further to supply energy to the body, and some are used as starting materials for the synthesis of nonprotein compounds that the body needs, such as thyroxine (Section 8.7), adrenaline, and melanin (Section 18.6).

Not all proteins contain the same amino acids. Most proteins from meat and dairy products contain all the amino acids needed by the body. However, most proteins from vegetable sources are *incomplete* proteins; they contain too little of one or more essential amino acids to support human growth. For example, bean protein is deficient in methionine, and wheat protein is deficient in lysine. Therefore, a balanced diet must include proteins from different sources.

16.2 THE CONFIGURATION OF AMINO ACIDS

Naturally occurring monosaccharides have the D-configuration.

Naturally occurring amino acids have the L-configuration.

L-alanine
an amino acid

The *α-carbon* of all the naturally occurring amino acids except glycine is an asymmetric center. Therefore, 19 of the 20 amino acids listed in Table 16.2 can exist as enantiomers. The D and L notation used for monosaccharides (Section 15.2) is also used for amino acids. An amino acid drawn in a Fischer projection with the carboxyl group on the top and the R group on the bottom of the vertical axis is a **D-amino acid** if the amino group is on the right, and an **L-amino acid** if the amino group is on the left. Unlike monosaccharides, where the D isomer is the one found in nature, most amino acids found in nature have the L configuration.

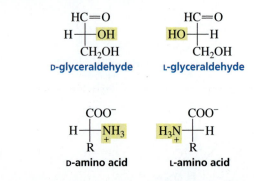

Why D-sugars and L-amino acids? Although it makes no difference which isomer nature "selected" to be synthesized, it is important that the same isomer is

synthesized by all organisms. For example, since mammals ended up having L-amino acids, L-amino acids must be the isomers synthesized by the organisms that mammals depend on for food.

AMINO ACIDS AND DISEASE

The Chamorro people of Guam have a high incidence of a syndrome that resembles amyotrophic lateral sclerosis (ALS or Lou Gehrig's disease) with elements of Parkinson's disease and dementia. This syndrome developed during World War II when, as a result of food shortages, the tribe ate large quantities of *Cycas circinalis* seeds. These seeds contain β-methylamino-L-alanine, an amino acid that binds to glutamate receptors. When monkeys are given β-methylamino-L-alanine, they develop some of the features of this syndrome. There is hope that, by studying this unusual amino acid, we may gain an understanding of how ALS and Parkinson's disease arise.

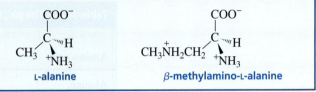

L-alanine   β-methylamino-L-alanine

A PEPTIDE ANTIBIOTIC

Gramicidin S, an antibiotic produced by a strain of bacteria, is a cyclic decapeptide. Notice that one of its residues is ornithine, an amino acid not listed in Table 16.2 because it occurs rarely in nature. Ornithine resembles lysine but has one less methylene group in its side chain. Notice also that the antibiotic contains two D-amino acids.

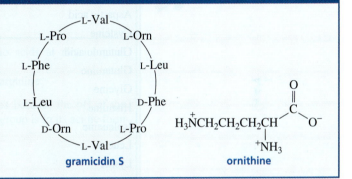

gramicidin S   ornithine

PROBLEM 1 ◆

Which isomer—(*R*)-alanine or (*S*)-alanine—is L-alanine?

PROBLEM 2   *SOLVED*

Threonine has two asymmetric centers and therefore has four stereoisomers.

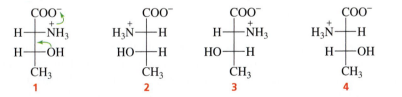

    1   2   3   4

Naturally occurring L-threonine is (2*S*,3*R*)-threonine. Which of the stereoisomers is L-threonine?

Solution Stereoisomer number 1 has the *R* configuration at both C-2 and C-3 because in both cases the arrow drawn from the highest to the next highest priority substituent is counterclockwise. In both cases, counterclockwise signifies *R* because the lowest-priority substituent (H) is on a horizontal bond (Section 15.2). Therefore, the configuration of (2*S*,3*R*)-threonine is the opposite of that in stereoisomer number 1 at C-2 and the same as that in stereoisomer number 1 at C-3. Thus, L-threonine is stereoisomer number 4. Notice that the $^+NH_3$ group is on the left, just as we would expect for the Fischer projection of an L-amino acid.

3-D Molecules:
Common naturally occurring amino acids.

PROBLEM 3 ◆

Do any other amino acids in Table 16.2 have more than one asymmetric center?

The partial double-bond character prevents free rotation about the peptide bond; therefore, the carbon and nitrogen atoms of the peptide bond and the two atoms to which each is attached are held rigidly in a plane (Figure 16.5).

▶ **Figure 16.5**
A segment of a polypeptide chain. Colored squares indicate the plane defined by each peptide bond. Notice that the R groups bonded to the α-carbons are on alternate sides of the peptide backbone.

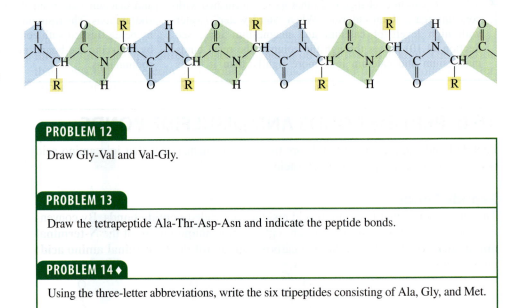

PROBLEM 12

Draw Gly-Val and Val-Gly.

PROBLEM 13

Draw the tetrapeptide Ala-Thr-Asp-Asn and indicate the peptide bonds.

PROBLEM 14 ◆

Using the three-letter abbreviations, write the six tripeptides consisting of Ala, Gly, and Met.

PROBLEM 15

Which bonds in the backbone of a peptide can rotate freely?

ENKEPHALINS

Enkephalins are pentapeptides that are synthesized by the body to control pain. They decrease the body's sensitivity to pain by binding to receptors in certain brain cells. Part of the three-dimensional structures of enkephalins must be similar to those of morphine and related painkillers such as Demerol because they bind to the same receptors (Section 21.6).

Tyr-Gly-Gly-Phe-Leu   Tyr-Gly-Gly-Phe-Met
leucine enkephalin   **methionine enkephalin**

Disulfide Bonds

When thiols are oxidized under mild conditions, they form disulfides. A **disulfide** is a compound with an S—S bond.

$$2\ R-SH \xrightarrow{\text{mild oxidation}} RS-SR$$
      a thiol   a disulfide

Disulfides are reduced to thiols.
Thiols are oxidized to disulfides.

Because thiols can be oxidized to disulfides, disulfides can be reduced to thiols.

$$RS-SR \xrightarrow{\text{reduction}} 2\ R-SH$$
    a disulfide   a thiol

Cysteine is an amino acid that contains a thiol group. Two cysteine molecules therefore can be oxidized to a disulfide. This disulfide is called cystine.

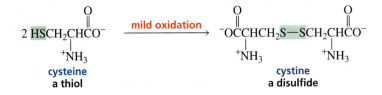

Two cysteine residues in a protein can be oxidized to a disulfide, creating a bond known as a **disulfide bridge**. Disulfide bridges contribute to the overall shape of a protein by linking cysteine residues found in different parts of the peptide backbone, as shown in Figure 16.6.

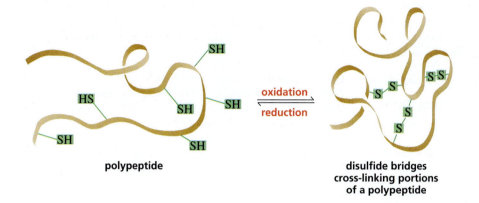

polypeptide

oxidation
reduction

disulfide bridges
cross-linking portions
of a polypeptide

◀ **Figure 16.6**
Disulfide bridges cross-linking portions of a peptide.

The hormone insulin, secreted by the pancreas, controls the level of glucose in the blood by regulating glucose metabolism. Insulin is a polypeptide with two peptide chains. It has three disulfide bridges, two of which hold the two chains together.

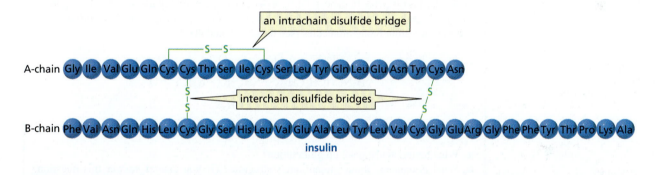

an intrachain disulfide bridge

A-chain Gly Ile Val Glu Gln Cys Cys Thr Ser Ile Cys Ser Leu Tyr Gln Leu Glu Asn Tyr Cys Asn

interchain disulfide bridges

B-chain Phe Val Asn Gln His Leu Cys Gly Ser His Leu Val Glu Ala Leu Tyr Leu Val Cys Gly Glu Arg Gly Phe Phe Tyr Thr Pro Lys Ala

insulin

HAIR: STRAIGHT OR CURLY?

Hair is made up of a protein called keratin that contains an unusually large number of cysteine residues. These furnish keratin with many disulfide bridges that preserve its three-dimensional structure. People can alter the structure of their hair (if they think it is either too straight or too curly) by changing the location of these disulfide bridges. This change is accomplished by first applying a reducing agent to the hair to reduce all the disulfide bridges in the protein strands. Then, after rearranging the hair into the desired shape (using curlers to curl it or combing it straight to uncurl it), an oxidizing agent is applied to form new disulfide bridges. The new disulfide bridges maintain the hair in its new shape. When this treatment is applied to straight hair, it is called a "permanent." When it is applied to curly hair, it is called "hair straightening."

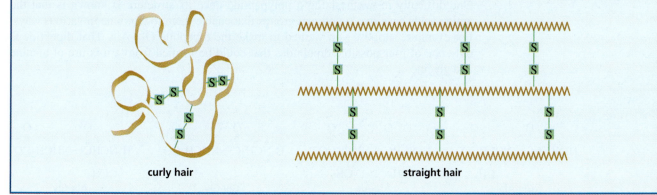

curly hair

straight hair

PEPTIDE HORMONES

Bradykinin, vasopressin, and oxytocin are peptide hormones. They are all nonapeptides. Bradykinin inhibits the inflammation of tissues. Vasopressin controls blood pressure by regulating the contraction of smooth muscle; it is also an antidiuretic. Oxytocin induces labor in pregnant women by stimulating the uterine muscle to contract, and it also stimulates milk production in nursing mothers. Vasopressin and oxytocin both have a disulfide bridge, and their C-terminal amino acids contain amide rather than carboxyl groups. Notice that the C-terminal amide group is indicated by writing "NH_2" after the name of the C-terminal amino acid. In spite of their very different physiological effects, vasopressin and oxytocin differ only by two amino acids.

bradykinin   Arg-Pro-Pro-Gly-Phe-Ser-Pro-Phe-Arg

vasopressin   Cys-Tyr-Phe-Gln-Asn-Cys-Pro-Arg-Gly-NH_2
                 S———————S

oxytocin   Cys-Tyr-Ile-Gln-Asn-Cys-Pro-Leu-Gly-NH_2
               S———————S

Oxytocin was the first small peptide to be synthesized. This was accomplished in 1953 by **Vincent du Vigneaud (1901–1978)**, *who later synthesized vasopressin. Du Vigneaud was born in Chicago and was a professor at George Washington University Medical School and later at Cornell University Medical College. For synthesizing these nonapeptides, he received the Nobel Prize in chemistry in 1955.*

3-D Molecules:
Glutathione;
Oxidized glutathione

PROBLEM 16 ◆

Glutathione is a tripeptide whose function is to destroy harmful oxidizing agents in the body. Oxidizing agents are thought to be responsible for some of the effects of aging and to play a causative role in cancer. Glutathione removes oxidizing agents by reducing them and as a result glutathione is oxidized, resulting in the formation of a disulfide bond between two glutathione molecules (see pages 448 and 460).

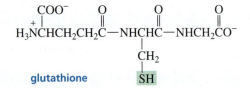

glutathione

a. What amino acids make up glutathione?
b. What is unusual about glutathione's structure? (If you cannot answer this question, draw the structure you would expect for the tripeptide, and compare your structure with the actual structure of glutathione.)

16.7 THE STRATEGY OF PEPTIDE BOND SYNTHESIS: N-PROTECTION AND C-ACTIVATION

One difficulty in synthesizing a polypeptide once its structure is known is that the amino acids have two functional groups that enable them to combine in various ways. For example, suppose you wanted to make the dipeptide Gly-Ala. That dipeptide is only one of four possible dipeptides that could be formed from a mixture of alanine and glycine.

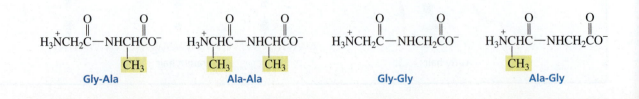

Physical Properties of Esters

Name	Structure	mp (°C)	bp (°C)
Methyl formate	$HCOOCH_3$	−100	32
Ethyl formate	$HCOOCH_2CH_3$	−80	54
Methyl acetate	CH_3COOCH_3	−98	57.5
Ethyl acetate	$CH_3COOCH_2CH_3$	−84	77
Propyl acetate	$CH_3COO(CH_2)_2CH_3$	−92	102
Methyl propionate	$CH_3CH_2COOCH_3$	−87.5	80
Ethyl propionate	$CH_3CH_2COOCH_2CH_3$	−74	99
Methyl butyrate	$CH_3CH_2CH_2COOCH_3$	−84.8	102.3
Ethyl butyrate	$CH_3CH_2CH_2COOCH_2CH_3$	−93	121

Physical Properties of Amides

Name	Structure	mp (°C)	bp (°C)
Formamide	$HCONH_2$	3	200 d*
Acetamide	CH_3CONH_2	82	221
Propanamide	$CH_3CH_2CONH_2$	80	213
Butanamide	$CH_3(CH_2)_2CONH_2$	116	216
Pentanamide	$CH_3(CH_2)_3CONH_2$	106	232

*d means the substance decomposes.

Physical Properties of Aldehydes

Name	Structure	mp (°C)	bp (°C)	Solubility (g/100 g H_2O at 25 °C)
Formaldehyde	HCHO	−92	−21	v. sol.
Acetaldehyde	CH_3CHO	−121	21	∞
Propionaldehyde	CH_3CH_2CHO	−81	49	16
Butyraldehyde	$CH_3(CH_2)_2CHO$	−96	75	7
Pentanal	$CH_3(CH_2)_3CHO$	−92	103	s. sol.
Hexanal	$CH_3(CH_2)_4CHO$	−56	131	s. sol.
Heptanal	$CH_3(CH_2)_5CHO$	−43	153	0.1
Octanal	$CH_3(CH_2)_6CHO$		171	insol.
Nonanal	$CH_3(CH_2)_7CHO$		192	insol.
Decanal	$CH_3(CH_2)_8CHO$	−5	209	insol.
Benzaldehyde	C_6H_5CHO	−26	178	0.3

Physical Properties of Ketones

Name	Structure	mp (°C)	bp (°C)	Solubility (g/100 g H_2O at 25 °C)
Acetone	CH_3COCH_3	−95	56	∞
2-Butanone	$CH_3COCH_2CH_3$	−86	80	25.6
2-Pentanone	$CH_3CO(CH_2)_2CH_3$	−78	102	5.5
2-Hexanone	$CH_3CO(CH_2)_3CH_3$	−57	127	1.6
2-Heptanone	$CH_3 CO(CH_2)_4CH_3$	−36	151	0.4
2-Octanone	$CH_3CO(CH_2)_5CH_3$	−16	173	insol.
2-Nonanone	$CH_3CO(CH_2)_6CH_3$	−7	195	insol.
2-Decanone	$CH_3CO(CH_2)_7CH_3$	14	210	insol.
3-Pentanone	$CH_3CH_2COCH_2CH_3$	−40	102	4.8
3-Hexanone	$CH_3CH_2CO(CH_2)_2CH_3$		123	1.5
3-Heptanone	$CH_3CH_2CO(CH_2)_3CH_3$	−39	149	0.3
Acetophenone	$CH_3COC_6H_5$	19	202	insol.
Propiophenone	$CH_3CH_2COC_6H_5$	18	218	insol.

Appendix II

pKa Values

Compound	pK_a	Compound	pK_a	Compound	pK_a
$CH_3C\overset{+}{\equiv}NH$	−10.1	$O_2N\text{—}C_6H_4\text{—}\overset{+}{N}H_3$	1.0	$CH_3\text{—}C_6H_4\text{—}COH$ (O)	4.3
HI	−10	pyrimidine $\overset{+}{N}H$	1.0		
HBr	−9			$CH_3O\text{—}C_6H_4\text{—}COH$ (O)	4.5
$CH_3\overset{+OH}{CH}$	−8	Cl_2CHCOH (O)	1.3	$C_6H_5\text{—}\overset{+}{N}H_3$	4.6
$CH_3\overset{+OH}{CCH_3}$	−7.3	HSO_4^-	2.0		
		H_3PO_4	2.1	CH_3COH (O)	4.8
HCl	−7	purine $H\overset{+}{N}$	2.5		
$C_6H_5\text{—}SO_3H$	−6.5			quinoline $\overset{+}{N}H$	4.9
$CH_3\overset{+OH}{COCH_3}$	−6.5	FCH_2COH (O)	2.7		
		$ClCH_2COH$ (O)	2.8	$CH_3\text{—}C_6H_4\text{—}\overset{+}{N}H_3$	5.1
$CH_3\overset{+OH}{COH}$	−6.1	$BrCH_2COH$ (O)	2.9	pyridine $\overset{+}{N}H$	5.2
H_2SO_4	−5	ICH_2COH (O)	3.2		
pyrrole $\overset{+}{N}H$	−3.8	HF	3.2	$CH_3O\text{—}C_6H_4\text{—}\overset{+}{N}H_3$	5.3
$CH_3CH_2\overset{+H}{O}CH_2CH_3$	−3.6	HNO_2	3.4	$CH_3C\overset{+}{=}NHCH_3$ \n CH_3	5.5
$CH_3CH_2\overset{+H}{O}H$	−2.4	$O_2N\text{—}C_6H_4\text{—}COH$ (O)	3.4		
$CH_3\overset{+H}{O}H$	−2.5	$HCOH$ (O)	3.8	CH_3CCH_2CH (O)(O)	5.9
H_3O^+	−1.7			$HO\overset{+}{N}H_3$	6.0
HNO_3	−1.3	$Br\text{—}C_6H_4\text{—}\overset{+}{N}H_3$	3.9	H_2CO_3	6.4
CH_3SO_3H	−1.2			imidazole $H\overset{+}{N}\text{—}NH$	6.8
$CH_3\overset{+OH}{CNH_2}$	0.0	$Br\text{—}C_6H_4\text{—}COH$ (O)	4.0	H_2S	7.0
F_3CCOH (O)	0.2	nicotinic COH (O)	4.2	$O_2N\text{—}C_6H_4\text{—}OH$	7.1
Cl_3CCOH (O)	0.64			$H_2PO_4^-$	7.2
$\overset{+}{N}\text{—}OH$ pyridine	0.79			$C_6H_5\text{—}SH$	7.8

[a] pK_a values are for the red H in each structure

pK_a Values (continued)

Compound	pK_a	Compound	pK_a	Compound	pK_a

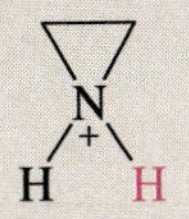

Column 1

Compound	pK_a
aziridinium (N with H, H)	8.0
$H_2N\overset{+}{N}H_3$	8.1
CH_3COOH (with =O)	8.2
$CH_3CH_2NO_2$	8.6
$CH_3\overset{O}{\overset{\|}{C}}CH_2\overset{O}{\overset{\|}{C}}CH_3$	8.9
$HC\equiv N$	9.1
morpholinium (O ring, N$^+$H H)	9.3
Cl—phenyl—OH	9.4
$\overset{+}{N}H_4$	9.4
$HOCH_2CH_2\overset{+}{N}H_3$	9.5
$H_3\overset{+}{N}CH_2\overset{O}{\overset{\|}{C}}O^-$	9.8
phenyl—OH	10.0
CH_3—phenyl—OH	10.2
HCO_3^-	10.2
CH_3NO_2	10.2
H_2N—phenyl—OH	10.3
CH_3CH_2SH	10.5
$(CH_3)_3\overset{+}{N}H$	10.6
$CH_3\overset{O}{\overset{\|}{C}}CH_2\overset{O}{\overset{\|}{C}}OCH_2CH_3$	10.7
$CH_3\overset{+}{N}H_3$	10.7

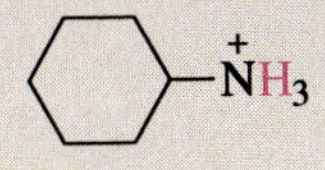

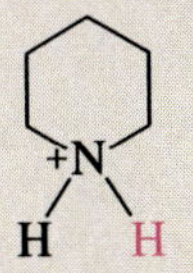

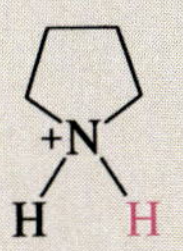

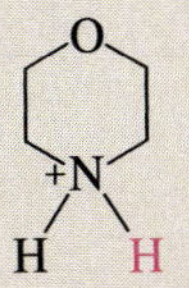

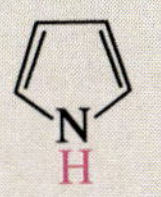

Column 2

Compound	pK_a
cyclohexyl—$\overset{+}{N}H_3$	10.7
$(CH_3)_2\overset{+}{N}H_2$	10.7
piperidinium (N$^+$H H ring)	11.1
$CH_3CH_2\overset{+}{N}H_3$	11.0
pyrrolidinium (N$^+$H H ring)	11.3
$HOOH$	11.6
HPO_4^{2-}	12.3
CF_3CH_2OH	12.4
$CH_3CH_2O\overset{O}{\overset{\|}{C}}CH_2\overset{O}{\overset{\|}{C}}OCH_2CH_3$	13.3
$HC\equiv CCH_2OH$	13.5
$H_2N\overset{O}{\overset{\|}{C}}NH_2$	13.7
$CH_3\overset{\overset{\displaystyle CH_3}{\|}}{\underset{\underset{\displaystyle CH_3}{\|}}{\overset{+}{N}}}CH_2CH_2OH$	13.9
imidazole (N=, NH)	14.4
CH_3OH	15.5
H_2O	15.7
CH_3CH_2OH	16.0
$CH_3\overset{O}{\overset{\|}{C}}NH_2$	16
phenyl—$\overset{O}{\overset{\|}{C}}CH_3$	16.0
pyrrole (NH)	~17

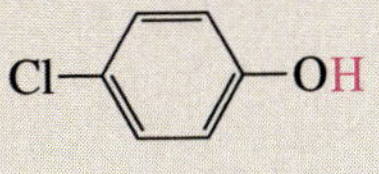

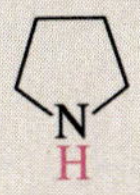

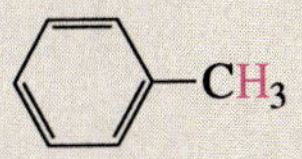

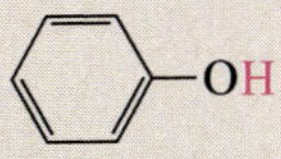

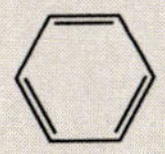

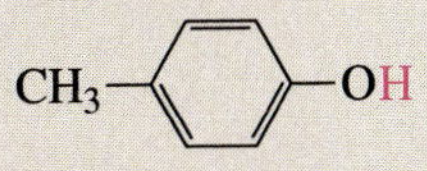

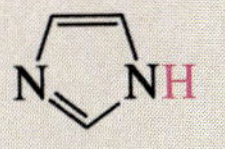

Column 3

Compound	pK_a
$CH_3\overset{O}{\overset{\|}{C}}H$	17
$(CH_3)_3COH$	18
$CH_3\overset{O}{\overset{\|}{C}}CH_3$	20
$CH_3\overset{O}{\overset{\|}{C}}CH_2CH_3$	24.5
$HC\equiv CH$	25
$CH_3C\equiv N$	25
$CH_3\overset{O}{\overset{\|}{C}}N(CH_3)_2$	30
NH_3	36
pyrrolidine (NH)	36
CH_3NH_2	40
phenyl—CH_3	41
benzene	43
$CH_2=CHCH_3$	43
$CH_2=CH_2$	44
cyclopropane	46
CH_4	60
CH_3CH_3	> 60

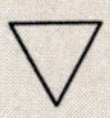

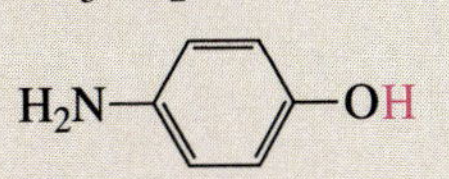

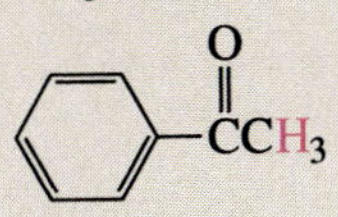

Appendix III

¹H NMR Chemical Shifts

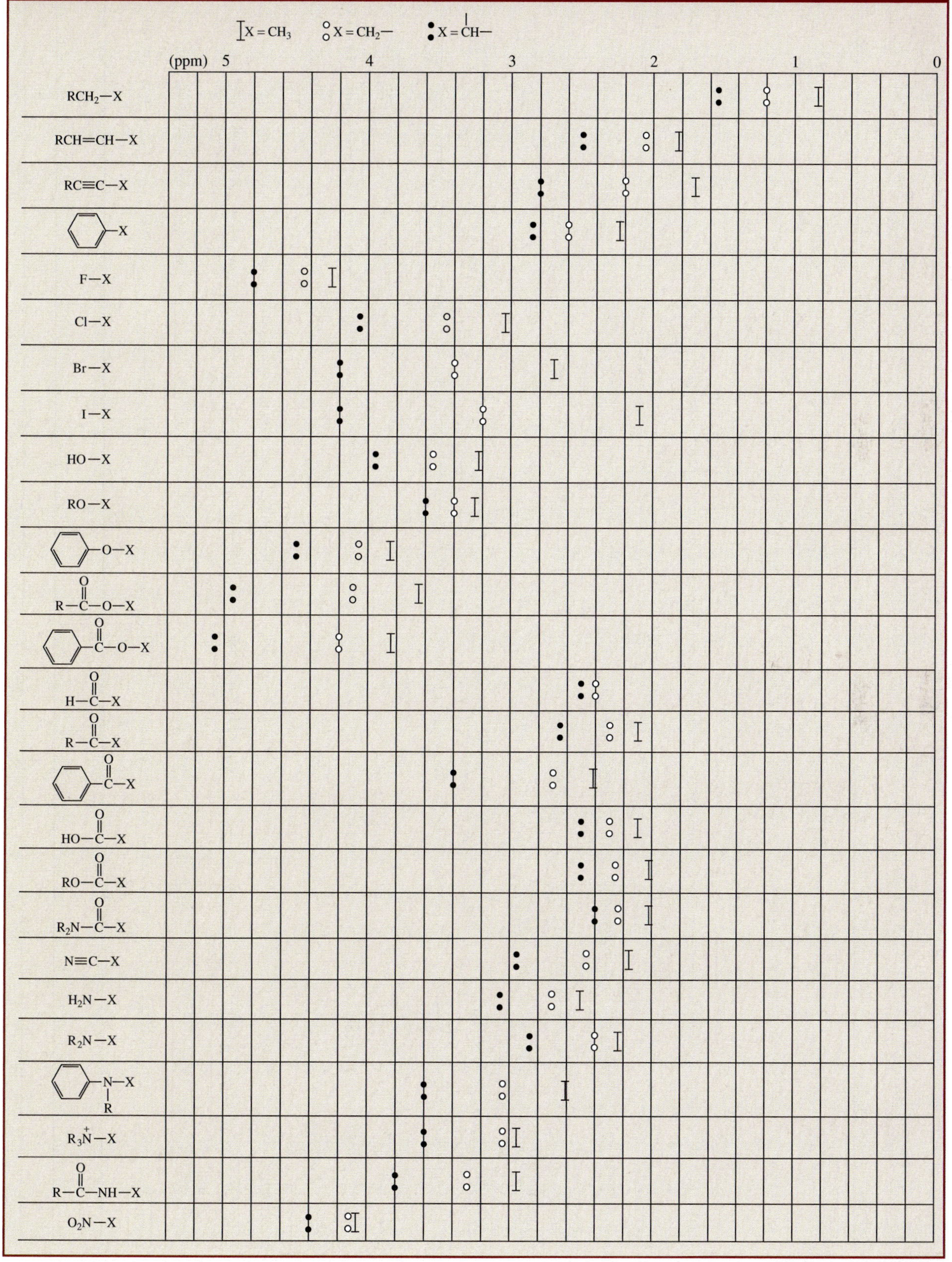

Legend: | X = CH₃   ○ X = CH₂—   ● X = CH—

(ppm) scale: 5 — 4 — 3 — 2 — 1 — 0

Rows (substrate types):

- RCH₂—X
- RCH=CH—X
- RC≡C—X
- ⟨benzene⟩—X
- F—X
- Cl—X
- Br—X
- I—X
- HO—X
- RO—X
- ⟨benzene⟩—O—X
- R—C(=O)—O—X
- ⟨benzene⟩—C(=O)—O—X
- H—C(=O)—X
- R—C(=O)—X
- ⟨benzene⟩—C(=O)—X
- HO—C(=O)—X
- RO—C(=O)—X
- R₂N—C(=O)—X
- N≡C—X
- H₂N—X
- R₂N—X
- ⟨benzene⟩—N(R)—X
- R₃N⁺—X
- R—C(=O)—NH—X
- O₂N—X

A-10

Characteristic Infrared Group Frequencies (S = strong, M = medium, W = weak). (Courtesy of N.B. Colthup, Stamford Research Laboratories, American Cyanamid Company, and the editor of the *Journal of the Optical Society*.) Overtone bands are marked 2ν.

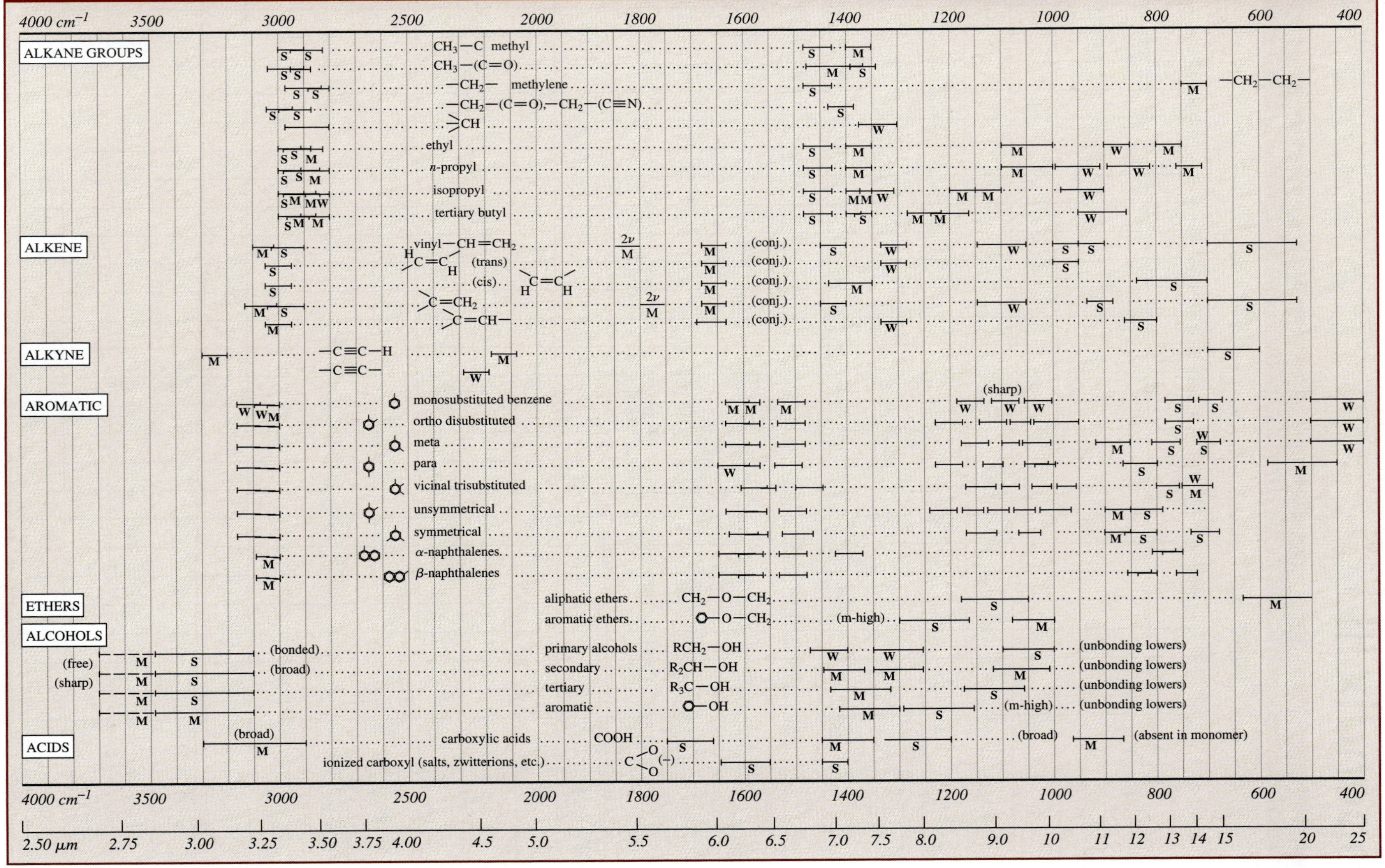

Characteristic Infrared Group Frequencies (continued)

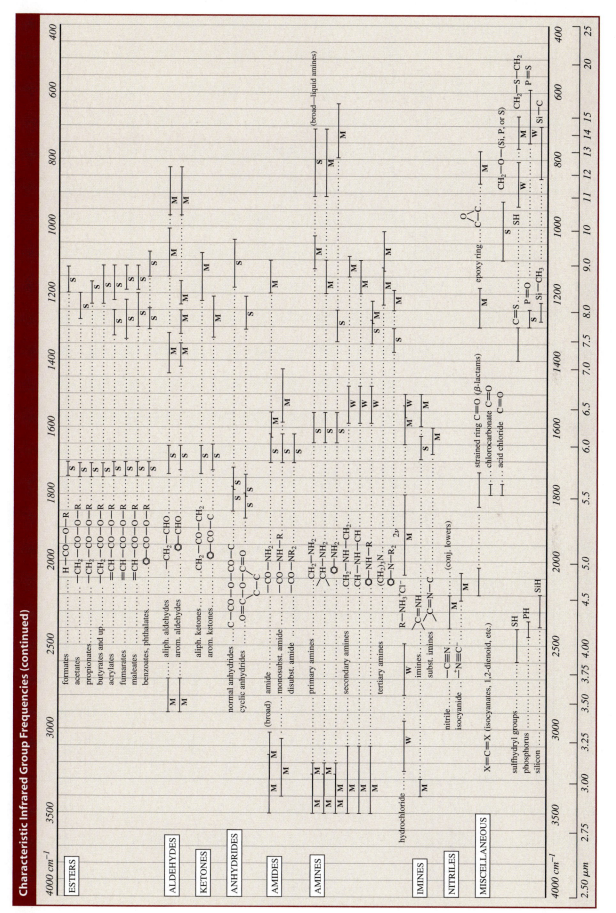

Characteristic Infrared Group Frequencies (continued)

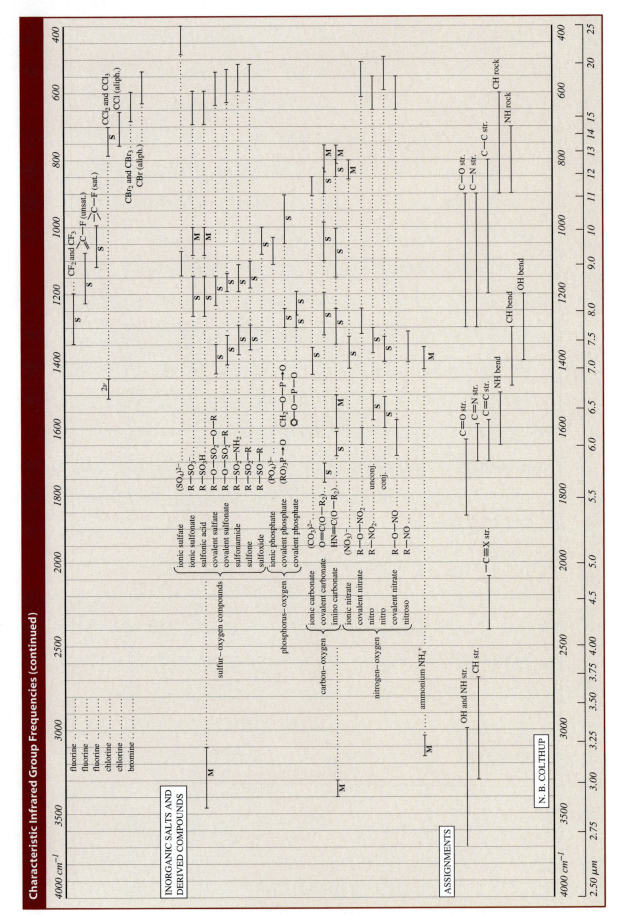

Answers to Selected Problems

CHAPTER 1

1. mn = 16 = 8 p + 8n; mn = 17 = 8 p + 9 n; mn = 18 = 8 p + 10 n   **2. a.** 4 **b.** 5 **c.** 6 **d.** 7   **3.** 1   **4. a.** N = 2 core, 5 valence; P = 10 core, 5 valence **b.** O = 2 core, 6 valence; S = 10 core, 6 valence   **5.** 7   **6. a.** Cl—CH₃ **b.** H—OH **c.** H—F **d.** Cl—CH₃   **7. a.** KCl **b.** Cl₂

8. a. HŌ—H̄ **b.** H₃C—NH₂ **c.** HO—Br **d.** I—Cl

9. a. LiH and HF **b.** HF   **10. a.** oxygen **b.** oxygen **c.** oxygen **d.** hydrogen

11. a. 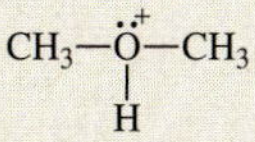 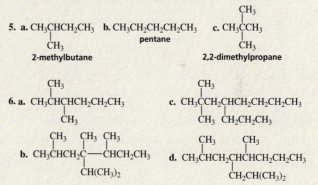 (structure) **b.** 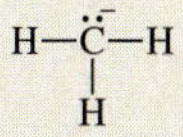 **c.** **d.**

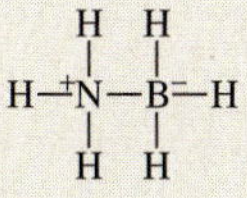

12. a. 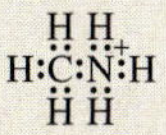 **b.** 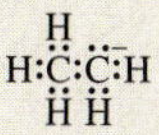 **c.** 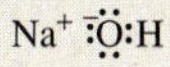 Na⁺ **d.**

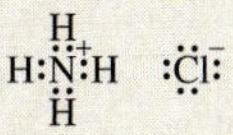

13. a. CH₃CH₂N̈H₂ **c.** CH₃CH₂ÖH **e.** CH₃CH₂C̈l: **b.** CH₃N̈HCH₃ **d.** CH₃ÖCH₃ **f.** HÖN̈H₂

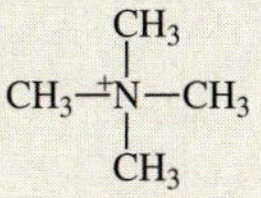

14. a. CH₃CH₂CH₂Cl **c.** CH₃COCH₂CH₃ (with O) **b.** CH₃CH₂C≡N **d.** 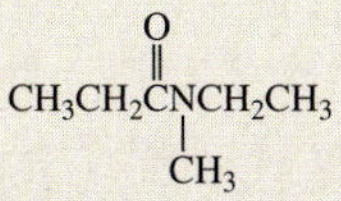 CH₃CH₂CNCH₂CH₃ (with O and CH₃)

15. a. Cl **b.** O **c.** N **d.** C and H   **16.** C₂H₇, C₃H₉   **19.** 2 C—C bonds from $sp^3 - sp^3$ overlap; 8 C—H bonds from $sp^3 - s$ overlap   **21.** greater than 104.5° and less than 109.5°   **22.** the hydrogens   **23.** Water is the most polar. Methane is the least polar.   **24.** ~ 107.3°   **25. a.** relative lengths: Br₂ > Cl₂; relative strengths: Cl₂ > Br₂ **b.** relative lengths: HBr > HCl > HF; relative strengths: HF > HCl > HBr   **26. a. 1.** C—Br **2.** C—C **3.** H—Cl **b. 1.** C—Cl **2.** C—H **3.** H—H   **27.** σ bond

28. a. CH₃CHCH=CHCH₂C≡CCH₃ (with CH₃, labeled sp^3, sp^2, sp, sp^3)

29. a. 109.5° **b.** 107.3° **c.** 109.5° **d.** 104.5°

CHAPTER 2

1. a. 1. ⁺NH₄ **2.** HCl **3.** H₂O **4.** H₃O⁺ **b. 1.** ⁻NH₂ **2.** Br⁻ **3.** NO₃⁻ **4.** HO⁻   **3. a.** compound with pKₐ = 5.2 **b.** compound with dissociation constant = 3.4 × 10⁻³   **4.** Kₐ = 1.51 10⁻⁵; weaker   **6. a.** basic **b.** acidic **c.** basic   **7. a.** CH₃COO⁻ **b.** ⁻NH₂ **c.** H₂O

8. CH₃NH⁻ > CH₃O⁻ > CH₃NH₂ > CH₃CO⁻ (with O) > CH₃OH

9. CH₃O⁻ + CH₃N⁺H₃

11. a. HBr **b.** CH₃CH₂CH₂O⁺H₂ **c.** structure on the right   **12. a.** F⁻ **b.** I⁻   **13. a.** oxygen **b.** H₂S **c.** CH₃SH **d.** CH₃C(=O)SH   **14. a.** HO⁻ **b.** NH₃ **c.** CH₃O⁻ **d.** CH₃O⁻   **15. a.** CH₃COO⁻ **b.** CH₃CH₂N⁺H₃ **c.** H₂O **d.** Br⁻

e. ⁺NH₄ **f.** HC≡N **g.** NO₂⁻ **h.** NO₃⁻   **16. a. 1.** neutral **2.** neutral **3.** charged **4.** charged **5.** charged **6.** charged **b. 1.** charged **2.** charged **3.** charged **4.** charged **5.** neutral **6.** neutral **c. 1.** neutral **2.** neutral **3.** neutral **4.** neutral **5.** neutral **6.** neutral

17. a. 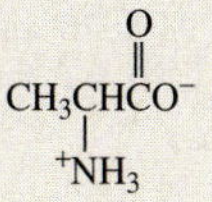 CH₃CHCO⁻ (with O and ⁺NH₃) **b.** no

CHAPTER 3

1. a. *n*-propyl alcohol or propyl alcohol **b.** dimethyl ether **c.** *n*-propylamine or propylamine

3. a. 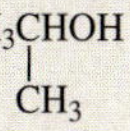 CH₃CHOH (with CH₃) **d.** 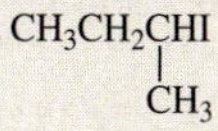 CH₃CH₂CHI (with CH₃) **b.** 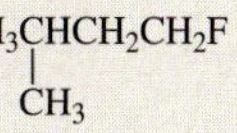 CH₃CHCH₂CH₂F (with CH₃) **e.** 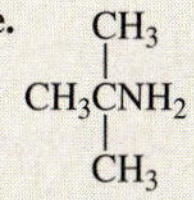 CH₃CNH₂ (with CH₃, CH₃) **c.** CH₃CH₂OCH₂CH₂CH₃ **f.** CH₃CH₂CH₂CH₂CH₂CH₂CH₂Br

4. a. ethyl methyl ether **b.** methyl propyl ether **c.** *sec*-butylamine **d.** butyl alcohol **e.** isobutyl bromide **f.** *sec*-butyl chloride

5. a. CH₃CHCH₂CH₃ (with CH₃) **b.** CH₃CH₂CH₂CH₂CH₃ pentane **c.** CH₃CCH₃ (with CH₃, CH₃) 2,2-dimethylpropane
2-methylbutane

6. a. CH₃CHCHCH₂CH₂CH₃ (with CH₃, CH₃) **c.** CH₃CCH₂CHCH₂CH₂CH₃ (with CH₃, CH₂CH₂CH₃) **b.** CH₃CHCH₂C—CHCH₂CH₃ (with CH₃, CH₃, CH(CH₃)₂) **d.** CH₃CHCH₂CHCHCH₂CH₂CH₃ (with CH₃, CH₃, CH₂CH(CH₃)₂)

8. a. 2,2,4-trimethylhexane **b.** 2,2-dimethylbutane **c.** 3,3-diethylhexane **d.** 2,5-dimethylheptane **e.** 4-isopropyloctane **f.** 4-ethyl-2,2,3-trimethylhexane

10. a. 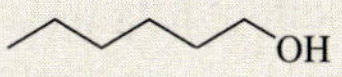 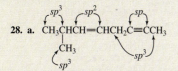 (structure with OH) **c.** 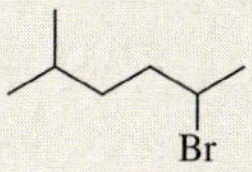 (structure with Br) **b.** 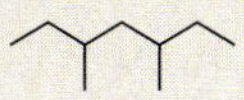 (structure) **d.**  (structure with O)

11. a. C₁₀H₂₀O **b.** C₁₀H₂₀O₂   **12. a.** 1-ethyl-2-methylcyclopentane **b.** ethylcyclobutane **c.** 3,6-dimethyldecane **d.** 5-isopropylnonane   **13. a.** same **b.** same   **14. a.** *sec*-butyl chloride, 2-chlorobutane **b.** isohexyl chloride, 1-chloro-4-methylpentane **c.** cyclohexyl bromide, bromocyclohexane **d.** isopropyl fluoride, 2-fluoropropane   **15. a.** a tertiary alkyl bromide **b.** a tertiary alcohol **c.** a primary amine   **16. a.** methylpropylamine; secondary **b.** trimethylamine; tertiary **c.** diethylamine; secondary **d.** butyldimethylamine; tertiary

17. a. 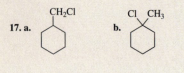 **b.**

c.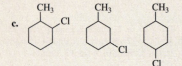

18. a. ~ 104.5° **b.** ~ 107.3° **c.** ~ 104.5°　**19.** pentane　**20. a.** O—H hydrogen bond is longer. **b.** O—H covalent bond is stronger.　**21. a.** 1, 4, and 5 **b.** 1, 2, 4, 5, and 6

22.

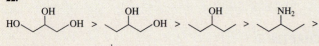

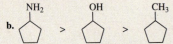

24. even number
25. a. $HOCH_2CH_2CH_2OH > CH_3CH_2CH_2OH > CH_3CH_2CH_2CH_2OH > CH_3CH_2CH_2CH_2Cl$

b.

26. ethanol　**27.** hexethal　**31.** isopropylcyclohexane　**32. a.** cis **b.** cis **c.** trans **d.** trans

CHAPTER 4

1. a. C_5H_8 **b.** C_4H_6 **c.** $C_{10}H_{16}$　**2. a.** 3 **b.** 4 **c.** 1　**4. a.** 4-methyl-2-pentene **b.** 2-chloro-3,4-dimethyl-3-hexene **c.** 1-bromo-4-methyl-3-hexene
d. 1,5-dimethylcyclohexene　**5. a.** 5 **b.** 4 **c.** 4 **d.** 6　**6. a.** 1 and 3
7. a. —I > —Br > —OH > —CH$_3$
b. —OH > —CH$_2$Cl > —CH=CH$_2$ > —CH$_2$CH$_2$OH
9. a. Z **b.** E

10.

11. a. middle one **b.** one on the right　**12.** *cis*-3,4-dimethyl-3-hexene > *trans*-3-hexene > *cis*-3-hexene > *cis*-2,5-dimethyl-3-hexene　**14.** nucleophiles:
H$^-$, CH$_3$O$^-$, CH$_3$C≡CH, NH$_3$; electrophiles: CH$_3$CHCH$_3$　**18. a.** 2 **b.** B **c.** 3
d. the first one **e.** products **f.** C to D **g.** B **h.** Yes, since the products are more stable than the reactants.

CHAPTER 5

2. ethyl cation

3. $CH_3CH_2\overset{+}{C}CH_3$ > $CH_3CH_2\overset{+}{C}HCH_3$ > $CH_3CH_2CH_2\overset{+}{C}H_2$

4. a. $CH_3CH_2CHCH_3$ (Br)　**c.**　**e.**

b. $CH_3CH_2\overset{CH_3}{\underset{Br}{C}}CH_3$　**d.** $CH_3\overset{CH_3}{\underset{Br}{C}}CH_2CH_2CH_3$　**f.** $CH_3CH_2CHCH_3$ (Br)

5. a. $CH_2=CCH_3$　**c.**

b.　**d.**

6. a. CH$_3$CH$_2$CH$_2$CHCH$_3$ (OH)　**b.**

c. CH$_3$CH$_2$CH$_2$CH$_2$CHCH$_3$ (OH) and CH$_3$CH$_2$CH$_2$CHCH$_2$CH$_3$ (OH)

d.

9. a. H$_2$O is the solvent:　CH$_3$CH$_2$CHCH$_3$ (OH) **major**

b. CH$_3$OH is the solvent:　CH$_3$CH$_2$CHCH$_3$ (OCH$_3$) **major**

11. C_nH_{2n-4}　**12.** $C_{14}H_{20}$
13. a. ClCH$_2$CH$_2$C≡CCH$_2$CH$_3$

b. CH$_3$C≡CCHCH$_3$ (Br)　**c.** HC≡CCH$_2$CCH$_3$ (CH$_3$)(CH$_3$)

14. a. 4-methyl-1-pentyne **b.** 2-hexyne　**16. a.** 5-bromo-2-pentyne
b. 6-bromo-2-chloro-4-octyne **c.** 5-methyl-3-heptyne **d.** 3-ethyl-1-hexyne
17. a. $sp^2–sp^2$ **b.** $sp^2–sp^3$ **c.** $sp–sp^2$ **d.** $sp–sp^3$ **e.** $sp–sp$
f. $sp^2–sp^2$ **g.** $sp^2–sp^3$ **h.** $sp–sp^3$ **i.** $sp^2–sp$

18. a. CH$_2$=CCH$_3$ (Br) **b.** CH$_3$CCH$_3$ (Br)(Br) **c.** CH$_3$CH$_2$CCH$_3$ (Br)(Br)

d. CH$_3$CCH$_2$CH$_2$CH$_3$ (Br)(Br)　+　CH$_3$CH$_2$CCH$_2$CH$_3$ (Br)(Br)

19. CH$_3$CH$_2$CCH$_2$CH$_2$CH$_2$CH$_3$ (O)　and　CH$_3$CH$_2$CH$_2$CCH$_2$CH$_2$CH$_3$ (O)

20. a. CH$_3$C≡CH **b.** CH$_3$CH$_2$C≡CCH$_2$CH$_3$ **c.** HC≡C—

21. CH$_3$CH=CCH$_2$CH$_2$CH(CH$_3$)$_2$ (OH)　and　CH$_3$CH$_2$C=CHCH$_2$CH(CH$_3$)$_2$ (OH)

22. a. 2-butyne + H$_2$ + Lindlar cat. **b.** 1-hexyne + H$_2$ + Lindlar cat.
23. a. 1-butene, *cis*-2-butene, *trans*-2-butene **b.** 1-pentene, *cis*-2-pentene, or *trans*-2-pentene **c.** 1-methylcyclopentene, 3-methylcyclopentene, 4-methylcyclopentene, methylenecyclopentene　**24.** The carbanion that would be formed is a stronger base than the amide ion.
25. a. CH$_3$CH$_2$CH$_2\overline{C}$H$_2$ > CH$_3$CH$_2$CH=$\overline{C}$H > CH$_3$CH$_2$C≡$\overline{C}$
　b. $^-$NH$_2$ > CH$_3$C≡C$^-$ > CH$_3$CH$_2$O$^-$ > F$^-$
26. a. CH$_3\overset{+}{C}$H$_2$　**b.** H$_2$C=$\overset{+}{C}$H
30. a. CH$_2$=CHCl **b.** CH$_2$=CCH$_3$ **c.** CF$_2$=CF$_2$ (COCH$_3$)(O)
32. beach balls

CHAPTER 6

1. a. CH$_3$CH$_2$CH$_2$OH　CH$_3$CHOH (CH$_3$)　CH$_3$CH$_2$OCH$_3$　**b.** 7
3. a. F, G, J, L, N, P, Q, R, S, Z **b.** A, C, D, H, I, M, O, T, U, V, W, X, Y
4. a, c, and **f**　**6. a, c,** and **f**

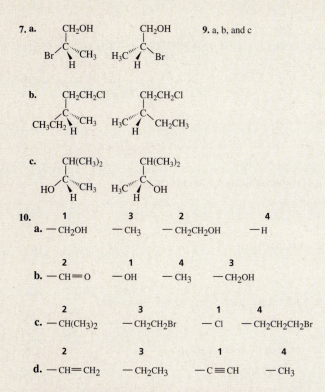

7. a.

CH₂OH structures (Br, CH₃, H) and (H₃C, Br, H)   **9.** a, b, and c

b. CH₂CH₂Cl structures (CH₃CH₂, CH₃, H) and (H₃C, CH₂CH₃, H)

c. CH(CH₃)₂ structures (HO, CH₃, H) and (H₃C, OH, H)

10.

	1	3	2	4
a.	—CH₂OH	—CH₃	—CH₂CH₂OH	—H

	2	1	4	3
b.	—CH=O	—OH	—CH₃	—CH₂OH

	2	3	1	4
c.	—CH(CH₃)₂	—CH₂CH₂Br	—Cl	—CH₂CH₂CH₂Br

	2	3	1	4
d.	—CH=CH₂	—CH₂CH₃	—C≡CH	—CH₃

11. a. *R* **b.** *R*   **12. a.** (*R*)-2-bromobutane **b.** (*R*)-1,3-dichlorobutane   **13. a.** enantiomers **b.** enantiomers   **15. a.** levorotatory **b.** dextrorotatory   **16. a.** *S* **b.** *R* **c.** *R* **d.** *S*   **17.** +168   **18. a.** −24 **b.** 0   **19.** From the data given, you cannot determine the configuration.   **20. a.** enantiomers **b.** identical compounds (Therefore, they are not isomers.) **c.** diastereomers   **21. a.** 8 **b.** $2^8 = 256$ **23. a.** diastereomers **b.** enantiomers **c.** identical **d.** constitutional isomers   **24.** B   **26.** the one on the left

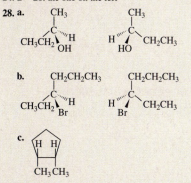

28. a. CH₃ structures (CH₃CH₂, H, OH) and (H, CH₂CH₃, HO)

b. CH₂CH₂CH₃ structures (CH₃CH₂, H, Br) and (H, CH₂CH₃, Br)

c. (H H, CH₃ CH₃)

29. a. (*R*)-malate and (*S*)-malate **b.** (*R*)-malate and (*S*)-malate

CHAPTER 7

1. a, b, e, f, and g have delocalized electrons.

4. a. CH₃C⁺≡CH≡CHCH₃ (with CH₃) **b.** cyclohexenone structure **c.** CH₃CH≡CH≡CHCH₃

5. a. All have the same length. **b.** 2/3 of a negative charge

6. (⁻O—C(=O)—O⁻) > (H—C(=O)—O⁻) > (H—C(=O)—OH)   **7.** (H₃C—C(=O)—O⁻)

8. ˙CH₂—CH=CH—CH=CH₂ ⟷ CH₂=CH—˙CH—CH=CH₂ ⟷ CH₂=CH—CH=CH—˙CH₂

9. 6   **10. a.** CH₃CH=CHCHCH₃ **b.** phenyl—⁺CCH₃ (with CH₃)

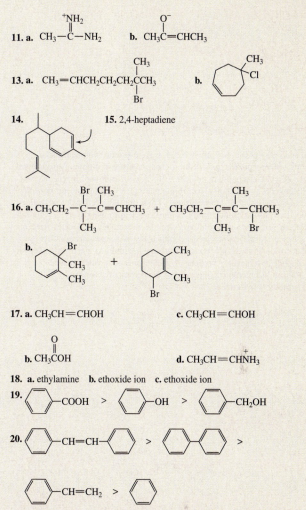

11. a. CH₃—C⁺(=NH₂)—NH₂ **b.** CH₃C(—O⁻)=CHCH₃

13. a. CH₃=CHCH₂CH₂CH₂CCH₃ (with CH₃ and Br) **b.** cycloheptene with CH₃ and Cl

14.   **15.** 2,4-heptadiene

16. a. CH₃CH₂—C(Br)(CH₃)—C(CH₃)=CHCH₃ + CH₃CH₂—C(CH₃)=C(CH₃)—CHCH₃ (with Br)

b. cyclohexene with Br, CH₃, CH₃ + cyclohexene with CH₃, CH₃, Br

17. a. CH₃CH=CHOH **c.** CH₃CH=CHOH

b. CH₃COH (with =O) **d.** CH₃CH=CHNH₃⁺

18. a. ethylamine **b.** ethoxide ion **c.** ethoxide ion

19. (benzene—COOH) > (benzene—OH) > (benzene—CH₂OH)

20. (benzene—CH=CH—benzene) > (benzene—benzene) >

(benzene—CH=CH₂) > (cyclohexane)

21. a. The compound on the right is blue. **b.** They will be the same color.

CHAPTER 8

3. the cycloheptatrienyl cation   **4.** only b   **7. a.** The nitrogen donates electrons by resonance into the ring. **b.** The nitrogen is the most electronegative atom in the molecule. **c.** The nitrogen atom withdraws electrons inductively from the ring.

8. a. CH₃CHCH₂CH₂CH₂CH₃ (with phenyl) **c.** CH₃CH₂CHCH₂CH₃ (with CH₂ phenyl)

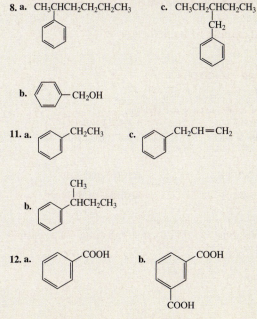

b. benzene—CH₂OH

11. a. benzene—CH₂CH₃ **c.** benzene—CH₂CH=CH₂

b. benzene—CHCH₂CH₃ (with CH₃)

12. a. benzene with COOH **b.** benzene with COOH and COOH

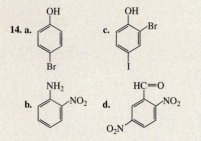

14. a. 4-bromophenol **b.** 2-nitroaniline **c.** 2-bromo-4-iodophenol **d.** 2,4-dinitrobenzaldehyde

15. a. *meta*-ethylphenol or 3-ethylphenol **b.** *meta*-bromochlorobenzene or 1-bromo-3-chlorobenzene **c.** *para*-bromobenzaldehyde or 4-bromo-benzaldehyde **d.** *ortho*-ethyltoluene or 2-ethyltoluene   **16. a.** 1,3,5-tribromobenzene **b.** *meta*-nitrophenol or 3-nitrophenol **c.** *para*-bromotoluene or 4-bromotoluene **d.** *ortho*-dichlorobenzene or 1,2-dichlorobenzene   **17. a.** donates electrons by resonance and withdraws electrons inductively **b.** donates electrons inductively **c.** withdraws electrons by resonance and withdraws electrons inductively **d.** donates electrons by resonance and withdraws electrons inductively **e.** donates electrons by resonance and withdraws electrons inductively **f.** withdraws electrons inductively
18. a. phenol > toluene > benzene > bromobenzene > nitrobenzene
   b. toluene > chloromethylbenzene > dichloromethylbenzene > difluoromethylbenzene

20. a.

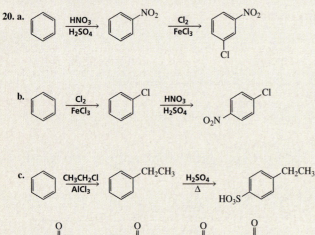

22. a. ClCH₂COH **b.** H₃ṄCH₂COH **c.** FCH₂COH **d.** HCOH

CHAPTER 9

1. a. It is tripled. **b.** It is half the original rate.

2. CH₃CH₂CH₂CH₂CH₂Br > CH₃CHCH₂CH₂Br >
$$\underset{CH_3}{|}$$

CH₃CH₂CHCH₂Br > CH₃CH₂CBr
with CH₃ and CH₃ substituents

3. b. (*S*)-2-butanol **c.** (*R*)-3-hexanol **d.** 3-pentanol
5. a. CH₃CH₂Br + HO⁻   **c.** CH₃CH₂Br + I⁻
   b. CH₃CHCH₂Br + HO⁻   **d.** CH₃CH₂CH₂I + HO⁻
   (with CH₃)

6. a. CH₃CH₂CH₂OCH₂CH₃   **c.** (CH₃)₃ṄCH₂CH₃
   b. CH₃C≡CCH₂CH₃   **d.** CH₃CH₂SCH₂CH₃

7. CH₃CBr > CH₃CHBr > CH₃CH₂CH₂Br > CH₃Br   **8. a.** 2 **b.** 1
(with CH₃ substituents)

9. CH₃CH₂CCH₂CH₃ > CH₃CHCH₂CH₂CH₃ > CH₃CHCH₂CH₂CH₃ >
with Br, Br, Cl substituents
ClCH₂CH₂CH₂CH₂CH₃

10. a. CH₃CH₂CH₂Br   **c.** cyclohexyl bromide
   b. CH₃CHCH₂CHCH₃   **d.** CH₃CCH₂CH₂Cl
   (with CH₃, Br)   (with CH₃, CH₃)

11. a. CH₃CH₂CHCH₃   **c.** cyclohexyl bromide
   (with Br)
   b. CH₃CH₂CH₂CCH₃   **d.** equally unreactive
   (with CH₃, Br)

13. a. CH₃CHBr   **b.** CH₃CHBr
   (with CH₃)   (with CH₃)

14. a. 2 is more reactive than 1. **b.** 2 is more reactive than 1. **c.** 2 is more reactive than 1. **d.** 1 is more reactive than 2.

15. b. CH₃CH₂CH=CCH₃   **d.** CH₃CCH=CH₂
   (with CH₃)   (with CH₃, CH₃)

   c. CH₃CH=CHCHCH₃
   (with CH₃)

16. CH₃C=CCH₂CH₃ > CH₃CHC=CHCH₃ > CH₃CHCCH₂CH₃
with CH₃ substituents (CH₂)

18. a. CH₃CH₂CHCH₃   **c.** cyclohexyl bromide
   (with Br)
   b. CH₃CH₂CH₂CCH₃   **d.** CH₃CCH₂CH₂Cl
   (with CH₃, Br)   (with CH₃, CH₃)

19. a. CH₃CH₂CHCH₃   **c.** cyclohexyl bromide
   (with Br)
   b. CH₃CH₂CH₂CCH₃
   (with CH₃, Br)

22. a. primarily substitution **b.** substitution and elimination **c.** substitution and elimination **d.** only elimination   **24. a.** no reaction **b.** no reaction **c.** substitution and elimination **d.** substitution and elimination   **25.** a and b

CHAPTER 10

2. a. 1-pentanol **primary b.** 4-methylcyclohexanol **secondary c.** 5-chloro-2-methyl-2-pentanol **tertiary d.** 2-ethyl-1-pentanol **primary e.** 5-methyl-3-hexanol **secondary f.** 2,6-dimethyl-4-octanol **secondary**

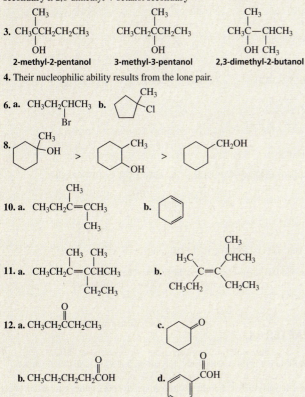

3.

CH₃CCH₂CH₂CH₃ | OH — 2-methyl-2-pentanol

CH₃CH₂CCH₂CH₃ | OH — 3-methyl-3-pentanol

CH₃C—CHCH₃ | OH CH₃ — 2,3-dimethyl-2-butanol

4. Their nucleophilic ability results from the lone pair.

6. a. CH₃CH₂CHCH₃ | Br **b.** (cyclopentane with CH₃ and Cl)

8. (cyclohexane-OH with CH₃) > (cyclohexane-OH with CH₃) > (cyclohexane-CH₂OH)

10. a. CH₃CH₂C=CCH₃ (with two CH₃) **b.** (cyclohexene)

11. a. CH₃CH₂C=CCHCH₃ (CH₃, CH₃, CH₂CH₃) **b.** (alkene with H₃C, CHCH₃, CH₃CH₂, CH₂CH₃)

12. a. CH₃CH₂CCH₂CH₃ (ketone) **c.** (cyclohexanone)

b. CH₃CH₂CH₂CH₂COH **d.** (benzoic acid, COH)

13. a. 2-butanol **b.** cyclohexanol **c.** 1-butanol **15. a. 1.** methoxyethane **2.** ethoxyethane **3.** 4-methoxyoctane **4.** 1-propoxybutane **b. no c. 1.** ethyl methyl ether **2.** diethyl ether **3.** no common name **4.** butyl propyl ether

17. a. CH₃CHCH₂OH + CH₃CH₂I | CH₃ **18. a.** (cyclohexane epoxide)

b. HOCH₂CH₂CH₂CH₂CH₂I

b. H₃C, H₃C / O \ CH₂CH₃ (epoxide)

c. CH₃C—I + CH₃CH₂OH | CH₃ with CH₃

d. HOCH₂CH₂CH₂CCH₃ | I with CH₃

19. a. HOCH₂CCH₃ | OCH₃ with CH₃ **c.** HOCH—CCH₃ | OCH₃ CH₃ CH₃

b. CH₃OCH₂CCH₃ | CH₃ **d.** CH₃OCH—CCH₃ | CH₃ CH₃

20. noncyclic ether

21. (bicyclic epoxide structure)

CHAPTER 11

1. a. propanamide, propionamide **b.** isobutyl butanoate, isobutyl butyrate **c.** potassium butanoate, potassium butyrate **d.** pentanoyl chloride, valeryl chloride **e.** N,N-dimethylhexanamide **f.** N-ethyl-2-methylbutanamide

2. a. CH₃CO— (phenyl ester)

d. CH₃CH₂CH₂CHCOCH₂CH₃ | Cl

b. CH₃CO⁻Na⁺

e. CH₃CHCH₂CNH₂ | Br

c. CH₃CNHCH₂— (phenyl)

f. CH₃CH₂CCl

3. The carbon–oxygen bond in an alcohol. **4.** The bond between oxygen and the methyl group is the longest; the bond between carbon and the carbonyl oxygen is the shortest. **6. a.** a new carboxylic acid derivative **b.** no reaction **c.** a mixture of two carboxylic acid derivatives **7. a.** acetate ion **b.** no reaction **10. a.** CH₃CH₂CH₂OH **b.** CH₃CH₂NH₂ **c.** (CH₃)₂NH

d. —NH₂ (phenyl) **e.** H₂O **f.** HO—(phenyl)—NO₂ **13.** phenyl acetate

14. a. (phenyl)COH + CH₃CH₂OH **b.** CH₃CH₂CH₂COH + CH₃OH

15. HOCH₂CH₂CH₂CH₂COH

16. CH₃CH₂COCH₂CH₂CH₂CH₃ + CH₃OH

19. a. CH₃CH₂CH₂CCl $\xrightarrow{CH_3OH}$ CH₃CH₂CH₂COCH₃

b. CH₃CCl $\xrightarrow{CH_3(CH_2)_7OH}$ CH₃COCH₂CH₂CH₂CH₂CH₂CH₂CH₂CH₃

22. a. CH₃CH₂CH₂CCl + 2 CH₃CH₂NH₂

b. (phenyl)CCl + 2 CH₃NH | CH₃

23. 2 and 4 **24. a. 1.** SOCl₂ **2.** (phenyl)—OH **b. 1.** SOCl₂ **2.** 2 CH₃CH₂NH₂

25. a. CH₃CH₂CH₂Br **b.** CH₃CHCH₂CH₂Br | CH₃

CHAPTER 12

1. If the ketone functional groups were anywhere else in these compounds, they would not be ketones and would not have the "one" suffix. **2. a.** 4-heptanone **b.** 4-phenylbutanal **3. a.** 3-methylpentanal, β-methylvaleraldehyde **b.** 2-methyl-4-heptanone, isobutyl propyl ketone **c.** 4-ethylhexanal, γ-ethylcaproaldehyde **d.** 6-methyl-3-heptanone, ethyl isopentyl ketone **4. a.** 2-heptanone **b.** 5-methyl-3-hexanone **5. a.** CH₃CH₃ + HO⁻ **b.** CH₃CH₃ + CH₃O⁻ **c.** CH₃CH₃ + CH₃N̄H

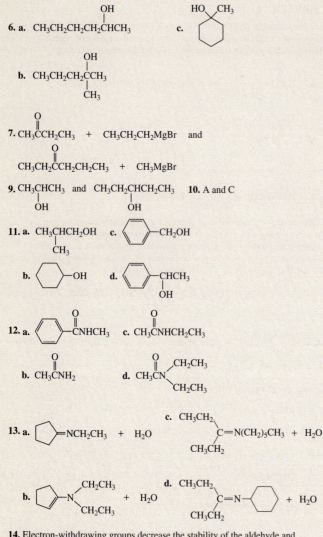

6. a. CH₃CH₂CH₂CH₂CHCH₃ (OH)   c. [structure]

b. CH₃CH₂CH₂CCH₃ (OH, CH₃)

7. CH₃CCH₂CH₃ (O) + CH₃CH₂CH₂MgBr   and

CH₃CH₂CCH₂CH₂CH₃ (O) + CH₃MgBr

9. CH₃CHCH₃ (OH) and CH₃CH₂CHCH₂CH₃ (OH)   10. A and C

11. a. CH₃CHCH₂OH (CH₃)   c. [benzene]—CH₂OH

b. [cyclohexane]—OH   d. [benzene]—CHCH₃ (OH)

12. a. [benzene]—CNHCH₃ (O)   c. CH₃CNHCH₂CH₃ (O)

b. CH₃CNH₂ (O)   d. CH₃CN (O) with N(CH₂CH₃)(CH₂CH₃)

13. a. [cyclopentane]=NCH₂CH₃ + H₂O   c. CH₃CH₂,CH₃CH₂ C=N(CH₂)₅CH₃ + H₂O

b. [cyclopentane]—N(CH₂CH₃)(CH₂CH₃) + H₂O   d. CH₃CH₂,CH₃CH₂ C=N—[cyclohexane] + H₂O

14. Electron-withdrawing groups decrease the stability of the aldehyde and increase the stability of the hydrate.

15. O₂N—[benzene]—C(O)—[benzene]—NO₂   **17. a.** hemiacetals: 7 **b.** acetals: 2, 3 **c.** hemiketals: 1, 8 **d.** ketals: 5 **e.** hydrates: 4, 6

18. a. [decalin structure with Br, O]   **b.** [decalin structure with OH, CH₃]

CHAPTER 13

2. a. the ketone **b.** the ester   **3.** The electrons left behind when the proton is removed are readily delocalized onto two oxygen atoms.   **4.** They have a hydrogen bonded to the nitrogen, which is more acidic than the hydrogen attached to the α-carbon.   **5.** It is easier for hydroxide ion to attack the reactive carbonyl group than to remove a hydrogen from an α-carbon.

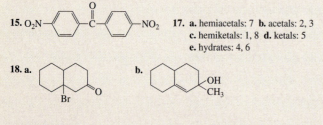

6. a. CH₃CH=CCH₂CH₃ (OH) **b.** [benzene]—C=CH₂ (OH) **c.** [cyclohexene with OH]

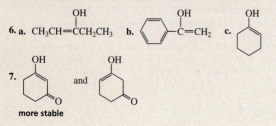

7. [structure] and [structure]

more stable

9. a. CH₃CH₂CCHCH₃ (O, CH₂CH₃)   b. [cyclohexanone with CH₂CH₃]

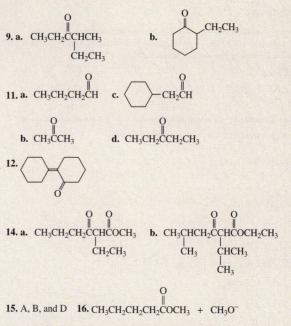

11. a. CH₃CH₂CH₂CH (O)   c. [cyclohexane]—CH₂CH (O)

b. CH₃CCH₃ (O)   d. CH₃CH₂CCH₂CH₃ (O)

12. [bicyclic ketone structure]

14. a. CH₃CH₂CH₂CCHCOCH₃ (O O, CH₂CH₃)   b. CH₃CHCH₂CCHCOCH₂CH₃ (O O, CH₃, CHCH₃, CH₃)

15. A, B, and D   **16.** CH₃CH₂CH₂CH₂COCH₃ (O) + CH₃O⁻

17. A and D   **18. a.** methyl bromide **b.** benzyl bromide **c.** isobutyl bromide
20. a. ethyl bromide **b.** pentyl bromide **c.** benzyl bromide   **22.** 7
23. a. 3 **b.** 7

CHAPTER 14

1. 2, 3, 5   **2.** m/z = 57   **4.** 1-bromopropane   **6.** 2-pentanone   **8.** C₆H₁₄
9. a. IR **b.** UV   **10. a.** 2000 cm⁻¹ **b.** 8 μm **c.** 2 μm   **11. a.** C≡C stretch
b. C—H stretch **c.** C≡N stretch **d.** C=O stretch   **13. a.** the
carbon–oxygen stretch of phenol **b.** the carbon–oxygen double-bond stretch of a
ketone **c.** the C—O stretch   **14.** one bonded to an sp³ carbon   **15.** C=O
17. a. a ketone **b.** a tertiary amine   **20. a.** 2 **b.** 1 **c.** 4 **d.** 3 **e.** 3 **f.** 3   **21. A** = 2
signals, **B** = 1 signal, **C** = 3 signals   **22. a.** 2 ppm **b.** 2 ppm   **23.** 1.5 ppm
24. upfield from the TMS peak   **26.** first spectrum = 1-iodopropane

27. a. CH₃CHCHBr (Br Br)   **b.** CH₃CHOCH₃ (CH₃)   **c.** CH₃CH₂CHCH₃ (Cl)

28. a. CH₃CH₂CHCH₃ (Cl)   **b.** CH₃CH₂CH₂Cl   **31.** B

32. a. CH₃CHCOH (O, Cl)   **b.** ClCH₂CH₂COH (O)

37. CH₃O—[benzene]—CH₃   **38. a.** 1. 3 2. 3 3. 4 4. 3 5. 2 6. 3

40. CH₃CH₂CH₂CH₂CH₂CCH₂CH₂CH₂CH₃ (O)

CHAPTER 15

1. D-Ribose is an aldopentose. D-Sedoheptulose is a ketoheptose. D-Mannose is an
aldohexose.   **3. a.** D-ribose **b.** L-talose   **4.** D-psicose   **5. a.** 16 **b.** 32 **c.** none
6. D-tagatose, D-galactose, and D-talose   **8. a.** D-iditol **b.** D-iditol and D-gulitol
9. L-galactose   **10.** D-tagatose   **11. a.** L-gulose **b.** L-gularic acid **c.** D-allose and
L-allose, D-altrose and D-talose, L-altrose and L-talose, D-galactose and
L-galactose   **12. a.** D-gulose and D-idose **b.** L-xylose and L-lyxose   **14. b.** C-2,
C-3, and C-4 **c.** C-1 and C-3   **15. b.** methyl α-D-galactoside; nonreducing

CHAPTER 16

1. (*S*)-alanine   **3.** isoleucine

6. b. $H_2NCCH_2CH_2CHCO^-$ (with $\overset{O}{\overset{\|}{}}$ ketone, $\overset{}{\underset{{}^+NH_3}{}}$)   **c.** $NH_2CNHCH_2CH_2CH_2CHCO^-$ (with $\overset{{}^+NH_2}{\overset{\|}{}}$, $\overset{O}{\overset{\|}{}}$, $\overset{}{\underset{{}^+NH_3}{}}$)

8. a. 5.43 **b.** 10.76 **c.** 5.68 **d.** 2.98   **9. a.** Asp **b.** Arg   **10.** 2-methylpropanal
14. A-G-M A-M-G M-G-A M-A-G G-A-M G-M-A   **16. a.** glutamate,
cysteine, and glycine   **17.** Leu-Val and Val-Val   **18.** 5.8%   **20.** Gly-Arg-Trp-
Ala-Glu-Leu-Met-Pro-Val-Asp   **21. a.** His-Lys, Leu-Val-Glu-Pro-Arg, Ala-Gly-
Ala **b.** Leu-Gly-Ser-Met-Phe-Pro-Tyr, Gly-Val   **23.** Leu-Tyr-Lys-Arg-Met-Phe-
Arg-Ser   **25. a.** cigar-shaped protein **b.** subunit of a hexamer

CHAPTER 17

2. 2 and 3   **3.** Only **2** can form an imine with the substrate.

4. $^-OCCCHCH_2CO^-$ + NADH + H^+ (with COO$^-$, two O, and O below)   **5.** CH_3CH-CO^- + NAD^+ (with OH and O)

6. a. 7   **b.** 3 and a separate group of 2

7. ring with S—S, $CH_2CH_2CH_2CH_2CO^-$ + $FADH_2$ (with O)

8. a. CH_3C- (with O)   **b.** CH_3C- (with O)   **9. a.** $^-OC-CH_2C-CO^-$ (with O, O, O)   **10.** 9

12. a. CH_3CHCO^- (with O and $^+NH_3$)   **b.** $^-OCCH_2CHCO^-$ (with O, O, and $^+NH_3$)

13. $CH_3CHCHOH$ (with H arrow and OH)   **14.** It is obtained by adding four hydrogens to folate.

15. $HSCH_2CH_2CHCO^-$ (with O and $^+NH_3$) + N^5-methyl-THF $\longrightarrow$ $CH_3SCH_2CH_2CHCO^-$ (with O and $^+NH_3$) + THF

16. the methylene group of N^5,N^{10}-methylene-THF

CHAPTER 18

4. The β-carbon has a partial positive charge.   **5.** eight   **6.** seven
9. two   **10.** a ketone   **11.** thiamine pyrophosphate   **12.** an aldehyde
14. pyridoxal phosphate   **15.** pyruvate   **16.** a secondary alcohol   **17.** two
18. citrate and isocitrate   **19.** 11

CHAPTER 19

2. glyceryl tripalmitate

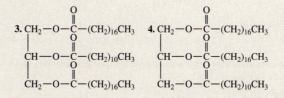

6. integral proteins have a higher percentage of nonpolar amino acids.
7. The bacteria could synthesize phosphoglycerides with more saturated fatty
acids.   **9. b.** a monoterpene   **11.** two ketone groups, a double bond, an aldehyde
group, a primary alcohol, a secondary alcohol

CHAPTER 20

2. a. 3′—C—C—T—G—T—T—A—G—A—C—G— 5′

b. guanine   **5.** Met-Asp-Pro-Val-Ile-Lys-His   **6.** Met-Asp-Pro-Leu-Leu-Asn

8. 5′—G—C—A—T—G—G—A—C—C—C—C—G—T—T—
A—T—T—A—A—A—C—A—C— 3′

9. a

Glossary

absorption band a peak in a spectrum that occurs as a result of the absorption of energy.

acetal R—C—H with OR groups

acetoacetic ester synthesis synthesis of a methyl ketone, using ethyl acetoacetate as the starting material.

acetylide ion $RC{\equiv}C^-$

achiral (optically inactive) an achiral molecule is identical to (that is, superimposable upon) its mirror image.

acid a substance that donates a proton.

acid anhydride

acid–base reaction a reaction in which an acid donates a proton to a base or accepts a share in a base's electrons.

acid catalyst a catalyst that increases the rate of a reaction by donating a proton.

acid-catalyzed reaction a reaction catalyzed by an acid.

acid dissociation constant a measure of the degree to which an acid dissociates in solution.

activating substituent a substituent that increases the reactivity of an aromatic ring. Electron-donating substituents activate aromatic rings toward electrophilic attack, and electron-withdrawing substituents activate aromatic rings toward nucleophilic attack.

active site a pocket or cleft in an enzyme where the substrate is bound.

acyl chloride

acyl group a carbonyl group bonded to an alkyl group or to an aryl group.

1,2-addition addition to the 1- and 2-positions of a conjugated system.

1,4-addition addition to the 1- and 4-positions of a conjugated system.

addition reaction a reaction in which atoms or groups are added to the reactant.

alcohol a compound with an OH group in place of one of the hydrogens of an alkane: ROH.

alcoholysis reaction with an alcohol.

aldaric acid a dicarboxylic acid with an OH group bonded to each carbon. Obtained by oxidizing the aldehyde and primary alcohol groups of an aldose.

aldehyde

alditol a compound with an OH group bonded to each carbon. Obtained by reducing an aldose or a ketose.

aldol addition a reaction between two molecules of an aldehyde (or two molecules of a ketone) that connects the α-carbon of one with the carbonyl carbon of the other.

aldol condensation an aldol addition followed by the elimination of water.

aldonic acid a carboxylic acid with an OH group bonded to each carbon. Obtained by oxidizing the aldehyde group of an aldose.

aldose a polyhydroxyaldehyde.

aliphatic a nonaromatic organic compound.

alkaloid a natural product, with one or more nitrogen heteroatoms, found in the leaves, bark, or seeds of plants.

alkane a hydrocarbon that contains only single bonds.

alkene a hydrocarbon that contains a double bond.

alkyl halide a compound with a halogen in place of one of the hydrogens of an alkane.

alkyl substituent (alkyl group) a substituent formed by removing a hydrogen from an alkane.

alkyne a hydrocarbon that contains a triple bond.

allyl group $CH_2{=}CHCH_2{-}$

allylic carbon an sp^3 carbon adjacent to a vinylic carbon.

allylic cation a species with a positive charge on an allylic carbon.

amide

amine a compound with alkyl groups in place of one or more of the hydrogens of ammonia: RNH_2, R_2NH, R_3N.

amino acid an α-aminocarboxylic acid. Naturally occurring amino acids have the L configuration.

amino acid analyzer an instrument that automates the ion-exchange separation of amino acids.

amino acid residue a monomeric unit of a peptide or protein.

amino acid side chain the substituent attached to the α-carbon of an amino acid.

aminolysis reaction with an amine.

amino sugar a sugar in which one of the OH groups is replaced by an NH_2 group.

anabolic reaction a reaction that a living organism carries out in order to synthesize complex molecules from simple precursor molecules.

anabolism reactions that living organisms carry out in order to synthesize complex molecules from simple precursor molecules.

angle strain the strain introduced into a molecule as a result of its bond angles being distorted from their ideal values.

angstrom unit of length; 100 picometers = 10^{-8} cm = 1 angstrom.

anomeric carbon the carbon in a cyclic sugar that is the carbonyl carbon in the open-chain form.

anomers two cyclic sugars that differ in configuration only at the carbon that is the carbonyl carbon in the open-chain form.

antibiotic a compound that interferes with the growth of a microorganism.

antibodies compounds that recognize foreign particles in the body.

anticodon the three bases at the bottom of the middle loop in tRNA.

antigens compounds that can generate a response from the immune system.

antiviral drug a drug that interferes with DNA or RNA synthesis in order to prevent a virus from replicating.

applied magnetic field the externally applied magnetic field.

aprotic solvent a solvent that does not have a hydrogen bonded to an oxygen or to a nitrogen.

arene oxide an aromatic compound that has had one of its double bonds converted to an epoxide.

aromatic a cyclic and planar compound with an uninterrupted ring of p orbital-bearing atoms containing an odd number of pairs of π electrons.

aryl group a benzene or a substituted-benzene group.

asymmetric center a carbon bonded to four different atoms or groups.

atomic number the number of protons (or electrons) that the neutral atom has.

atomic orbital an orbital associated with an atom.

atomic weight the average mass of the atoms in the naturally occurring element.

autoradiograph the exposed photographic plate obtained in autoradiography.

axial bond a bond of the chair conformation of cyclohexane that is perpendicular to the plane in which the chair is drawn (an up–down bond).

back-side attack nucleophilic attack on the side of the carbon opposite the side bonded to the leaving group.

bactericidal drug a drug that kills bacteria.

bacteriostatic drug a drug that inhibits the further growth of bacteria.

base[1] a substance that accepts a proton.

base[2] a heterocyclic compound (a purine or a pyrimidine) in DNA and RNA.

base catalyst a catalyst that increases the rate of a reaction by removing a proton.

basicity the tendency of a compound to share its electrons with a proton.

bending vibration a vibration that does not occur along the line of the bond. It results in changing bond angles.

benzyl group ⟨benzene ring⟩—CH_2—

benzylic carbon an sp^3 carbon bonded to a benzene ring.

benzylic cation a compound with a positive charge on a benzylic carbon.

bimolecular reaction (second-order reaction) a reaction whose rate depends on the concentration of two reactants.

biochemistry (biological chemistry) the chemistry of biological systems.

biodegradable polymer a polymer that can be broken into small segments by an enzyme-catalyzed reaction.

bioorganic compound an organic compound found in biological systems.

biopolymer a polymer that is synthesized in nature.

biosynthesis synthesis in a biological system.

biotin the coenzyme required by enzymes that catalyze carboxylation of a carbon adjacent to an ester or a keto group.

blind screen (random screen) the search for a pharmacologically active compound without any information about which chemical structures might show activity.

boiling point the temperature at which the vapor pressure equals the atmospheric pressure.

bond an attractive force between two atoms.

bond dissociation energy the energy required to break a bond, or the amount of energy released when a bond is formed.

bond length the internuclear distance between two atoms at minimum energy (maximum stability).

bond strength the energy required to break a bond.

brand name name that identifies a commercial product and distinguishes it from other products. It can be used only by the owner of the registered trademark.

buffer a weak acid and its conjugate base.

carbanion a compound containing a negatively charged carbon.

carbocation a species containing a positively charged carbon.

carbohydrate a sugar or a saccharide. Naturally occurring carbohydrates have the D configuration.

α-carbon a carbon bonded to a carbonyl carbon or to a leaving group.

β-carbon a carbon adjacent to an α-carbon carbonyl.

carbonyl carbon the carbon of a carbonyl group.

carbonyl compound a compound that contains a carbonyl group.

carbonyl group a carbon doubly bonded to an oxygen.

carbonyl oxygen the oxygen of a carbonyl group.

carboxyl group COOH

carboxylic acid

$$R-\overset{\overset{\displaystyle O}{\|}}{C}-OH$$

carboxylic acid derivative a compound that is hydrolyzed to a carboxylic acid.

carboxyl oxygen the single-bonded oxygen of a carboxlic acid or an ester.

carotenoid a class of compounds (a tetraterpene) responsible for the red and orange colors of fruits, vegetables, and fall leaves.

catabolic reaction a reaction that a living organism carries out in order to break down complex molecules into simple molecules and energy.

catabolism reactions that living organisms carry out in order to break down complex molecules into simple molecules and energy.

catalyst a species that increases the rate at which a reaction occurs without being consumed in the reaction. Because it does not change the equilibrium constant of the reaction, it does not change the amount of product that is formed.

catalytic hydrogenation the addition of hydrogen to a double or a triple bond with the aid of a metal catalyst.

cation-exchange resin a negatively charged resin used in ion-exchange chromatography.

chain-growth polymer a polymer made by adding monomers to the growing end of a chain.

chain reaction a reaction in which propagating steps are repeated over and over.

chair conformation the conformation of cyclohexane that roughly resembles a chair. It is the most stable conformation of cyclohexane.

chemically equivalent protons protons with the same connectivity relationship to the rest of the molecule.

chemical shift the location of a signal in an NMR spectrum. It is measured downfield from a reference compound (most often, TMS).

chiral (optically active) a chiral molecule has a nonsuperimposable mirror image.

cholesterol a steroid that is the precursor of all other animal steroids.

chromatography a separation technique in which the mixture to be separated is dissolved in a solvent and the solvent is passed through a column packed with an absorbent stationary phase.

cis fused two cyclohexane rings fused together such that if the second ring were considered to be two substituents of the first ring, one substituent would be in an axial position and the other would be in an equatorial position.

cis isomer the isomer with identical substituents on the same side of the double bond or on the same side of a cyclic structure.

cis–trans isomers geometric isomers.

citric acid cycle (Krebs cycle) a series of reactions that converts the acetyl group of acetyl-CoA into two molecules of CO_2.

Claisen condensation a reaction between two molecules of an ester that connects the α-carbon of one with the carbonyl carbon of the other and eliminates an alkoxide ion.

codon a sequence of three bases in mRNA that specifies the amino acid to be incorporated into a protein.

coenzyme a cofactor that is an organic molecule.

coenzyme A a thiol used by biological organisms to form thioesters.

coenzyme B_{12} the coenzyme required by enzymes that catalyze certain rearrangement reactions.

cofactor an organic molecule or a metal ion that certain enzymes need to catalyze a reaction.

common name nonsystematic nomenclature.

complex carbohydrate a carbohydrate containing two or more sugar molecules linked together.

condensation reaction a reaction combining two molecules while removing a small molecule (usually water or an alcohol).

configuration the three-dimensional structure of a particular atom in a compound. The configuration is designated by R or S.

conformation the three-dimensional shape of a molecule at a given instant that can change as a result of rotations about σ bonds.

conformers different conformations of a molecule.

conjugate acid a species accepts a proton to form its conjugate acid.

conjugate addition 1,4-addition.

conjugate base a species loses a proton to form its conjugate base.

conjugated diene a hydrocarbon with two conjugated double bonds.

conjugated double bonds double bonds separated by one single bond.

constitutional isomers molecules that have the same molecular formula but differ in the way their atoms are connected.

core electrons electrons in inner shells.

coupled protons protons that split each other. Coupled protons have the same coupling constant.

coupling constant the distance (in hertz) between two adjacent peaks of a split NMR signal.

covalent bond a bond created as a result of sharing electrons.

crown ether a cyclic molecule that contains several ether linkages.

crown–guest complex the complex formed when a crown ether binds a substrate.

C-terminal amino acid the terminal amino acid of a peptide (or protein) that has a free carboxyl group.

cycloalkane an alkane with its carbon chain arranged in a closed ring.

deactivating substituent a substituent that decreases the reactivity of an aromatic ring. Electron-withdrawing substituents deactivate aromatic rings toward electrophilic attack, and electron-donating substituents deactivate aromatic rings toward nucleophilic attack.

deamination loss of ammonia.

decarboxylation loss of carbon dioxide.

dehydration loss of water.

dehydrogenase an enzyme that carries out an oxidation reaction by removing hydrogen from the substrate.

delocalization energy (resonance energy) the extra stability a compound achieves as a result of having delocalized electrons.

delocalized electrons electrons that are shared by more than two atoms.

denaturation destruction of the highly organized tertiary structure of a protein.

deoxyribonucleic acid (DNA) a polymer of deoxyribonucleotides.

deoxyribonucleotide a nucleotide in which the sugar component is D-2-deoxyribose.

deoxy sugar a sugar in which one of the OH groups has been replaced by an H.

dextrorotatory the enantiomer that rotates polarized light in a clockwise direction.

diastereomer a configurational stereoisomer that is not an enantiomer.

diene a hydrocarbon with two double bonds.

dinucleotide two nucleotides linked by phosphodiester bonds.

dipeptide two amino acids linked by an amide bond.

dipole–dipole interaction an interaction between the dipole of one molecule and the dipole of another.

disaccharide a compound containing two sugar molecules linked together.

disulfide R—S—S—R

disulfide bridge a disulfide (—S—S—) bond in a peptide or protein.

DNA (deoxyribonucleic acid) a polymer of deoxyribonucleotides.

donation of electrons by resonance donation of electrons through p orbital overlap with neighboring π bonds.

double bond a σ bond and a π bond between two atoms.

double helix the secondary structure of DNA.

doublet an NMR signal split into two peaks.

downfield at a higher frequency in an NMR spectrum.

drug a compound that reacts with a biological molecule, triggering a physiological effect.

drug resistance biological resistance to a particular drug.

eclipsed conformer a conformer in which the bonds on adjacent carbons are aligned as viewed looking down the carbon–carbon bond.

***E* conformation** the conformation of a carboxylic acid or carboxylic acid derivative in which the carbonyl oxygen and the substituent bonded to the carboxyl oxygen or nitrogen are on opposite sides of the single bond.

Edman's reagent phenyl isothiocyanate. A reagent used to determine the N-terminal amino acid of a polypeptide.

effective magnetic field the magnetic field that a proton "senses" through the surrounding cloud of electrons.

***E* isomer** the isomer with the high-priority groups on opposite sides of the double bond.

electronegative element an element that readily acquires an electron.

electronegativity tendency of an atom to pull electrons toward itself.

electronic configuration description of the orbitals that the electrons in an atom occupy.

electrophile an electron-deficient atom or molecule.

electrophilic addition reaction an addition reaction in which the first species that adds to the reactant is an electrophile.

electrophilic aromatic substitution a reaction in which an electrophile substitutes for a hydrogen of an aromatic ring.

electrophoresis a technique that separates amino acids on the basis of their pI values.

electrostatic attraction attractive force between opposite charges.

electrostatic potential map a model that shows the charge distribution in a species.

elimination reaction a reaction that involves the elimination of atoms (or molecules) from the reactant.

enamine an α,β-unsaturated tertiary amine.

enantiomers nonsuperimposable mirror-image molecules.

enkephalins pentapeptides synthesized by the body to control pain.

enol an α,β-unsaturated alcohol.

enolization keto–enol interconversion.

enzyme a protein that is a catalyst.

epimers monosaccharides that differ in configuration at only one carbon.

epoxide an ether in which the oxygen is incorporated into a three-membered ring.

epoxy resin substance formed by mixing a low-molecular-weight prepolymer with a compound that forms a cross-linked polymer.

equatorial bond a bond of the chair conformer of cyclohexane that juts out from the ring in approximately the same plane that contains the chair.

equilibrium constant the ratio of products to reactants at equilibrium or the ratio of the rate constants for the forward and reverse reactions.

E1 reaction a first-order elimination reaction.

E2 reaction a second-order elimination reaction.

essential amino acid an amino acid that humans must obtain from their diet because they cannot synthesize it at all or cannot synthesize it in adequate amounts.

ester

$$\underset{\text{ester}}{R}\overset{\displaystyle\overset{O}{\|}}{-C-}OR$$

ether a compound containing an oxygen bonded to two carbons (ROR).

fat a triester of glycerol that exists as a solid at room temperature.

fatty acid a long-chain carboxylic acid.

favorable reaction a reaction in which the concentration of products is greater than the concentration of reactants at equilibrium.

fibrous protein a water-insoluble protein in which the polypeptide chains are arranged in bundles.

fingerprint region the right-hand third of an IR spectrum where the absorption bands are characteristic of the compound as a whole.

Fischer esterification reaction the reaction of a carboxylic acid with alcohol in the presence of an acid catalyst to form an ester.

Fischer projection a method of representing the spatial arrangement of groups bonded to a chirality center. The chirality center is the point of intersection of two perpendicular lines; the horizontal lines represent bonds that project out of the plane of the paper toward the viewer, and the vertical lines represent bonds that point back from the plane of the paper away from the viewer.

flavin adenine dinucleotide (FAD) a coenzyme required in certain oxidation reactions. It is reduced to $FADH_2$, which can act as a reducing agent in another reaction.

formal charge the number of valence electrons − (the number of nonbonding electrons +1/2 the number of bonding electrons).

free energy of activation ($\Delta G^{\ddagger}$) the true energy barrier to a reaction.

free radical a species with an unpaired electron.

frequency the velocity of a wave divided by its wavelength (in units of cycles/s).

Friedel–Crafts acylation an electrophilic substitution reaction that puts an acyl group on a benzene ring.

functional group the center of reactivity in a molecule.

functional group interconversion the conversion of one functional group into another functional group.

functional group region the left-hand two-thirds of an IR spectrum where most functional groups show absorption bands.

furanose a five-membered-ring sugar.

furanoside a five-membered-ring glycoside.

gene a segment of DNA.

generic name a commercially nonrestricted name for a drug.

genetic code the amino acid specified by each three-base sequence of mRNA.

geometric isomers cis–trans (or *E,Z*) isomers.

globular protein a water-soluble protein that tends to have a roughly spherical shape.

gluconeogenesis the synthesis of D-glucose from pyruvate.

glycolysis (glycolytic pathway) the series of reactions that converts D-glucose into two molecules of pyruvate.

glycoprotein a protein that is covalently bonded to a polysaccharide.

glycoside the acetal of a sugar.

glycosidic bond the bond between the anomeric carbon and the alcohol in a glycoside.

α-1,4'-glycosidic linkage a glycosidic linkage between the C-1 oxygen of one sugar and the C-4 of a second sugar with the oxygen atom of the glycosidic linkage in the axial position.

α-1,6'-glycosidic linkage   a glycosidic linkage between the C-1 oxygen of one sugar and the C-6 of a second sugar with the oxygen atom of the glycosidic linkage in the axial position.

β-1,4'-glycosidic linkage   a glycosidic linkage between the C-1 oxygen of one sugar and the C-4 of a second sugar with the oxygen atom of the glycosidic linkage in the equatorial position.

Grignard reagent   the compound that results when magnesium is inserted between the carbon and halogen of an alkyl halide (RMgBr, RMgCl).

halogenation   reaction with halogen (Br_2, Cl_2, I_2).

α-helix   the backbone of a polypeptide coiled in a right-handed spiral with hydrogen bonding occurring within the helix.

$$\textbf{hemiacetal} \quad R-\underset{\underset{\displaystyle OR}{|}}{\overset{\overset{\displaystyle OH}{|}}{C}}-H$$

$$\textbf{hemiketal} \quad R-\underset{\underset{\displaystyle OR}{|}}{\overset{\overset{\displaystyle OH}{|}}{C}}-R$$

heptose   a monosaccharide with seven carbons.

heteroatom   an atom other than carbon or hydrogen.

heterocyclic compound (heterocycle)   a cyclic compound in which one or more of the atoms of the ring are heteroatoms.

hexose   a monosaccharide with six carbons.

hormone   an organic compound synthesized in a gland and delivered by the bloodstream to its target tissue.

human genome   the total DNA of a human cell.

hybrid orbital   an orbital formed by mixing (hybridizing) orbitals.

$$\textbf{hydrate} \quad R-\underset{\underset{\displaystyle OH}{|}}{\overset{\overset{\displaystyle OH}{|}}{C}}-R \quad (H)$$

hydrated   water has been added to a compound.

hydration   addition of water to a compound.

hydride ion   a negatively charged hydrogen (a hydrogen atom with an extra electron).

hydrocarbon   a compound that contains only carbon and hydrogen.

α-hydrogen   usually, a hydrogen bonded to the carbon adjacent to a carbonyl carbon.

hydrogenation   addition of hydrogen.

hydrogen bond   an unusually strong dipole–dipole attraction (5 kcal/mol) between a hydrogen bonded to O, N, or F and the nonbonding electrons of an O, N, or F of another molecule.

hydrogen ion (proton)   a positively charged hydrogen (a hydrogen atom without its electron).

hydrolysis   reaction with water.

hydrophobic interactions   interactions between nonpolar groups. These interactions increase stability by decreasing the amount of structured water (increasing entropy).

imine $R_2C=NR$

induced-dipole–induced-dipole interaction   an interaction between a temporary dipole in one molecule and the dipole the temporary dipole induces in another molecule.

induced-fit model   a model that describes the specificity of an enzyme for its substrate: The shape of the active site does not become completely complementary to the shape of the substrate until after the enzyme binds the substrate.

inductive electron donation   donation of electrons through σ bond(s).

inductive electron withdrawal   withdrawal of electrons through σ bond(s).

infrared radiation   electromagnetic radiation familiar to us as heat.

infrared spectroscopy   uses infrared energy to provide knowledge of the functional groups in a compound.

infrared (IR) spectrum   a plot of percent transmission versus wave number (or wavelength) of infrared radiation.

initiation step   the step in which radicals are created, or the step in which the radical needed for the first propagation step is created.

intermediate   a species that is formed during a reaction and that is not the final product of the reaction.

intermolecular reaction   a reaction that takes place between two molecules.

internal alkyne   an alkyne with the triple bond not at the end of the carbon chain.

intramolecular reaction   a reaction that takes place within a molecule.

inversion of configuration   turning the configuration of a carbon inside out like an umbrella in a windstorm, so that the resulting product has a configuration opposite that of the reactant.

ion–dipole interaction   the interaction between an ion and the dipole of a molecule.

ion-exchange chromatography   a technique that uses a column packed with an insoluble resin to separate compounds on the basis of their charges and polarities.

ionic bond   a bond formed through the attraction of two ions of opposite charges.

isoelectric point (pI)   the pH at which there is no net charge on an amino acid.

isolated diene   a hydrocarbon containing two isolated double bonds.

isolated double bonds   double bonds separated by more than one single bond.

isomers   nonidentical compounds with the same molecular formula.

isotopes   atoms with the same number of protons but different numbers of neutrons.

IUPAC nomenclature   systematic nomenclature of chemical compounds.

Kekulé structure   a model that represents the bonds between atoms as lines.

$$\textbf{ketal} \quad R-\underset{\underset{\displaystyle OR}{|}}{\overset{\overset{\displaystyle OR}{|}}{C}}-R$$

keto–enol tautomerism (keto–enol interconversion)   interconversion of keto and enol tautomers.

keto–enol tautomers   a ketone and its isomeric α,β-unsaturated alcohol.

β-keto ester   an ester with a second carbonyl group at the β-position.

$$\textbf{ketone} \quad \underset{R}{\overset{\overset{\displaystyle O}{\|}}{\diagdown C \diagup}} R$$

ketose   a polyhydroxyketone.

Kiliani–Fischer synthesis   a method used to increase the number of carbons in an aldose by one, resulting in the formation of a pair of C-2 epimers.

λ_{max}   the wavelength at which there is maximum UV or Vis absorbance.

lead compound   the prototype in a search for other biologically active compounds.

leaving group   the group that is displaced in a nucleophilic substitution reaction.

levorotatory   the enantiomer that rotates polarized light in a counterclockwise direction.

Lewis acid   a substance that accepts an electron pair.

Lewis base   a substance that donates an electron pair.

Lewis structure   a model that represents the bonds between atoms as lines or dots and the valence electrons as dots.

ligation   sharing of nonbonding electrons with a metal ion.

lipid   a water-insoluble compound found in a living system.

lipid bilayer   two layers of phosphoacylglycerols arranged so that their polar heads are on the outside and their nonpolar fatty acid chains are on the inside.

localized electrons   electrons that are restricted to a particular locality.

lone-pair electrons (nonbonding electrons)   valence electrons not used in bonding.

magnetic resonance imaging (MRI)   NMR used in medicine. The difference in the way water is bound in different tissues produces a variation in signal between organs as well as between healthy and diseased tissue.

major groove   the wider and deeper of the two alternating grooves in DNA.

malonic ester synthesis   the synthesis of a carboxylic acid, using diethyl malonate as the starting material.

mass number   the number of protons plus the number of neutrons in an atom.

mechanism of a reaction   a description of the step-by-step process by which reactants are changed into products.

melting point   the temperature at which a solid becomes a liquid.

membrane   the material that surrounds a cell in order to isolate its contents.

meso compound   a compound that contains chirality centers and a plane of symmetry.

metabolism   reactions that living organisms carry out in order to obtain the energy and to synthesize the compounds they require.

meta director   a substituent that directs an incoming substituent meta to an existing substituent.

methine hydrogen   a tertiary hydrogen.

methylene group   a CH_2 group.

micelle   a spherical aggregation of molecules, each with a long hydrophobic tail and a polar head, arranged so that the polar head points to the outside of the sphere.

minor groove   the narrower and more shallow of the two alternating grooves in DNA.

mixed triglyceride   a triacylglycerol in which the fatty-acid components are different.

molecular modification   changing the structure of a lead compound.

molecular recognition   the recognition of one molecule by another as a result of specific interactions; for example, the specificity of an enzyme for its substrate.

molecular weight   the average weighted mass of the atoms in a molecule.

monomer   a repeating unit in a polymer.

monosaccharide (simple carbohydrate)   a single sugar molecule.

monoterpene   a terpene that contains 10 carbons.

MRI scanner   an NMR spectrometer used in medicine for whole-body NMR.

multiplet   an NMR signal split into more than seven peaks.

multiplicity   the number of peaks in an NMR signal.

multistep synthesis   preparation of a compound by a route that requires several steps.

mutarotation   a slow change in optical rotation to an equilibrium value.

$N + 1$ rule   an ^1H NMR signal for a hydrogen with N equivalent hydrogens bonded to an adjacent carbon is split into $N + 1$ peaks. A ^{13}C NMR signal for a carbon bonded to N hydrogens is split into $N + 1$ peaks.

natural-abundance atomic weight   the average mass of the atoms in the naturally occurring element.

natural product   a product synthesized in nature.

nicotinamide adenine dinucleotide (NAD^+)   a coenzyme required in certain oxidation reactions. It is reduced to NADH, which can act as a reducing agent in another reaction.

nitration   substitution of a nitro group (NO_2) for a hydrogen of a benzene ring.

nitrile   a compound that contains a carbon–nitrogen triple bond ($RC\equiv N$).

NMR spectroscopy   the absorption of electromagnetic radiation to determine the structural features of an organic compound. In the case of NMR spectroscopy, it determines the carbon–hydrogen framework.

nonbonding electrons (lone-pair electrons)   valence electrons not used in bonding.

nonpolar covalent bond   a bond formed between two atoms that share the bonding electrons equally.

nonpolar molecule   a molecule with no charge or partial charge on any of its atoms.

nonreducing sugar   a sugar that cannot be oxidized by reagents such as Ag^+ and Cu^+. Nonreducing sugars are not in equilibrium with the open-chain aldose or ketose.

N-terminal amino acid   the terminal amino acid of a peptide (or protein) that has a free amino group.

nucleic acid   the two kinds of nucleic acid are DNA and RNA.

nucleophile   an electron-rich atom or molecule.

nucleophilic acyl substitution reaction   a reaction in which a group bonded to an acyl or aryl group is substituted by another group.

nucleophilic addition reaction   a reaction that involves the addition of a nucleophile to a reagent.

nucleophilicity   a measure of how readily an atom or a molecule with a pair of nonbonding electrons attacks an atom.

nucleophilic substitution reaction   a reaction in which a nucleophile substitutes for an atom or a group.

nucleoside   a heterocyclic base (a purine or a pyrimidine) bonded to the anomeric carbon of a sugar (D-ribose or D-2-deoxyribose).

nucleotide   a heterocycle attached in the β-position to a phosphorylated ribose or deoxyribose.

observed rotation   the amount of rotation observed in a polarimeter.

octet rule   states that an atom will give up, accept, or share electrons in order to achieve a filled shell. Because a filled second shell contains eight electrons, this is known as the octet rule.

oil   a triester of glycerol that exists as a liquid at room temperature.

oligonucleotide   3 to 10 nucleotides linked by phosphodiester bonds.

oligopeptide   3 to 10 amino acids linked by amide bonds.

oligosaccharide   3 to 10 sugar molecules linked by glycosidic bonds.

operating frequency   the frequency at which an NMR spectrometer operates.

optically active   rotates the plane of polarized light.

optically inactive   does not rotate the plane of polarized light.

orbital   the volume of space around the nucleus in which an electron is most likely to be found.

orbital hybridization   mixing of orbitals.

organic compound   a compound that contains carbon.

organic synthesis   preparation of organic compounds from other organic compounds.

organometallic compound   a compound containing a carbon–metal bond.

orphan drugs   drugs for diseases or conditions that affect fewer than 200,000 people.

ortho-para-director   a substituent that directs an incoming substituent ortho and para to an existing substituent.

oxidation reaction   a reaction in which the number of C—H bonds decreases.

oxidation–reduction reaction (redox reaction)   a reaction that involves the transfer of electrons from one species to another.

oxidative cleavage   an oxidation reaction that cuts the reactant into two or more pieces.

packing   the fitting of individual molecules into a frozen crystal lattice.

paraffin   an alkane.

parent hydrocarbon   the longest continuous carbon chain in a molecule.

partial hydrolysis   a technique that hydrolyzes only some of the peptide bonds in a polypeptide.

pentose   a monosaccharide with five carbons.

peptide   polymer of amino acids linked together by amide bonds. A peptide contains fewer amino acid residues than a protein does.

peptide bond   the amide bond that links the amino acids in a peptide or protein.

peroxyacid   a carboxylic acid with an OOH group instead of an OH group.

perspective formula   a method of representing the spatial arrangement of groups bonded to a chirality center. Two bonds are drawn in the plane of the paper; a solid wedge is used to depict a bond that projects out of the plane of the paper toward the viewer, and a hatched wedge is used to represent a bond that projects back from the plane of the paper away from the viewer.

pH   the pH scale is used to describe the acidity of a solution ($pH = -\log[H^+]$).

phenyl group

pheromone   a compound secreted by an animal that stimulates a physiological or behavioral response from a member of the same species.

phosphoacylglycerol (phosphoglyceride)   a compound formed when two OH groups of glycerol form esters with fatty acids and the terminal OH group forms a phosphate ester.

phosphoanhydride bond   the bond holding two phosphoric acid molecules together.

phospholipid   a lipid that contains a phosphate group.

phosphoryl transfer reaction   the transfer of a phosphate group from one compound to another.

pi (π) bond   a bond formed as a result of side-to-side overlap of p orbitals.

pK_a   describes the tendency of a compound to lose a proton ($pK_a = -\log K_a$, where K_a is the acid dissociation constant).

plane polarized light   light that oscillates only in one plane.

plane of symmetry   an imaginary plane that bisects a molecule into mirror images.

β-pleated sheet   the backbone of a polypeptide that is extended in a zigzag structure with hydrogen bonding between neighboring chains.

polar covalent bond   a covalent bond between atoms with different electronegativities.

polarimeter   an instrument that measures the rotation of polarized light.

polymer   a large molecule made by linking monomers together.

polymerization   the process of linking up monomers to form a polymer.

polynucleotide   many nucleotides linked by phosphodiester bonds.

polypeptide   many amino acids linked by amide bonds.

polysaccharide   a compound containing more than 10 sugar molecules linked together.

polyunsaturated fatty acid   a fatty acid with more than one double bond.

porphyrin ring system   consists of four pyrrole rings joined by one-carbon bridges.

primary alcohol   an alcohol in which the OH group is bonded to a primary carbon.

primary alkyl halide   an alkyl halide in which the halogen is bonded to a primary carbon.

primary alkyl radical   a radical with the unpaired electron on a primary carbon.

primary amine   an amine with one alkyl group bonded to the nitrogen.

primary carbocation   a carbocation with the positive charge on a primary carbon.

primary carbon   a carbon bonded to only one other carbon.

primary hydrogen   a hydrogen bonded to a primary carbon.

primary structure (of a nucleic acid)   the sequence of bases in a nucleic acid.

primary structure (of a protein)   the sequence of amino acids in a protein.

propagating site   the reactive end of a chain-growth polymer.

propagation step   in the first of a pair of propagation steps, a radical (or an electrophile or a nucleophile) reacts to produce another radical (or an electrophile or a nucleophile) that reacts in the second step to produce the radical (or the electrophile or the nucleophile) that was the reactant in the first propagation step.

protein   a polymer containing 40 to 4000 amino acids linked by amide bonds.

proton   a positively charged hydrogen (hydrogen ion); a positively charged particle in an atomic nucleus.

proton-coupled ^{13}C NMR spectrum   a ^{13}C NMR spectrum in which each signal is split by the hydrogens bonded to the C that produced the spectrum.

pyranose   a six-membered-ring sugar.

pyranoside   a six-membered-ring glycoside.

pyridoxal phosphate   the coenzyme required by enzymes that catalyze certain transformations of amino acids.

quartet   an NMR signal split into four peaks.

quaternary structure   a description of the way the individual polypeptide chains of a protein are arranged with respect to each other.

racemic mixture (racemate, racemic modification)   a mixture of equal amounts of a pair of enantiomers.

radical   an atom or a molecule with an unpaired electron.

radical chain reaction   a reaction in which radicals are formed and react in repeating propagating steps.

radical inhibitor   a compound that traps radicals.

radical substitution reaction   a substitution reaction that has a radical intermediate.

random screen (blind screen)   the search for a pharmacologically active compound without any information about what chemical structures might show activity.

rate constant   a measure of how easy or difficult it is to reach the transition state of a reaction (to get over the energy barrier to the reaction).

rate law   the relationship between the rate of a reaction and the concentration of the reactants.

rate-determining step (rate-limiting step)   the step in a reaction that has the transition state with the highest energy.

_R_ configuration   after assigning relative priorities to the four groups bonded to a chirality center, if the lowest priority group is on a vertical axis in a Fischer projection (or pointing away from the viewer in a perspective formula), an arrow drawn from the highest priority group to the next-highest-priority group goes in a clockwise direction.

reaction coordinate diagram   describes the energy changes that take place during the course of a reaction.

reactivity–selectivity principle   states that the greater the reactivity of a species, the less selective it will be.

receptor   a site on a cell at which a drug binds in order to exert its physiological effect.

reducing sugar   a sugar that can be oxidized by reagents such as Ag^+ or Br_2. Reducing sugars are in equilibrium with the open-chain aldose or ketose.

reduction reaction   a reaction in which the number of C—H bonds increases.

reference compound   a compound added to a sample whose NMR spectrum is to be taken. The positions of the signals in the NMR spectrum are measured from the position of the signal given by the reference compound.

regioselective reaction   a reaction that leads to the preferential formation of one constitutional isomer over another.

replication   the synthesis of identical copies of DNA.

resonance   a compound with delocalized electrons is said to have resonance.

resonance contributor   a structure with localized electrons that approximates the true structure of a compound with delocalized electrons.

resonance electron donation   donation of electrons through _p_ orbital overlap with neighboring π bonds.

resonance electron withdrawal   withdrawal of electrons through _p_ orbital overlap with neighboring π bonds.

resonance hybrid   the actual structure of a compound with delocalized electrons; it is represented by two or more structures with localized electrons.

resonance stabilization   the extra stability associated with a compound as a result of its having delocalized electrons.

restriction endonuclease   an enzyme that cleaves DNA at a specific base sequence.

restriction fragment   a fragment that is formed when DNA is cleaved by a restriction endonuclease.

ribonucleic acid (RNA)   a polymer of ribonucleotides.

ribonucleotide   a nucleotide in which the sugar component is D-ribose.

RNA (ribonucleic acid)   a polymer of ribonucleotides.

saponification   hydrolysis of an ester (such as a fat) under basic conditions.

saturated hydrocarbon   a hydrocarbon that is completely saturated (that is, contains no double or triple bonds) with hydrogen.

_S_ configuration   after assigning relative priorities to the four groups bonded to a chirality center, if the lowest priority group is on a vertical axis in a Fischer projection (or pointing away from the viewer in a perspective formula), an arrow drawn from the highest priority group to the next-highest priority group goes in a counterclockwise direction.

secondary alcohol   an alcohol in which the OH group is bonded to a secondary carbon.

secondary alkyl halide   an alkyl halide in which the halogen is bonded to a secondary carbon.

secondary alkyl radical   a radical with the unpaired electron on a secondary carbon.

secondary amine   an amine with two alkyl groups bonded to the nitrogen.

secondary carbocation   a carbocation with the positive charge on a secondary carbon.

secondary carbon   a carbon bonded to two other carbons.

secondary hydrogen   a hydrogen bonded to a secondary carbon.

secondary structure   a description of the conformation of the backbone of a protein.

semiconservative replication   the mode of replication that results in a daughter molecule of DNA having one of the original DNA strands plus a newly synthesized strand.

sense strand   the strand in DNA that is not read during transcription; it has the same sequence of bases as the synthesized mRNA strand (with a U, T difference).

separated charges   a positive and a negative charge that can be neutralized by the movement of electrons.

sesquiterpene   a terpene that contains 15 carbons.

shielding   phenomenon caused by electron donation to the environment of a proton. The electrons shield the proton from the full effect of the applied magnetic field. The more a proton is shielded, the farther to the right its signal appears in an NMR spectrum.

sigma (σ) bond   a bond with a cylindrically symmetrical distribution of electrons.

simple carbohydrate (monosaccharide)   a single sugar molecule.

simple triglyceride   a triacylglycerol in which the fatty acid components are the same.

single bond   a σ bond.

singlet   an unsplit NMR signal.

skeletal structure   shows the carbon–carbon bonds as lines and does not show the carbon–hydrogen bonds.

S_N1 reaction   a unimolecular nucleophilic substitution reaction.

S$_N$2 reaction   a bimolecular nucleophilic substitution reaction.

soap   a sodium or potassium salt of a fatty acid.

solvation   the interaction between a solvent and another molecule (or ion).

specific rotation   the amount of rotation that will be caused by a compound with a concentration of 1.0 g/mL in a sample tube 1.0 dm long.

spectroscopy   study of the interaction of matter and electromagnetic radiation.

sphingolipid   a lipid that contains sphingosine.

α-spin state   nuclei in this spin state have their magnetic moments oriented in the same direction as the applied magnetic field.

β-spin state   nuclei in this spin state have their magnetic moments oriented opposite the direction of the applied magnetic field.

squalene   a triterpene that is a precursor of steroid molecules.

staggered conformer   a conformer in which the bonds on one carbon bisect the bond angle on the adjacent carbon when viewed looking down the carbon–carbon bond.

stereochemistry   the field of chemistry that deals with the structures of molecules in three dimensions.

stereoisomers   isomers that differ in the way their atoms are arranged in space.

steric effects   effects due to the fact that groups occupy a certain volume of space.

steric hindrance   refers to bulky groups at the site of a reaction that make it difficult for the reactants to approach each other.

steric strain (van der Waals strain, van der Waals repulsion)   the repulsion between the electron cloud of an atom or a group of atoms and the electron cloud of another atom or group of atoms.

steroid   a class of compounds that contains a steroid ring system.

straight-chain alkane   an alkane in which the carbons form a continuous chain with no branches.

stretching frequency   the frequency at which a stretching vibration occurs.

substitution reaction   a reaction in which an atom or group substitutes for another atom or group.

substrate   the reactant of an enzyme-catalyzed reaction.

subunit   an individual chain of an oligomer.

sulfonation   substitution of a hydrogen of a benzene ring by a sulfonic acid group (— SO$_3$H).

synthetic polymer   a polymer that is not synthesized in nature.

systematic nomenclature   nomenclature based on structure.

tautomerism   interconversion of tautomers.

tautomers   rapidly equilibrating isomers that differ in the location of their bonding electrons.

template strand (antisense strand)   the strand in DNA that is read during transcription.

terminal alkyne   an alkyne with the triple bond at the end of the carbon chain.

termination step   when two radicals combine to produce a molecule in which all the electrons are paired.

terpene   a lipid, isolated from a plant, that contains carbon atoms in multiples of five.

tertiary alcohol   an alcohol in which the OH group is bonded to a tertiary carbon.

tertiary alkyl halide   an alkyl halide in which the halogen is bonded to a tertiary carbon.

tertiary alkyl radical   a radical with the unpaired electron on a tertiary carbon.

tertiary amine   an amine with three alkyl groups bonded to the nitrogen.

tertiary carbocation   a carbocation with the positive charge on a tertiary carbon.

tertiary carbon   a carbon bonded to three other carbons.

tertiary hydrogen   a hydrogen bonded to a tertiary carbon.

tertiary structure   a description of the three-dimensional arrangement of all the atoms in a protein.

tetrahedral bond angle   the bond angle (109.5°) formed by adjacent bonds of an sp^3 carbon.

tetrahedral carbon   an sp^3 carbon; a carbon that forms covalent bonds by using four sp^3 hybridized orbitals.

tetrahedral intermediate   the intermediate formed in a nucleophilic acyl substitution reaction.

tetrahydrofolate (THF)   the coenzyme required by enzymes that catalyze reactions that donate a group containing a single carbon to their substrates.

tetraterpene   a terpene that contains 40 carbons.

tetrose   a monosaccharide with four carbons.

thiamine pyrophosphate (TPP)   the coenzyme required by enzymes, which catalyze a reaction that transfers a two-carbon fragment to a substrate.

thioester   the sulfur analog of an ester.

$$\underset{R}{\overset{\displaystyle O \atop \displaystyle \|}{\diagdown}}\,\underset{}{\overset{C}{}}\,\diagup SR$$

thiol   the sulfur analog of an alcohol (RSH).

trademark   a registered name, symbol, or picture.

transamination   a reaction in which an amino group is transferred from one compound to another.

transcription   the synthesis of mRNA from a DNA blueprint.

transesterification reaction   the reaction of an ester with an alcohol to form a different ester.

trans fused   two cyclohexane rings fused together such that if the second ring were considered to be two substituents of the first ring, both substituents would be in equatorial positions.

trans isomer   the isomer with identical substituents on opposite sides of the double bond or on opposite sides of a cyclic structure.

transition state   the highest point on a hill in a reaction coordinate diagram. In the transition state, bonds in the reactant that will break are partially broken and bonds in the product that will form are partially formed.

translation   the synthesis of a protein from an mRNA blueprint.

triglyceride   the compound formed when the three OH groups of glycerol are esterified with fatty acids.

triose   a monosaccharide with three carbons.

tripeptide   three amino acids linked by amide bonds.

triple bond   a σ bond plus two π bonds.

triplet   an NMR signal split into three peaks.

triterpene   a terpene that contains 30 carbons.

ultraviolet light   electromagnetic radiation with wavelengths ranging from 180 to 400 nm.

upfield   toward lower frequency in an NMR spectrum.

unimolecular reaction (first-order reaction)   a reaction whose rate depends on the concentration of one reactant.

unsaturated hydrocarbon   a hydrocarbon that contains one or more double or triple bonds.

UV/Vis spectroscopy   the absorption of electromagnetic radiation in the ultraviolet and visible regions of the spectrum; used to determine information about conjugated systems.

valence electron   an electron in an unfilled shell.

van der Waals forces   induced-dipole–induced-dipole interactions.

vinyl group   CH$_2$= CH—

vinylic carbon   a carbon in a carbon–carbon double bond.

vinylic cation   a compound with a positive charge on a vinylic carbon.

vinylic radical   a compound with an unpaired electron on a vinylic carbon.

visible light   electromagnetic radiation with wavelengths ranging from 400 to 780 nm.

vitamin   a substance needed in small amounts for normal body function that the body cannot synthesize at all or cannot synthesize in adequate amounts.

vitamin KH$_2$ the coenzyme required by the enzyme that catalyzes the carboxylation of glutamate side chains.

wavelength   distance from any point on one wave to the corresponding point on the next wave (usually in units of μm or nm).

wavenumber   the number of waves in 1 cm.

wax   an ester formed from a long-chain carboxylic acid and a long-chain alcohol.

withdraw electrons by resonance   withdrawal of electrons through p orbital overlap with neighboring bonds.

Z isomer   the isomer with the high-priority groups on the same side of the double bond.

zwitterion   a compound with a negative charge and a positive charge on nonadjacent atoms.

Photo Credits

Index

Periodic Table of the Elements

Main groups

Transition metals

Period	1A 1	2A 2	3B 3	4B 4	5B 5	6B 6	7B 7	8B 8	8B 9	8B 10	1B 11	2B 12	3A 13	4A 14	5A 15	6A 16	7A 17	8A 18
1	1 **H** 1.00794																	2 **He** 4.002602
2	3 **Li** 6.941	4 **Be** 9.012182											5 **B** 10.811	6 **C** 12.0107	7 **N** 14.0067	8 **O** 15.9994	9 **F** 18.998403	10 **Ne** 20.1797
3	11 **Na** 22.989770	12 **Mg** 24.3050											13 **Al** 26.981538	14 **Si** 28.0855	15 **P** 30.973761	16 **S** 32.065	17 **Cl** 35.453	18 **Ar** 39.948
4	19 **K** 39.0983	20 **Ca** 40.078	21 **Sc** 44.955910	22 **Ti** 47.867	23 **V** 50.9415	24 **Cr** 51.9961	25 **Mn** 54.938049	26 **Fe** 55.845	27 **Co** 58.933200	28 **Ni** 58.6934	29 **Cu** 63.546	30 **Zn** 65.39	31 **Ga** 69.723	32 **Ge** 72.64	33 **As** 74.92160	34 **Se** 78.96	35 **Br** 79.904	36 **Kr** 83.80
5	37 **Rb** 85.4678	38 **Sr** 87.62	39 **Y** 88.90585	40 **Zr** 91.224	41 **Nb** 92.90638	42 **Mo** 95.94	43 **Tc** [98]	44 **Ru** 101.07	45 **Rh** 102.90550	46 **Pd** 106.42	47 **Ag** 107.8682	48 **Cd** 112.411	49 **In** 114.818	50 **Sn** 118.710	51 **Sb** 121.760	52 **Te** 127.60	53 **I** 126.90447	54 **Xe** 131.293
6	55 **Cs** 132.90545	56 **Ba** 137.327	71 **Lu** 174.967	72 **Hf** 178.49	73 **Ta** 180.9479	74 **W** 183.84	75 **Re** 186.207	76 **Os** 190.23	77 **Ir** 192.217	78 **Pt** 195.078	79 **Au** 196.96655	80 **Hg** 200.59	81 **Tl** 204.3833	82 **Pb** 207.2	83 **Bi** 208.98038	84 **Po** [208.98]	85 **At** [209.99]	86 **Rn** [222.02]
7	87 **Fr** [223.02]	88 **Ra** [226.03]	103 **Lr** [262.11]	104 **Rf** [261.11]	105 **Db** [262.11]	106 **Sg** [266.12]	107 **Bh** [264.12]	108 **Hs** [269.13]	109 **Mt** [268.14]	110 [271.15]	111 [272.15]	112 [277]		114 [285]		116 [289]		

*Lanthanide series

57 ***La** 138.9055	58 **Ce** 140.116	59 **Pr** 140.90765	60 **Nd** 144.24	61 **Pm** [145]	62 **Sm** 150.36	63 **Eu** 151.964	64 **Gd** 157.25	65 **Tb** 158.92534	66 **Dy** 162.50	67 **Ho** 164.93032	68 **Er** 167.259	69 **Tm** 168.93421	70 **Yb** 173.04

†Actinide series

89 **†Ac** [227.03]	90 **Th** 232.0381	91 **Pa** 231.03588	92 **U** 238.02891	93 **Np** [237.05]	94 **Pu** [244.06]	95 **Am** [243.06]	96 **Cm** [247.07]	97 **Bk** [247.07]	98 **Cf** [251.08]	99 **Es** [252.08]	100 **Fm** [257.10]	101 **Md** [258.10]	102 **No** [259.10]

[a]The labels on top (1A, 2A, etc.) are common American usage. The labels below these (1, 2, etc.) are those recommended by the International Union of Pure and Applied Chemistry.

The names and symbols for elements 110 and above have not yet been decided.

Atomic weights in brackets are the masses of the longest-lived or most important isotope of radioactive elements.

Further information is available at *http://www.shef.ac.uk/chemistry/web-elements/*.

The production of element 116 was reported in May 1999 by scientists at Lawrence Berkeley National Laboratory.

Common Functional Groups

Alkane RCH_3

Aniline

Alkene
internal   terminal

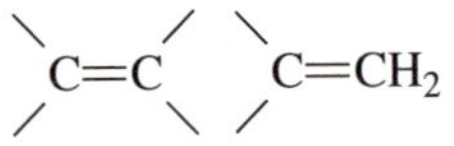

Phenol

Alkyne $RC\equiv CR$ $RC\equiv CH$
internal   terminal

Carboxylic acid

Nitrile $RC\equiv N$

Acyl chloride

Ether $R-O-R$

Acid anhydride

Thiol RCH_2-SH

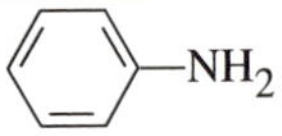

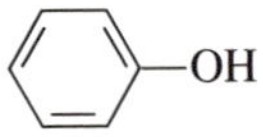

Ester

Disulfide $R-S-S-R$

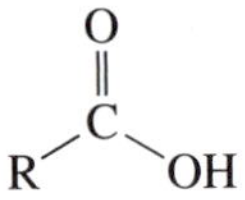

Amide

Epoxide

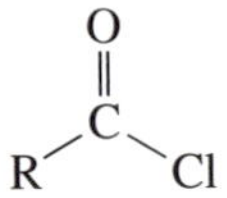

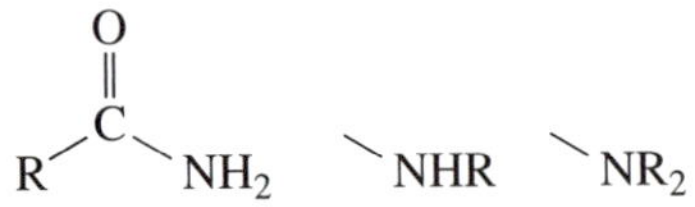

Aldehyde

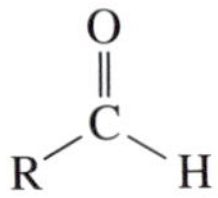

Ketone

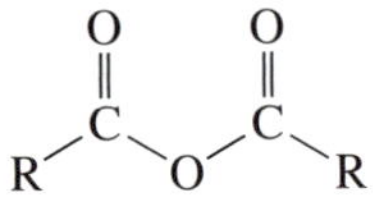

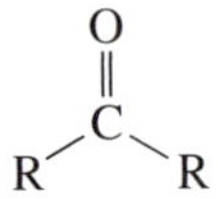

	primary	secondary	tertiary
Alkyl halide	$R-CH_2-X$ $X = F, Cl, Br, or I$	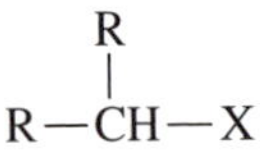 $R-\overset{R}{\underset{}{CH}}-X$	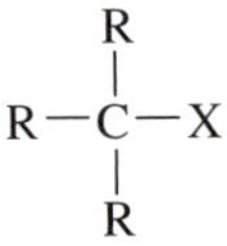 $R-\overset{R}{\underset{R}{C}}-X$
Alcohol	$R-CH_2-OH$	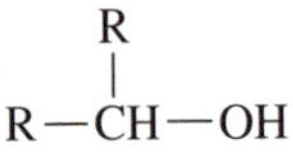 $R-\overset{R}{\underset{}{CH}}-OH$	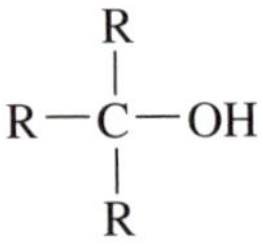 $R-\overset{R}{\underset{R}{C}}-OH$
Amine	$R-NH_2$	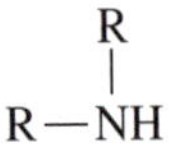 $R-\overset{R}{\underset{}{NH}}$	 $R-\overset{R}{\underset{R}{N}}$

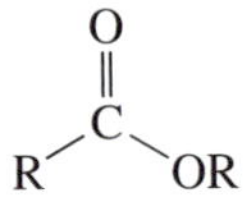